LINEAR ALGEBRA WITH APPLICATIONS

LINEAR ALGEBRA WITH APPLICATIONS

Roger Baker
Kenneth Kuttler
Brigham Young University, USA

NEW JERSEY • LONDON • SINGAPORE • BEIJING • SHANGHAI • HONG KONG • TAIPEI • CHENNAI

Published by

World Scientific Publishing Co. Pte. Ltd.
5 Toh Tuck Link, Singapore 596224
USA office: 27 Warren Street, Suite 401-402, Hackensack, NJ 07601
UK office: 57 Shelton Street, Covent Garden, London WC2H 9HE

Library of Congress Cataloging-in-Publication Data
Baker, R. C. (Roger Clive), 1947– author.
Linear algebra with applications / by Roger Baker (Brigham Young University, USA) & Kenneth Kuttler (Brigham Young University, USA).
pages cm
Includes bibliographical references and index.
ISBN 978-9814590532 (hardcover : alk. paper)
1. Algebras, Linear. I. Kuttler, Kenneth, author. II. Title.
QA184.2.B354 2014
512'.5--dc23
2014002876

British Library Cataloguing-in-Publication Data
A catalogue record for this book is available from the British Library.

Printed in Singapore

Preface

Linear algebra is the most widely studied branch of mathematics after calculus. There are good reasons for this. If you, the reader, pursue the study of mathematics, you will find linear algebra used everywhere from number theory to PDEs (partial differential equations). If you apply mathematics to concrete problems, then again linear algebra is a powerful tool. Obviously you must start by grasping the principles of linear algebra, and that is the aim of this book. The readership we had in mind includes majors in disciplines as widely separated as mathematics, engineering and social sciences. The notes at the end of the book on further reading may be useful, and allow for many different directions for further studies. Suggestions from readers on this or any other aspect of the book should be sent to {baker@math.byu.edu} or {klkuttle@math.byu.edu}.

We have used Maple for longer computations (an appendix at the back addresses Maple for beginners). There are, of course, other options.

Occasionally, a sequence of adjacent exercises will have each exercise depending on the preceding exercise. We will indicate such dependence by a small vertical arrow. An asterisk means that the labeled material is difficult.

Contents

Chapter 1

Numbers, vectors and fields

Chapter summary

This chapter is on fundamental notation concerning sets and gives a brief introduction to fields, one of the ingredients which is important in the study of linear algebra. It contains a specific example of a vector space $\mathbb{R}^3$ along with some of the geometric ideas concerning vector addition and the dot and cross product. It also contains a description of the most important examples of fields. These include the real numbers, complex numbers, and the field of residue classes.

1.1 Functions and sets

A set is a collection of things called elements of the set. For example, one speaks of the set of integers, whole numbers such as 1,2,−4, and so on. This set, whose existence will be assumed, is denoted by $\mathbb{Z}$. The symbol $\mathbb{R}$ will denote the set of real numbers which is usually thought of as points on the number line. Other sets could be the set of people in a family or the set of donuts in a display case at the store. Sometimes parentheses, { } specify a set by listing the things which are in the set between the parentheses. For example, the set of integers between -1 and 2, including these numbers, could be denoted as $\{-1, 0, 1, 2\}$. The notation signifying x is an element of a set S, is written as $x \in S$. We write $x \notin S$ for 'x is not an element of S'. Thus, $1 \in \{-1, 0, 1, 2, 3\}$ and $7 \notin \{-1, 0, 1, 2, 3\}$. Here are some axioms about sets. Axioms are statements which are accepted, not proved.

(1) Two sets are equal if and only if they have the same elements.
(2) To every set A, and to every condition $S(x)$ there corresponds a set B, whose elements are exactly those elements x of A for which $S(x)$ holds.
(3) For every collection of sets there exists a set that contains all the elements that belong to at least one set of the given collection. This set is called the union of the sets.

Example 1.1. As an illustration of these axioms, $\{1, 2, 3\} = \{3, 2, 1\}$ because they have the same elements. If you consider $\mathbb{Z}$ the set of integers, let $S(x)$ be the condition that x is even. Then the set B consisting of all elements x of $\mathbb{Z}$ such that $S(x)$ is true, specifies the even integers. We write this set as follows:

$$\{x \in \mathbb{Z} : S(x)\}$$

Next let $A = \{1, 2, 4\}, B = \{2, 5, 4, 0\}$. Then the union of these two sets is the set $\{1, 2, 4, 5, 0\}$. We denote this union as $A \cup B$.

The following is the definition of a function.

Definition 1.1. Let X, Y be nonempty sets. A function f is a rule which yields a unique $y \in Y$ for a given $x \in X$. It is customary to write $f(x)$ for this element of Y. It is also customary to write

$$f : X \to Y$$

to indicate that f is a function defined on X which gives an element of Y for each $x \in X$.

Example 1.2. Let $X = \mathbb{R}$ and $f(x) = 2x$.

The following is another general consideration.

Definition 1.2. Let $f : X \to Y$ where f is a function. Then f is said to be one-to-one (injective), if whenever $x_1 \neq x_2$, it follows that $f(x_1) \neq f(x_2)$. The function is said to be onto, (surjective), if whenever $y \in Y$, there exists an $x \in X$ such that $f(x) = y$. The function is bijective if the function is both one-to-one and onto.

Example 1.3. The function $f(x) = 2x$ is one-to-one and onto from $\mathbb{R}$ to $\mathbb{R}$. The function $f : \mathbb{R} \to \mathbb{R}$ given by the formula $f(x) = x^2$ is neither one-to-one nor onto.

Notation 1.1. There are certain standard notations which pertain to sets. We write

$$A \subseteq B \text{ or } A \subset B$$

to mean that every element of A is an element of B. We say that A is a **subset** of B. We also say that the set A is contained in the set B.

1.2 Three dimensional space

We use $\mathbb{R}$ to denote the set of all real numbers. Let $\mathbb{R}^3$ be the set of all ordered triples (x_1, x_2, x_3) of real numbers. ('Ordered' in the sense that, for example, $(1, 2, 3)$ and $(2, 1, 3)$ are different triples.) We can describe the three-dimensional space in which we move around very neatly by specifying an origin of coordinates $\mathbf{0}$ and three axes at right angles to each other. Then we assign a point of $\mathbb{R}^3$ to each point in space according to perpendicular distances from these axes, with signs showing the directions from these axes. If we think of the axes as pointing east, north and vertically (Fig. 1), then $(2, 6, -1)$'is' the point with distances 2 (units) from the N axis in direction E, 6 from the E axis in direction N, and 1 vertically below the plane containing the E and N axes. (There is, of course, a similar identification of $\mathbb{R}^2$ with the set of points in a plane.)

The triples (x_1, x_2, x_3) are called **vectors**. You can mentally identify a vector with a point in space, or with a directed segment from $\mathbf{0} = (0, 0, 0)$ to that point.

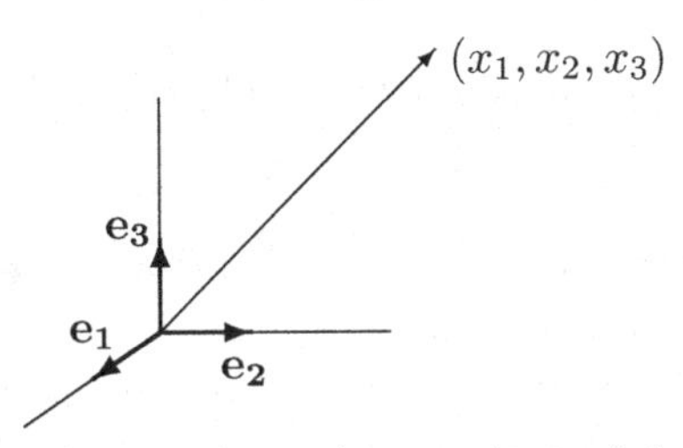

Let us write $\mathbf{e}_1 = (1, 0, 0), \mathbf{e}_2 = (0, 1, 0), \mathbf{e}_3 = (0, 0, 1)$ for the unit vectors along the axes.

Vectors $(x_1, x_2, x_3), (y_1, y_2, y_3)$ are 'abbreviated' using the same letter in bold face:

$$\mathbf{x} = (x_1, x_2, x_3), \ \mathbf{y} = (y_1, y_2, y_3).$$

(Similarly in $\mathbb{R}^n$; see below.) The following algebraic operations in $\mathbb{R}^3$ are helpful.

Definition 1.3. The sum of $\mathbf{x}$ and $\mathbf{y}$ is

$$\mathbf{x} + \mathbf{y} = (x_1 + y_1, x_2 + y_2, x_3 + y_3).$$

For $c \in \mathbb{R}$, the **scalar multiple** $c\mathbf{x}$ is

$$c\mathbf{x} = (cx_1, cx_2, cx_3).$$

There are algebraic rules that govern these operations, such as

$$c(\mathbf{x} + \mathbf{y}) = c\mathbf{x} + c\mathbf{y}.$$

We defer the list of rules until Section 1.6, when we treat a more general situation; see Axioms **1**, ..., **8**.

Definition 1.4. The **inner product**, sometimes called the **dot product** of $\mathbf{x}$ and $\mathbf{y}$ is

$$\langle \mathbf{x}, \mathbf{y} \rangle = x_1y_1 + x_2y_2 + x_3y_3.$$

Note that for this product that the 'outcome' $\langle \mathbf{x}, \mathbf{y} \rangle$ is a scalar, that is, a member of $\mathbb{R}$. The algebraic rules of the inner product (easily checked) are

$$\langle \mathbf{x}+\mathbf{z}, \mathbf{y} \rangle = \langle \mathbf{x}, \mathbf{y} \rangle + \langle \mathbf{z}, \mathbf{y} \rangle, \tag{1.1}$$

$$\langle c\mathbf{x}, \mathbf{y} \rangle = c\langle \mathbf{x}, \mathbf{y} \rangle, \tag{1.2}$$

$$\langle \mathbf{x}, \mathbf{y} \rangle = \langle \mathbf{y}, \mathbf{x} \rangle, \tag{1.3}$$

$$\langle \mathbf{x}, \mathbf{x} \rangle > 0 \text{ for } \mathbf{x} \neq \mathbf{0}. \tag{1.4}$$

Definition 1.5. The **cross product** of $\mathbf{x}$ and $\mathbf{y}$ is

$$\mathbf{x} \times \mathbf{y} = (x_2y_3 - x_3y_2, -x_1y_3 + x_3y_1, x_1y_2 - x_2y_1).$$

For this product, the 'outcome' $\mathbf{x}\times\mathbf{y}$ is a vector. Note that $\mathbf{x}\times\mathbf{y} = -\mathbf{y}\times\mathbf{x}$ and $\mathbf{x}\times\mathbf{x} = 0$, among other 'rules' that may be unexpected. To see the cross product in the context of classical dynamics and electrodynamics, you could look in Chapter 1 of Johns (1992). The above definitions first arose in describing the actions of forces on moving particles and bodies.

In describing our operations geometrically, we begin with length and distance. The **length** of $\mathbf{x}$ (distance from $\mathbf{0}$ to $\mathbf{x}$) is

$$|\mathbf{x}| = \sqrt{x_1^2 + x_2^2 + x_3^3} = \sqrt{\langle \mathbf{x}, \mathbf{x} \rangle}.$$

This follows from Pythagoras's theorem (see the Exercises). Note that the directed segment $c\mathbf{x}$ points in the same direction as $\mathbf{x}$ if $c > 0$; the opposite direction if $c < 0$. Moreover,

$$|c\mathbf{x}| = |c||\mathbf{x}|.$$

(multiplication by c 'scales' our vector).

The sum of $\mathbf{u}$ and $\mathbf{v}$ can be formed by 'translating' the directed segment $\mathbf{u}$, that is, moving it parallel to itself, so that its initial point is at $\mathbf{v}$. The terminal point is then at $\mathbf{u}+\mathbf{v}$, as shown in the following picture.

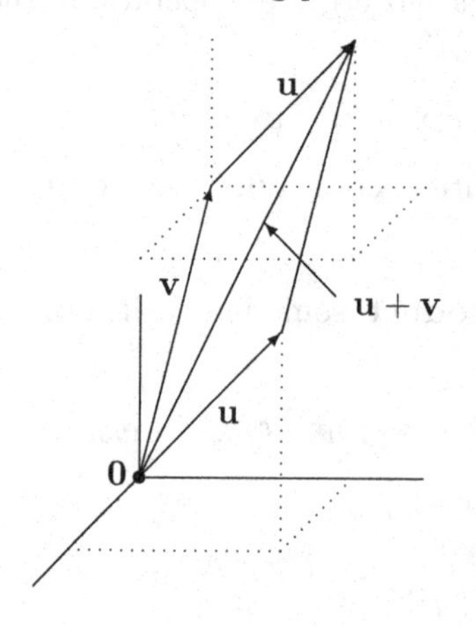

The points $\mathbf{0}, \mathbf{u}, \mathbf{v}, \mathbf{u}+\mathbf{v}$ are the vertices of a parallelogram.

Note that the directed segment from $\mathbf{x}$ to $\mathbf{y}$, which we write $[\mathbf{x}, \mathbf{y}]$, is obtained by 'translating' $\mathbf{y}-\mathbf{x}$, because

$$\mathbf{x} + (\mathbf{y} - \mathbf{x}) = \mathbf{y}.$$

Hence the distance $d(\mathbf{x}, \mathbf{y})$ from $\mathbf{x}$ to $\mathbf{y}$ is

$$d(\mathbf{x}, \mathbf{y}) = |\mathbf{y} - \mathbf{x}|.$$

Example 1.4. Find the coordinates of the point P, a fraction t $(0 < t \leq 1)$ of the way from $\mathbf{x}$ to $\mathbf{y}$ along $[\mathbf{x}, \mathbf{y}]$.

We observe that P is located at

$$\mathbf{x} + t(\mathbf{y} - \mathbf{x}) = t\mathbf{y} + (1-t)\mathbf{x}.$$

Example 1.5. The midpoint of $[\mathbf{x}, \mathbf{y}]$ is $\frac{1}{2}(\mathbf{x}+\mathbf{y})$ (take $t = 1/2$).

The **centroid** $\frac{\mathbf{a}+\mathbf{b}+\mathbf{c}}{3}$ of a triangle with vertices $\mathbf{a}, \mathbf{b}, \mathbf{c}$ lies on the line joining $\mathbf{a}$ to the midpoint $\frac{\mathbf{b}+\mathbf{c}}{2}$ of the opposite side. This is the point of intersection of the three lines in the following picture.

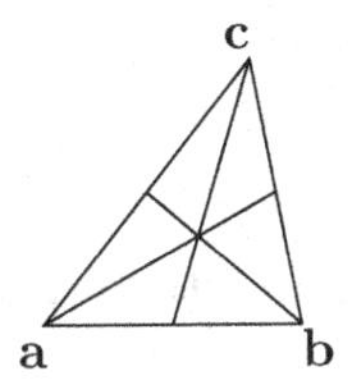

For

$$\frac{1}{3}\mathbf{a} + \frac{2}{3}\frac{(\mathbf{b}+\mathbf{c})}{2} = \frac{1}{3}(\mathbf{a}+\mathbf{b}+\mathbf{c}).$$

The same statement is true for the other vertices, giving three concurrent lines intersecting as shown.

Example 1.6. The set

$$L(\mathbf{a}, \mathbf{b}) = \{\mathbf{x} : \mathbf{x} = \mathbf{a} + t\mathbf{b},\ t \in \mathbb{R}\} \tag{1.5}$$

is a straight line through $\mathbf{a}$ pointing in the direction $\mathbf{b}$ (where $\mathbf{b} \neq \mathbf{0}$).

Example 1.7. Write in the form (1.5) the straight line through $(1, 0, -1)$ and $(4, 1, 2)$.

Solution. We can take $\mathbf{b} = (4, 1, 2) - (1, 0, -1) = (3, 1, 3)$. The line is $L(\mathbf{a}, \mathbf{b})$ with $\mathbf{a} = (1, 0, -1), \mathbf{b} = (3, 1, 3)$.

Observation 1.1. The inner product is geometrically interpreted as follows. Let θ be the angle between the directed segments $\mathbf{a}$ and $\mathbf{b}$, with $0 \leq \theta \leq \pi$. Then

$$\langle \mathbf{a}, \mathbf{b} \rangle = |\mathbf{a}||\mathbf{b}| \cos\theta. \tag{1.6}$$

To see this, we evaluate $l = |\mathbf{a} - \mathbf{b}|$ in two different ways.

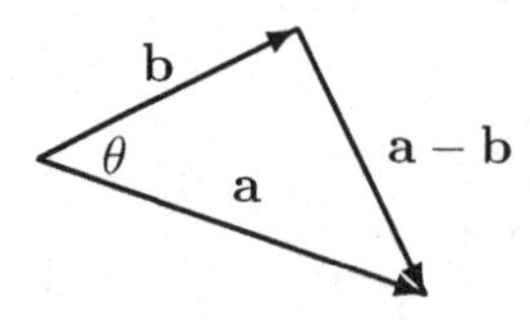

By a rule from trigonometry,

$$l^2 = |\mathbf{b} - \mathbf{a}|^2 = |\mathbf{a}|^2 + |\mathbf{b}|^2 - 2|\mathbf{a}||\mathbf{b}| \cos\theta. \tag{1.7}$$

On the other hand, recalling (1.1)–(1.3),

$$\begin{aligned} |\mathbf{b} - \mathbf{a}|^2 = \langle(\mathbf{b} - \mathbf{a}), (\mathbf{b} - \mathbf{a})\rangle &= \langle \mathbf{b}, \mathbf{b}\rangle - 2\langle \mathbf{a}, \mathbf{b}\rangle + \langle \mathbf{a}, \mathbf{a}\rangle \\ &= |\mathbf{b}|^2 - 2\langle \mathbf{a}, \mathbf{b}\rangle + |\mathbf{a}|^2. \end{aligned} \tag{1.8}$$

Comparing (1.7) and (1.8) yields (1.6).

Note that nonzero vectors $\mathbf{x}$ and $\mathbf{y}$ are perpendicular to each other, or **orthogonal**, if

$$\langle \mathbf{x}, \mathbf{y}\rangle = 0.$$

For then $\cos\theta = 0$, and $\theta = \pi/2$, in (1.6).

In particular, $\mathbf{x} \times \mathbf{y}$ is **orthogonal to** (perpendicular to) both $\mathbf{x}$ and $\mathbf{y}$. For example,

$$\langle \mathbf{x}, \mathbf{x} \times \mathbf{y}\rangle = x_1(x_2y_3 - x_3y_2) + x_2(-x_1y_3 + x_3y_1) + x_3(x_1y_2 - x_2y_1) = 0.$$

The length of $\mathbf{x} \times \mathbf{y}$ is

$$|\mathbf{x}|\ |\mathbf{y}| \sin\theta,$$

with θ as above. To see this,

$$|\mathbf{x} \times \mathbf{y}|^2 = (x_1y_2 - x_2y_1)^2 + (-x_1y_3 + x_3y_1)^2 + (x_1y_2 - x_2y_1)^2, \tag{1.9}$$

while

$$\begin{aligned} |\mathbf{x}|^2|\mathbf{y}|^2 \sin^2\theta &= |\mathbf{x}|^2|\mathbf{y}|^2 - |\mathbf{x}|^2|\mathbf{y}|^2 \cos^2\theta \\ [2mm] &= (x_1^2 + x_2^2 + x_3^2)(y_1^2 + y_2^2 + y_3^2) - (x_1y_1 + x_2y_2 + x_3y_3)^2, \end{aligned} \tag{1.10}$$

and the right-hand sides of (1.9), (1.10) are equal.

We close our short tour of geometry by considering a plane P through $\mathbf{a} = (a_1, a_2, a_3)$ with a **normal u**; that is, $\mathbf{u}$ is a vector perpendicular to P. Thus the vectors $\mathbf{x} - \mathbf{a}$ for $\mathbf{x} \in P$ are each perpendicular to the given normal vector $\mathbf{u}$ as indicated in the picture.

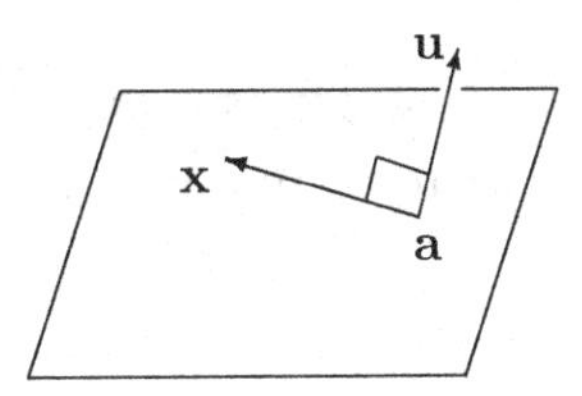

Thus the equation of the plane P is

$$\langle \mathbf{u}, \mathbf{x} - \mathbf{a} \rangle = u_1 (x_1 - a_1) + u_2 (x_2 - a_2) + u_3 (x_3 - a_3) = 0$$

Example 1.8. Find the equation of the plane P' through $\mathbf{a} = (1, 1, 2), \mathbf{b} = (2, 1, 3)$ and $\mathbf{c} = (5, -1, 5)$.

A normal to P' is

$$\mathbf{u} = (\mathbf{b} - \mathbf{a}) \times (\mathbf{c} - \mathbf{a}) = (2, 1, -2)$$

since $\mathbf{u}$ is perpendicular to the differences $\mathbf{b} - \mathbf{a}$ and $\mathbf{c} - \mathbf{a}$. Hence P' has equation

$$\langle (\mathbf{x} - \mathbf{a}), \mathbf{u} \rangle = 0,$$

reducing to

$$2x_1 + x_2 - 2x_3 = -1.$$

1.3 An n–dimensional setting

In the nineteenth century (and increasingly after 1900) mathematicians realized the value of abstraction. The idea is to treat a general class of algebraic or other mathematical structures 'all in one go'. A simple example is $\mathbb{R}^n$. This is the set of all ordered $n-$ tuples $(x_1, \ldots, x_n)$, with each $x_i \in \mathbb{R}$. We cannot illustrate $\mathbb{R}^n$on paper for $n \geq 4$, but even so, we can think of $\mathbb{R}$ as having a geometry. We call the $n-$ tuples **vectors**.

Definition 1.6. Let $\mathbf{x} = (x_1, \ldots, x_n)$ and $\mathbf{y} = (y_1, \ldots, y_n)$ be vectors in $\mathbb{R}^n$. We let

$$\mathbf{x} + \mathbf{y} = (x_1 + y_1, \ldots, x_n + y_n),$$
$$c\mathbf{x} = (cx_1, \ldots, cx_n) \text{ for } c \in \mathbb{R}.$$

Clearly $\mathbb{R}^1$ can be identified with $\mathbb{R}$. There is no simple way to extend the idea of cross product for $n > 3$. But we have no difficulty with inner product.

Definition 1.7. The inner product of $\mathbf{x}$ and $\mathbf{y}$ is

$$\langle \mathbf{x}, \mathbf{y} \rangle = x_1 y_1 + x_2 y_2 + \cdots + x_n y_n.$$

You can easily check that the rules (1.1)–(1.4) hold in this context. This product is also called the **dot product**.

We now define orthogonality.

Definition 1.8. Nonzero vectors $\mathbf{x}, \mathbf{y}$ in $\mathbb{R}^n$ are **orthogonal** if $\langle \mathbf{x}, \mathbf{y} \rangle = 0$. An **orthogonal set** is a set of nonzero vectors $\mathbf{x}_1, \ldots, \mathbf{x}_k$ with $\langle \mathbf{x}_i, \mathbf{x}_j \rangle = 0$ whenever $i \neq j$.

Definition 1.9. The **length** of $\mathbf{x}$ in $\mathbb{R}^n$ is

$$|\mathbf{x}| = \sqrt{x_1^2 + \cdots + x_n^2} = \sqrt{\langle \mathbf{x}, \mathbf{x} \rangle}.$$

The **distance** from $\mathbf{x}$ to $\mathbf{y}$ is $d(\mathbf{x}, \mathbf{y}) = |\mathbf{y} - \mathbf{x}|$.

Several questions arise at once. What is the largest number of vectors in an orthogonal set? (You should easily find an orthogonal set in $\mathbb{R}^n$ with n members.) Is the 'direct' route from $\mathbf{x}$ to $\mathbf{z}$ shorter than the route via $\mathbf{y}$, in the sense that

$$d(\mathbf{x}, \mathbf{z}) \leq d(\mathbf{x}, \mathbf{y}) + d(\mathbf{y}, \mathbf{z})? \tag{1.11}$$

Is it still true that

$$|\langle \mathbf{x}, \mathbf{y} \rangle| \leq |\mathbf{x}||\mathbf{y}| \tag{1.12}$$

(a fact obvious from (1.6) in $\mathbb{R}^3$)?

To show the power of abstraction, we hold off on the answers until we discuss a more general structure (inner product space) in Chapter 6. Why give detailed answers to these questions when we can get much more general results that include these answers? However, see the exercises for a preview of these topics.

1.4 Exercises

(1) Explain why the set $\left\{(x, y, z) \in \mathbb{R}^3 : (x-1)^2 + (y-2)^2 + z^2 = 4\right\}$, usually written simply as $(x-1)^2 + (y-2)^2 + z^2 = 4$, is a sphere of radius 2 which is centered at the point $(1, 2, 0)$.

(2) Given two points $(a, b, c), (x, y, z)$, show using the formula for distance between two points that for $t \in (0, 1)$,

$$\frac{|t(x-a, y-b, z-c)|}{|(x-a, y-b, z-c)|} = t.$$

Explain why this shows that the point on the line between these two points located at

$$(a, b, c) + t(x-a, y-b, z-c)$$

divides the segment joining the points in the ratio $t : 1-t$.

(3) Find the line joining the points $(1, 2, 3)$ and $(3, 5, 7)$. Next find the point on this line which is $1/3$ of the way from $(1, 2, 3)$ to $(3, 5, 7)$.

(4) A triangle has vertices $(1, 3), (3, 7), (5, 5)$. Find its centroid.

(5) There are many proofs of the Pythagorean theorem, which says the square of the hypotenuse equals the sum of the squares of the other two sides, $c^2 = a^2+b^2$ in the following right triangle.

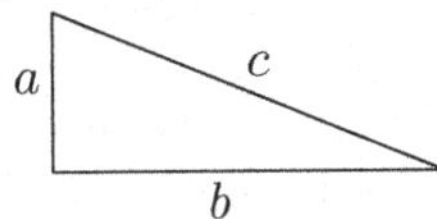

Here is a simple one.[1] It is based on writing the area of the following trapezoid two ways. Sum the areas of three triangles in the following picture or write the area of the trapezoid as $(a+b)\,a + \frac{1}{2}(a+b)(b-a)$, which is the sum of a triangle and a rectangle as shown. Do it both ways and see the pythagorean theorem appear.

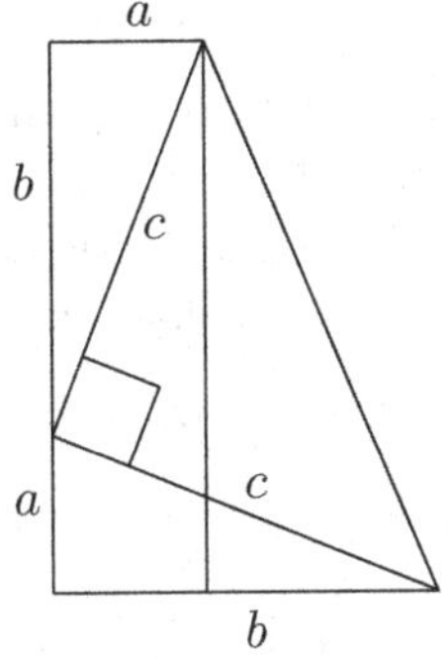

(6) Find the cosine of the angle between the two vectors $(2,1,3)$ and $(3,-1,2)$.

(7) Let (a,b,c) and (x,y,z) be two nonzero vectors in $\mathbb{R}^3$. Show from the properties of the dot product that

$$(a,b,c) - \frac{(a,b,c)\cdot(x,y,z)}{x^2+y^2+z^2}(x,y,z)$$

is perpendicular to (x,y,z).

(8) Prove the Cauchy-Schwarz inequality using the geometric description of the inner product in terms of the cosine of the included angle. This inequality states that $|\langle \mathbf{x},\mathbf{y}\rangle| \leq |\mathbf{x}|\,|\mathbf{y}|$. This geometrical argument is not the right way to prove this inequality but it is sometimes helpful to think of it this way.

(9) The right way to think of the Cauchy-Schwarz inequality for vectors $\mathbf{a}$ and $\mathbf{b}$ in $\mathbb{R}^n$ is in terms of the algebraic properties of the inner product. Using the properties of the inner product, (1.1)–(1.4) show that for a real number t,

$$0 \leq p(t) = \langle \mathbf{a}+t\mathbf{b}, \mathbf{a}+t\mathbf{b}\rangle = |\mathbf{a}|^2 + 2t\langle \mathbf{a},\mathbf{b}\rangle + t^2|\mathbf{b}|^2 .$$

Choose t in an auspicious manner to be the value which minimizes this nonnegative polynomial. Another way to get the Cauchy-Schwarz inequality is to note that this is a polynomial with either one real root or no real roots. Why? What does this observation say about the discriminant? (The discriminant is the b^2-4ac term under the radical in the quadratic formula.)

---

[1] This argument involving the area of a trapezoid is due to James Garfield, who was one of the presidents of the United States.

(10) Define a nonstandard inner product as follows.

$$\langle \mathbf{a}, \mathbf{b} \rangle = \sum_{j=1}^{n} w_j a_j b_j$$

where $w_j > 0$ for each j. Show that the properties of the inner product (1.1)–(1.4) continue to hold. Write down what the Cauchy-Schwarz inequality says for this example.

(11) The normal vector to a plane is $\mathbf{n} = (1, 2, 4)$ and a point on this plane is $(2, 1, 1)$. Find the equation of this plane.

(12) Find the equation of a plane which goes through the three points,

$$(1, 1, 2), (-1, 3, 0), (4, -2, 1).$$

(13) Find the equation of a plane which contains the line $(x, y, z) = (1, 1, 1) + t(2, 1, -1)$ and the point $(0, 0, 0)$.

(14) Let $\mathbf{a}, \mathbf{b}$ be two vectors in $\mathbb{R}^3$. Think of them as directed line segments which start at the same point. The parallelogram determined by these vectors is

$$P(\mathbf{a}, \mathbf{b}) = \left\{ \mathbf{x} \in \mathbb{R}^3 : \mathbf{x} = s_1 \mathbf{a} + s_2 \mathbf{b} \text{ where each } s_i \in (0, 1) \right\}.$$

Explain why the area of this parallelogram is $|\mathbf{a} \times \mathbf{b}|$.

(15) Let $\mathbf{a}, \mathbf{b}, \mathbf{c}$ be three vectors in $\mathbb{R}^3$. Think of them as directed line segments which start at $\mathbf{0}$. The parallelepiped determined by these three vectors is the following.

$$P(\mathbf{a}, \mathbf{b}, \mathbf{c}) = \left\{ \mathbf{x} \in \mathbb{R}^3 : \mathbf{x} = s_1 \mathbf{a} + s_2 \mathbf{b} + s_3 \mathbf{c} \text{ where each } s_i \in (0, 1) \right\}$$

A picture of a parallelepiped follows.

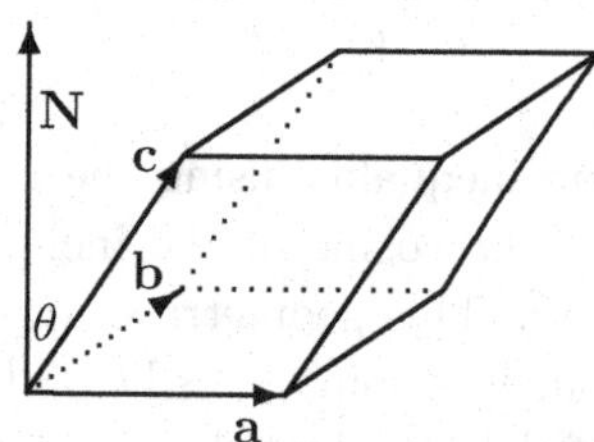

Note that the normal $\mathbf{N}$ is perpendicular to both $\mathbf{a}$ and $\mathbf{b}$. Explain why the volume of the parallelepiped determined by these vectors is $|(\mathbf{a} \times \mathbf{b}) \cdot \mathbf{c}|$.

(16) There is more to the cross product than what is mentioned earlier. The vectors $\mathbf{a}, \mathbf{b}, \mathbf{a} \times \mathbf{b}$ satisfies a right hand rule. This means that if you place the fingers of your **right hand** in the direction of $\mathbf{a}$ and wrap them towards $\mathbf{b}$, the thumb of your right hand points in the direction of $\mathbf{a} \times \mathbf{b}$ as shown in the following picture.

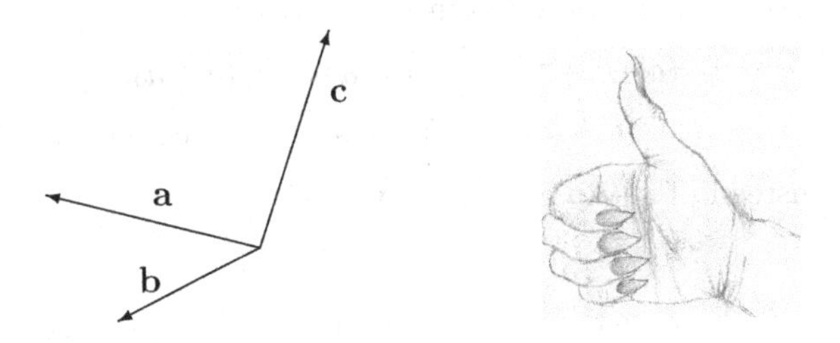

Show that with the definition of the cross product given above, $\mathbf{a}, \mathbf{b}, \mathbf{a} \times \mathbf{b}$ always satisfies the right hand rule whenever each of $\mathbf{a}, \mathbf{b}$ is one of the standard basis vectors $\mathbf{e}_1, \mathbf{e}_2, \mathbf{e}_3$.

(17) Explain why any relation of the form $\{(x, y, z) : ax + by + cz = d\}$ where $a^2 + b^2 + c^2 \neq 0$ is always a plane. Explain why $\langle a, b, c \rangle$ is normal to the plane. Usually we write the above set in the form $ax + by + cz = d$.

(18) Two planes in $\mathbb{R}^3$are said to be parallel if there exists a single vector $\mathbf{n}$ which is normal to both planes. Find a plane which is parallel to the plane $2x+3y-z = 7$ and contains the point $(1, 0, 2)$.

(19) Let $ax + by + cz = d$ be an equation of a plane. Thus $a^2 + b^2 + c^2 \neq 0$. Let $\mathbf{X}_0 = (x_0, y_0, z_0)$ be a point not on the plane. The point of the plane which is closest to the given point is obtained by taking the intersection of the line through (x_0, y_0, z_0) which is in the direction of the normal to the plane and finding the distance between this point of intersection and the given point. Show that this distance equals $\frac{|ax_0+by_0+cz_0-d|}{\sqrt{a^2+b^2+c^2}}$.

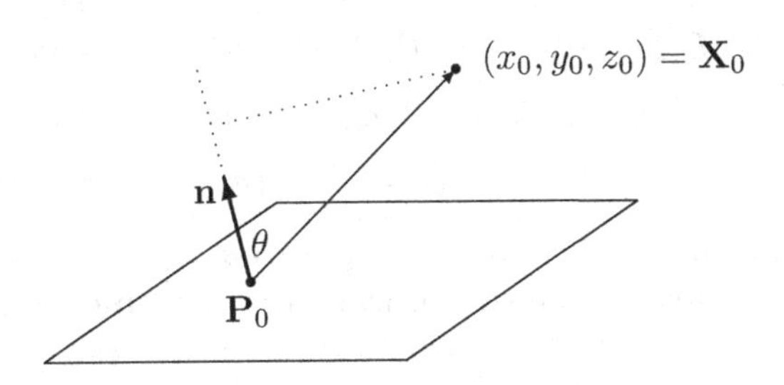

1.5 Complex numbers

It is very little work (and highly useful) to step up the generality of our studies by replacing real numbers by complex numbers. Many mathematical phenomena (in the theory of functions, for example) become clearer when we 'enlarge' $\mathbb{R}$. Let $\mathbb{C}$ be the set of all expressions $a + bi$, where i is an abstract symbol, $a \in \mathbb{R}, b \in \mathbb{R}$, with the following rules of addition and multiplication:

$$(a + bi) + (c + di) = a + c + (b + d)i,$$
$$(a + bi)(c + di) = ac - bd + (bc + ad)i.$$

We write a instead of $a + 0i$ and identify $\mathbb{R}$ with the set $\{a + 0i : a \in \mathbb{R}\}$, so that

$$\mathbb{R} \subset \mathbb{C}.$$

Note that if we ignore the multiplication, $\mathbb{C}$ has the same structure as $\mathbb{R}^2$, giving us a graphical representation of the set of **complex numbers** $\mathbb{C}$.

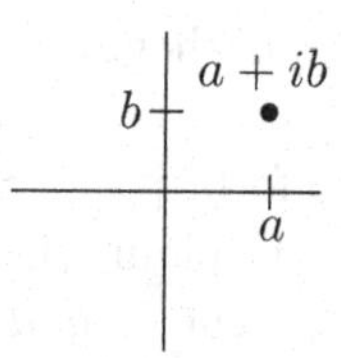

We write ci instead of $0 + ci$. Note that from the above definition of multiplication, $i^2 = -1$. Thus the equation

$$x^2 + 1 = 0, \tag{1.13}$$

which has no solution in $\mathbb{R}$, does have solutions $i, -i$ in $\mathbb{C}$. There is a corresponding factorization

$$x^2 + 1 = (x + i)(x - i).$$

In fact, it is shown in complex analysis that any polynomial of **degree** n, (written $\deg P = n$),

$$P(x) = a_0 x^n + a_1 x^{n-1} + \cdots + a_{n-1}x + a_n$$

with $a_0 \neq 0$, and all coefficients a_j in $\mathbb{C}$, can be factored into linear factors:

$$P(z) = a_0(z - z_1) \ldots (z - z_n)$$

for some $z_1, \ldots, z_n$ in $\mathbb{C}$. The z_j are the **zeros** of P, that is, the solutions of $P(z) = 0$. The existence of this factorization is the **fundamental theorem of algebra**. Maple or some other computer algebra system can sometimes be used to find $z_1, \ldots, z_n$ if no simple method presents itself.

Example 1.9. To find the zeros of a polynomial, say $z^3 - 1$, enter it as

$$\text{Solve } (x \wedge 3 - 1 = 0, x)$$

and type 'Enter'. We get

$$1, -1/2 + I\frac{1}{2}\sqrt{3}, -1/2 - I\frac{1}{2}\sqrt{3}.$$

(Maple uses I for i). If we type in $x^3 - 1$ and right click, select 'factor', we get

$$(x - 1)(x^2 + x + 1).$$

This leaves to you the simple step of solving a real quadratic equation.

To perform a division, for example,

$$\frac{a+bi}{c+di} = \frac{(a+bi)(c-di)}{(c+di)(c-di)} = \frac{ac+bd}{c^2+d^2} + i\left(\frac{bc-ad}{c^2+d^2}\right).$$

Here of course $z = z_1/z_2$ means that

$$z_2 z = z_1.$$

(We suppose $z_2 \neq 0$.) The **absolute value** $|z|$ of $z = a+bi$ is $\sqrt{a^2+b^2}$ (its length, if we identify $\mathbb{C}$ with $\mathbb{R}^2$). The **complex conjugate** of $z = a+ib$ is $\bar{z} = a - ib$. You can easily verify the properties

$$\overline{z_1+z_2} = \overline{z_1} + \overline{z_2}, \quad \overline{z_1 z_2} = \overline{z_1}\,\overline{z_2}, \quad z\bar{z} = |z|^2,$$

and writing

$$\Re(z) = \Re(a+ib) = a, \ \Im(z) = \Im(a+ib) = b,$$

$$\frac{1}{2}(z+\bar{z}) = \Re(z), \ \frac{1}{2i}(z-\bar{z}) = \Im(z),$$

$\Re(z)$ is the 'real part of z' $\Im(z)$ is the 'imaginary part of z'.

A simple consequence of the above is $|z_1 z_2| = |z_1||z_2|$. (The square of each side is $z_1\overline{z_1}z_2\overline{z_2}$.)

Now $\mathbb{C}^n$ is defined to be the set of

$$\mathbf{z} = (z_1, \ldots, z_n) \qquad (z_j \in \mathbb{C}).$$

You can easily guess the definition of addition and scalar multiplication for $\mathbb{C}^n$:

$$\begin{aligned} \mathbf{w} &= (w_1, \ldots, w_n), \\ \mathbf{z} + \mathbf{w} &= (z_1 + w_1, \ldots, z_n + w_n), \\ c\mathbf{z} &= (cz_1, \ldots, cz_n) \qquad (c \in \mathbb{C}). \end{aligned}$$

The definition of the inner product in $\mathbb{C}^n$ is less obvious, namely

$$\langle \mathbf{z}, \mathbf{w} \rangle = z_1\overline{w_1} + \cdots + z_n\overline{w_n}.$$

The reason for this definition is that it carries over the property that $\langle \mathbf{z}, \mathbf{z} \rangle > 0$. We have

$$\langle \mathbf{z}, \mathbf{z} \rangle > 0 \text{ if } \mathbf{z} \neq \mathbf{0},$$

because

$$\langle \mathbf{z}, \mathbf{z} \rangle = |z_1|^2 + \cdots + |z_n|^2.$$

We shall see in Chapter 6 that $\mathbb{C}^n$ is an example of a **complex inner product space**.

1.6 Fields

To avoid pointless repetition, we step up the level of abstraction again. Both $\mathbb{R}$ and $\mathbb{C}$ are examples of **fields**. Linear algebra can be done perfectly well in a setting where the scalars are the elements of some given field, say F. In this book, we will use the following notation:

Notation 1.2. The symbol $\square$, invented by Paul Halmos, shows the end of a proof.

Definition 1.10. A **field** is a set F containing at least two distinct members, which we write as 0 and 1. Moreover, there is a rule for forming the sum $a+b$ and the product ab whenever a, b are in F. We require further that, for any a, b, c in F,

(1) $a+b$ and ab are in F.
(2) $a+b=b+a$ and $ab=ba$.
(3) $a+(b+c)=(a+b)+c$ and $a(bc)=(ab)c$.
(4) $a(b+c)=ab+ac$
(5) $a+0=a$.
(6) $a1=a$.
(7) For each a in F, there is a member $-a$ of F such that

$$a+(-a)=0.$$

(8) For each a in $F, a \neq 0$, there is a member a^{-1} of F such that $aa^{-1}=1$.

It is well known that $F=\mathbb{R}$ has all these properties, and it is not difficult to check that $F=\mathbb{C}$ has them too.

The following proposition follows easily from the above axioms.

Proposition 1.1. The additive identity 0 is unique. That is if $0'+a=a$ for all a, then $0'=0$. The multiplicative identity 1 is unique. That is if $a1'=a$ for all a, then $1'=1$. Also if $a+b=0$ then $b=-a$, so the additive inverse is unique. Also if $a \neq 0$ and $ab=1$, then $b=a^{-1}$, so the multiplicative inverse is unique. Also $0a=0$ and $-a=(-1)a$.

Proof. (We will not cite the uses of (1)–(8).) First suppose $0'$ acts like 0. Then

$$0'=0'+0=0.$$

Next suppose $1'$ acts like 1. Then

$$1'=1'1=1.$$

If $a+b=0$ then add $-a$ to both sides. Then

$$b=(-a+a)+b=-a+(a+b)=-a.$$

If $ab=1$, then

$$a^{-1}=a^{-1}(ab)=\left(a^{-1}a\right)b=1b=b.$$

Next, it follows from the axioms that

$$0a = (0+0)\,a = 0a + 0a.$$

Adding $-(0a)$ to both sides, it follows that $0 = 0a$. Finally,

$$a + (-1)\,a = 1a + (-1)\,a = (1 + -1)\,a = 0a = 0,$$

and so from the uniqueness of the additive inverse, $-a = (-1)\,a$. This proves the proposition. □

We give one further example, which is useful in number theory and cryptography.

Example 1.10. Congruence classes.

Let m be a fixed positive integer. We introduce **congruence classes** $\bar{0}, \bar{1}$, $\bar{2}, \ldots, \overline{m-1}$ in the following way; $\bar{k}$ consists of all integers that leave remainder k on division by m. That is, $l \in \bar{k}$ means

$$l = mj + k$$

for some integer j. The congruence classes are also called **residue classes**. If h is an integer with $h \in \bar{k}$, we write $\bar{h} = \bar{k}$, e.g. $\overline{26} = \bar{2}$ if $m = 8$.

Thus if $m = 8$,

$$\bar{5} = \{\ldots, -11, -3, 5, 13, 21, \ldots\}.$$

To add or multiply congruence classes, let $\bar{a} + \bar{b} = \overline{a+b}$, $\bar{a}\bar{b} = \overline{ab}$.

For instance, if $m = 11$, then $\bar{9} + \bar{7} = \bar{5}, \bar{9}\bar{7} = \bar{8}$. It is not hard to see that the set $\mathbb{Z}_m = \{\bar{0}, \bar{1}, \ldots, \overline{m-1}\}$ obeys the rules (1) - (7) above. To get the rule (8), we must assume that m is prime. This is seen to be necessary, e.g. if $m = 10$, we will never have $\bar{5}\bar{k} = \bar{1}$ since $\bar{5}\bar{k} \in \{\bar{0}, \bar{5}\}$. For a prime p, we write F_p instead of $\mathbb{Z}_p$. See Herstein (1964) for a detailed proof that F_p is a field. See also the Exercises 25 - 27 on Page 22. We display in Figure 1.7 a multiplication table for F_7. It is clear from the table that each nonzero member of F_7 has an inverse.

	$\bar{1}$	$\bar{2}$	$\bar{3}$	$\bar{4}$	$\bar{5}$	$\bar{6}$
$\bar{1}$	1	2	3	4	5	6
$\bar{2}$	2	4	6	1	3	5
$\bar{3}$	3	6	2	5	1	4
$\bar{4}$	4	1	5	2	6	3
$\bar{5}$	5	3	1	6	4	2
$\bar{6}$	6	5	4	3	2	1

Fig. 1.7 Multiplication table for F_7.

It is convenient to define the dot product as in Definition 1.7 for any two vectors in F^n for F an arbitrary field. That is, for $\mathbf{x} = (x_1, \cdots, x_n)\,, \mathbf{y} = (y_1, \cdots, y_n)\,,$ the dot product $\mathbf{x} \cdot \mathbf{y}$ is given by

$$\mathbf{x} \cdot \mathbf{y} = x_1 y_1 + \cdots + x_n y_n$$

1.7 Ordered fields

The real numbers $\mathbb{R}$ are an example of an ordered field. More generally, here is a definition.

Definition 1.11. Let F be a field. It is an ordered field if there exists an order, $<$ which satisfies

(1) For any $x \neq y$, either $x < y$ or $y < x$.
(2) If $x < y$ and either $z < w$ or $z = w$, then, $x + z < y + w$.
(3) If $0 < x, 0 < y$, then $xy > 0$.

With this definition, the familiar properties of order can be proved. The following proposition lists many of these familiar properties. The relation '$a > b$' has the same meaning as '$b < a$'.

Proposition 1.2. The following are obtained.

(1) If $x < y$ and $y < z$, then $x < z$.
(2) If $x > 0$ and $y > 0$, then $x + y > 0$.
(3) If $x > 0$, then $-x < 0$.
(4) If $x \neq 0$, either x or $-x$ is > 0.
(5) If $x < y$, then $-x > -y$.
(6) If $x \neq 0$, then $x^2 > 0$.
(7) If $0 < x < y$ then $x^{-1} > y^{-1}$.

Proof. First consider (1), called the transitive law. Suppose that $x < y$ and $y < z$. Then from the axioms, $x + y < y + z$ and so, adding $-y$ to both sides, it follows that $x < z$.

Next consider (2). Suppose $x > 0$ and $y > 0$. Then from (2),

$$0 = 0 + 0 < x + y.$$

Next consider (3). It is assumed $x > 0$, so

$$0 = -x + x > 0 + (-x) = -x.$$

Now consider (4). If $x < 0$, then

$$0 = x + (-x) < 0 + (-x) = -x.$$

Consider (5). Since $x < y$, it follows from (2) that

$$0 = x + (-x) < y + (-x),$$

and so by (4) and Proposition 1.1,

$$(-1)(y + (-x)) < 0.$$

Also from Proposition 1.1 $(-1)(-x) = -(-x) = x$, and so

$$-y + x < 0.$$

Hence

$$-y < -x.$$

Consider (6). If $x > 0$, there is nothing to show. It follows from the definition. If $x < 0$, then by (4), $-x > 0$ and so by Proposition 1.1 and the definition of the order,

$$(-x)^2 = (-1)(-1)x^2 > 0.$$

By this proposition again, $(-1)(-1) = -(-1) = 1$, and so $x^2 > 0$ as claimed. Note that $1 > 0$ because $1 = 1^2$.

Finally, consider (7). First, if $x > 0$ then if $x^{-1} < 0$, it would follow that $(-1)x^{-1} > 0$, and so $x(-1)x^{-1} = (-1)1 = -1 > 0$. However, this would require

$$0 > 1 = 1^2 > 0$$

from what was just shown. Therefore, $x^{-1} > 0$. Now the assumption implies $y + (-1)x > 0$, and so multiplying by x^{-1},

$$yx^{-1} + (-1)xx^{-1} = yx^{-1} + (-1) > 0.$$

Now multiply by y^{-1}, which by the above satisfies $y^{-1} > 0$, to obtain

$$x^{-1} + (-1)y^{-1} > 0,$$

and so

$$x^{-1} > y^{-1}.$$

This proves the proposition. □

In an ordered field the symbols $\leq$ and $\geq$ have the usual meanings. Thus $a \leq b$ means $a < b$ or else $a = b$, etc.

1.8 Division

The **Archimedean property** says that for any $a > 0$, and $b \geq 0$ there exists a positive integer n such that $b < na$. We assume the Archimedean property of the real numbers. We also assume the **well ordering** property of the positive integers which says that if S is a nonempty set of positive integers, then it contains a unique smallest number. These two properties imply the following division theorem.

Theorem 1.1. *Suppose $0 < a$ and let $b \geq 0$. Then there exists a unique integer p and real number r such that $0 \leq r < a$ and $b = pa + r$.*

Proof. Let $S \equiv \{n \in \mathbb{N} : an > b\}$. By the Archimedian property this set is nonempty. Let $p + 1$ be the smallest element of S. Then $pa \leq b$ because $p + 1$ is the smallest in S. Therefore,

$$r \equiv b - pa \geq 0.$$

If $r \geq a$ then $b - pa \geq a$, and so $b \geq (p+1)\, a$ contradicting $p + 1 \in S$. Therefore, $r < a$ as desired.

To verify uniqueness of p and r, suppose p_i and r_i, $i = 1, 2$, both work and $r_2 > r_1$. Then a little algebra shows

$$p_1 - p_2 = \frac{r_2 - r_1}{a} \in (0, 1)\,.$$

Thus $p_1 - p_2$ is an integer between 0 and 1 and there are none of these. The case that $r_1 > r_2$ cannot occur either by similar reasoning. Thus $r_1 = r_2$ and it follows that $p_1 = p_2$. □

A geometric description of the conclusion of the above theorem is to say that the succession of disjoint intervals $[0, a), [a, 2a), [2a, 3a), \cdots$ includes all nonnegative real numbers. Here is a picture of some of these intervals.

$0a \quad 1a \quad 2a \quad 3a \quad 4a$

Thus an arbitrary nonnegative real number b is in exactly one such interval. $[pa, (p+1)\, a)$ where p is some nonnegative integer. Thus $b = pa + (b - pa)$ where $r = b - pa$ is less that a because $(p+1)a > b \geq pa$.

This theorem is called the Euclidean algorithm when a and b are integers. In this case, you have all the numbers are positive integers.

That which you can do for integers often can be modified and done to polynomials because polynomials behave a lot like integers. You add and multiply polynomials using the distributive law and then combine terms which have the same power of x using the standard rules of exponents. Thus

$$\left(2x^2 + 5x - 7\right) + \left(3x^3 + x^2 + 6\right) = 3x^2 + 5x - 1 + 3x^3,$$

and

$$\left(2x^2 + 5x - 7\right)\left(3x^3 + x^2 + 6\right) = 6x^5 + 17x^4 + 5x^2 - 16x^3 + 30x - 42.$$

The second assertion is established as follows. From the distributive law,

$$\begin{aligned} & \left(2x^2 + 5x - 7\right)\left(3x^3 + x^2 + 6\right) \\ = \; & \left(2x^2 + 5x - 7\right) 3x^3 + \left(2x^2 + 5x - 7\right) x^2 + \left(2x^2 + 5x - 7\right) 6 \\ = \; & 2x^2\left(3x^3\right) + 5x\left(3x^3\right) - 7\left(3x^3\right) + 2x^2\left(x^2\right) \\ & + 5x\left(x^2\right) - 7\left(x^2\right) + 12x^2 + 30x - 42 \end{aligned}$$

which simplifies to the claimed result. Note that $x^2x^3 = x^5$ because the left side simply says to multiply x by itself 5 times. Other axioms satisfied by the integers are also satisfied by polynomials and like integers, polynomials typically don't have multiplicative inverses which are polynomials. In this section the polynomials have coefficients which come from a field. This field is usually $\mathbb{R}$ in calculus but it doesn't have to be.

The following is the Euclidean algorithm for polynomials. This is a lot like the Euclidean algorithm for numbers, Theorem 1.1. Here is the definition of the degree of a polynomial.

Definition 1.12. Let $a_n x^n + \cdots + a_1 x + a_0$ be a polynomial. The **degree** of this polynomial is n if $a_n \neq 0$. The degree of a polynomial is the largest exponent on x provided the polynomial does not have all the $a_i = 0$. If each $a_i = 0$, we don't speak of the degree because it is not defined. In writing this, it is only assumed that the coefficients a_i are in some field such as the real numbers or the rational numbers. Two polynomials are defined to be equal when their degrees are the same and corresponding coefficients are the same.

Theorem 1.2. *Let $f(x)$ and $g(x) \neq 0$ be polynomials with coefficients in a some field. Then there exists a polynomial, $q(x)$ such that*

$$f(x) = q(x) g(x) + r(x)$$

where the degree of $r(x)$ is less than the degree of $g(x)$ or $r(x) = 0$. All these polynomials have coefficients in the same field. The two polynomials $q(x)$ and $r(x)$ are unique.

Proof. Consider the polynomials of the form $f(x) - g(x) l(x)$ and out of all these polynomials, pick one which has the smallest degree. This can be done because of the well ordering of the natural numbers. Let this take place when $l(x) = q_1(x)$ and let

$$r(x) = f(x) - g(x) q_1(x).$$

It is required to show that the degree of $r(x) <$ degree of $g(x)$ or else $r(x) = 0$.

Suppose $f(x) - g(x) l(x)$ is never equal to zero for any $l(x)$. Then $r(x) \neq 0$. It is required to show the degree of $r(x)$ is smaller than the degree of $g(x)$. If this doesn't happen, then the degree of $r \geq$ the degree of g. Let

$$\begin{aligned} r(x) &= b_m x^m + \cdots + b_1 x + b_0 \\ g(x) &= a_n x^n + \cdots + a_1 x + a_0 \end{aligned}$$

where $m \geq n$ and b_m and a_n are nonzero. Then let $r_1(x)$ be given by

$$r_1(x) = r(x) - \frac{x^{m-n} b_m}{a_n} g(x)$$

$$= (b_m x^m + \cdots + b_1 x + b_0) - \frac{x^{m-n} b_m}{a_n} (a_n x^n + \cdots + a_1 x + a_0)$$

which has smaller degree than m, the degree of $r(x)$. But

$$\begin{aligned} r_1(x) &= \overbrace{f(x) - g(x) q_1(x)}^{r(x)} - \frac{x^{m-n} b_m}{a_n} g(x) \\ &= f(x) - g(x) \left(q_1(x) + \frac{x^{m-n} b_m}{a_n} \right), \end{aligned}$$

and this is not zero by the assumption that $f(x) - g(x) l(x)$ is never equal to zero for any $l(x)$, yet has smaller degree than $r(x)$, which is a contradiction to the choice of $r(x)$.

It only remains to verify that the two polynomials $q(x)$ and $r(x)$ are unique. Suppose $q'(x)$ and $r'(x)$ satisfy the same conditions as $q(x)$ and $r(x)$. Then

$$(q(x) - q'(x)) g(x) = r'(x) - r(x)$$

If $q(x) \neq q'(x)$, then the degree of the left is greater than the degree of the right. Hence the equation cannot hold. It follows that $q'(x) = q(x)$ and $r'(x) = r(x)$. This proves the Theorem. □

1.9 Exercises

(1) Find the following. Write each in the form $a + ib$.
   (a) $(2+i)(4-2i)^{-1}$
   (b) $(1+i)(2+3i)^{-1} + 4 + 6i$
   (c) $(2+i)^2$

(2) Let $z = 5 + i9$. Find z^{-1}.

(3) Let $z = 2 + i7$ and let $w = 3 - i8$. Find $zw, z + w, z^2$, and w/z.

(4) Give the complete solution to $x^4 + 16 = 0$.

(5) Show that for z, w complex numbers,

$$\overline{zw} = \overline{z}\,\overline{w},\ \overline{z+w} = \overline{z} + \overline{w}.$$

Explain why this generalizes to any finite product or any finite sum.

(6) Suppose $p(x)$ is a polynomial with real coefficients. That is

$$p(x) = a_n x^n + a_{n-1} x^{n-1} + \cdots + a_1 x + a_0, \text{ each } a_k \text{ real.}$$

As mentioned above, such polynomials may have complex roots. Show that if $p(z) = 0$ for $z \in \mathbb{C}$, then $p(\overline{z}) = 0$ also. That is, the roots of a real polynomial come in conjugate pairs.

(7) Let $z = a + ib$ be a complex number. Show that there exists a complex number $w = c + id$ such that $|w| = 1$ and $wz = |z|$. **Hint:** Try something like $\overline{z}/|z|$ if $z \neq 0$.

(8) Show that there can be no order which makes $\mathbb{C}$ into an ordered field. What about F_3? **Hint:** Consider i. If there is an order for $\mathbb{C}$ then either $i > 0$ or $i < 0$.

(9) The lexicographic order on $\mathbb{C}$ is defined as follows. $a + ib < x + iy$ means that $a < x$ or if $a = x$, then $b < y$. Why does this not give an order for $\mathbb{C}$? What exactly goes wrong? If S consists of all real numbers in the open interval $(0, 1)$ show that each of $1 - in$ is an upper bound for S in this lexicographic order. Therefore, there is no least upper bound although there is an upper bound.

(10) Show that if $a+ib$ is a non zero complex number, then there exists a unique r and $\theta \in [0, 2\pi)$ such that

$$a+ib = r(\cos(\theta) + i\sin(\theta)).$$

Hint: Try $r = \sqrt{a^2+b^2}$ and observe that $(a/r, b/r)$ is a point on the unit circle.

(11) Show that if $z \in \mathbb{C}$, $z = r(\cos(\theta) + i\sin(\theta))$, then

$$\overline{z} = r(\cos(-\theta) + i\sin(-\theta)).$$

(12) Show that if $z = r_1(\cos(\theta) + i\sin(\theta))$ and $w = r_2(\cos(\alpha) + i\sin(\alpha))$ are two complex numbers, then

$$zw = r_1 r_2 (\cos(\theta+\alpha) + i\sin(\theta+\alpha)).$$

(13) Prove DeMoivre's theorem which says that for any integer n,

$$(\cos(\theta) + i\sin\theta)^n = \cos(n\theta) + i\sin(n\theta).$$

Thus $(r(\cos\theta + i\sin\theta))^n = r^n(\cos(n\theta) + i\sin(n\theta))$. In particular if z is on the circle of radius r, then z^n is on a circle of radius r^n.

(14) Suppose you have any polynomial in $\cos\theta$ and $\sin\theta$. By this we mean an expression of the form

$$\sum_{\alpha=0}^{m}\sum_{\beta=0}^{n} a_{\alpha\beta}\cos^{\alpha}\theta\sin^{\beta}\theta$$

where $a_{\alpha\beta} \in \mathbb{C}$. Can this always be written in the form

$$\sum_{\gamma=-(n+m)}^{m+n} b_{\gamma}\cos\gamma\theta + \sum_{\tau=-(n+m)}^{n+m} c_{\tau}\sin\tau\theta?$$

Explain.

(15) Using DeMoivre's theorem, show that every nonzero complex number has exactly k k^{th} roots. **Hint:** The given complex number is

$$r(\cos(\theta) + i\sin(\theta)).$$

where $r > 0$. Then if $\rho(\cos(\alpha) + i\sin(\alpha))$ is a kth root, then by DeMoivre's theorem,

$$\rho^k = r,\ \cos(k\alpha) + i\sin(k\alpha) = \cos(\theta) + i\sin(\theta).$$

What is the necessary relationship between $k\alpha$ and θ?

(16) Factor x^3+8 as a product of linear factors.

(17) Write x^3+27 in the form $(x+3)(x^2+ax+b)$ where x^2+ax+b cannot be factored any more using only real numbers.

(18) Completely factor x^4+16 as a product of linear factors.

(19) Factor x^4+16 as the product of two quadratic polynomials each of which cannot be factored further without using complex numbers.

(20) It is common to see i referred to as $\sqrt{-1}$. Let's use this definition. Then
$$-1 = i^2 = \sqrt{-1}\sqrt{-1} = \sqrt{1} = 1,$$
so adding 1 to both ends, it follows that $0 = 2$. This is a remarkable assertion, but is there something wrong here?

(21) Give the complete solution to $x^4 + 16 = 0$.

(22) Graph the complex cube roots of 8 in the complex plane. Do the same for the four fourth roots of 16.

(23) This problem is for people who have had a calculus course which mentions the completeness axiom, every nonempty set which is bounded above has a least upper bound and every nonempty set which is bounded below has a greatest lower bound. Using this, show that for every $b \in \mathbb{R}$ and $a > 0$ there exists $m \in \mathbb{N}$, the natural numbers $\{1, 2, \cdots\}$ such that $ma > b$. This says $\mathbb{R}$ is Archimedean. **Hint:** If this is not so, then b would be an upper bound for the set $S = \{ma : m \in \mathbb{N}\}$. Consider the least upper bound g and argue from the definition of least upper bound there exists $ma \in \left(g - \frac{a}{2}, g\right)$. Now what about $(m + 1)\,a$?

(24) Verify Lagrange's identity which says that
$$\left(\sum_{i=1}^{n} a_i^2\right)\left(\sum_{i=1}^{n} b_i^2\right) - \left|\sum_{i} a_i b_i\right|^2 = \sum_{i<j} (a_i b_j - a_j b_i)^2$$
Now use this to prove the Cauchy Schwarz inequality in $\mathbb{R}^n$.

(25) Show that the operations of $+$ and multiplication on congruence classes are well defined. The definition says
$$\overline{a}\overline{b} = \overline{ab}.$$
If $\overline{a} = \overline{a_1}$ and $\overline{b} = \overline{b_1}$, is it the case that
$$\overline{a_1 b_1} = \overline{ab}?$$
A similar question must be considered for addition of congruence classes. Also verify the field axioms (1) - (7).

(26) An integer a is said to divide b, written $a|b$, if for some integer m,
$$b = ma.$$
The greatest common divisor of two non zero integers a, b is the positive integer p which divides them both and has the property that every integer which does divide both also divides p. Show that this greatest common divisor p exists, and if a, b are any two positive integers, there exist integers m, n such that
$$p = ma + nb.$$
The greatest common divisor of a and b is often denoted as (a, b). **Hint:** Consider $S = \{m, n \in \mathbb{Z} : ma + nb > 0\}$. Then S is nonempty. Why? Let q be the smallest element of S with associated integers m, n. If q does not divide

a then, by Theorem 1.1, $a = qp + r$ where $0 < r < q$. Solve for r and show that $r \in S$, contradicting the property of q which says it is the smallest integer in S. Since $q = ma + nb$, argue that if some l divides both a, b then it must divide q. Thus $q = p$.

(27) A prime number is a postive integer, larger than 1 with the property that the only positive integers which divide it are 1 and itself. Explain why the property (8) holds for F_p with p a prime. **Hint:** Recall that $F_p = \left\{\bar{0}, \bar{1}, \cdots, \overline{p-1}\right\}$. If $\overline{m} \in F_p$, $\overline{m} \neq \bar{0}$, then $(m, p) = 1$. Why? Therefore, there exist integers s, t such that

$$1 = sm + tp.$$

Explain why this implies

$$\bar{1} = \bar{s}\overline{m}.$$

(28) ↑Prove Wilson's theorem. This theorem states that if p is a prime, then $(p-1)! + 1$ is divisible by p. Wilson's theorem was first proved by Lagrange in the 1770's. **Hint:** Check directly for $p = 2, 3$. Show that $\overline{p-1} = -\bar{1}$ and that if $a \in \{2, \cdots, p-2\}$, then $(\bar{a})^{-1} \neq \bar{a}$. Thus a residue class $\bar{a}$ and its multiplicative inverse for $a \in \{2, \cdots, p-2\}$ occur in pairs. Show that this implies that the residue class of $(p-1)!$ must be -1. From this, draw the conclusion.

(29) If $p(x)$ and $q(x)$ are two nonzero polynomials having coefficients in F a field of scalars, the greatest common divisor of $p(x)$ and $q(x)$ is a monic polynomial $l(x)$ (Of the form $x^n + a_{n-1}x^{n-1} + \cdots + a_1x + a_0$) which has the property that it divides both $p(x)$ and $q(x)$, and that if $r(x)$ is any polynomial which divides both $p(x)$ and $q(x)$, then $r(x)$ divides $l(x)$. Show that the greatest common divisor is unique and that if $l(x)$ is the greatest common divisor, then there exist polynomials $n(x)$ and $m(x)$ such that $l(x) = m(x)p(x) + n(x)q(x)$.

(30) Give an example of a nonzero polynomial having coefficients in the field F_2 which sends every residue class of F_2 to 0. Now describe all polynomials having coefficients in F_3 which send every residue class of F_3 to 0.

(31) Show that in the arithmetic of F_p, $(\bar{x} + \bar{y})^p = (\bar{x})^p + (\bar{y})^p$, a well known formula among students.

(32) Using Problem 27 above, consider $(\bar{a}) \in F_p$ for p a prime, and suppose $(\bar{a}) \neq \bar{1}, \bar{0}$. Fermat's little theorem says that $(\bar{a})^{p-1} = \bar{1}$. In other words $(a)^{p-1} - 1$ is divisible by p. Prove this. **Hint:** Show that there must exist $r \geq 1, r \leq p-1$ such that $(\bar{a})^r = \bar{1}$. To do so, consider $\bar{1}, (\bar{a}), (\bar{a})^2, \cdots$. Then these all have values in $\left\{\bar{1}, \bar{2}, \cdots, \overline{p-1}\right\}$, and so there must be a repeat in $\left\{\bar{1}, (\bar{a}), \cdots, (\bar{a})^{p-1}\right\}$, say $p - 1 \geq l > k$ and $(\bar{a})^l = (\bar{a})^k$. Then tell why $(\bar{a})^{l-k} - \bar{1} = 0$. Let r be the first positive integer such that $(\bar{a})^r = \bar{1}$. Let $G = \left\{\bar{1}, (\bar{a}), \cdots, (\bar{a})^{r-1}\right\}$. Show that every residue class in G has its multiplicative inverse in G. In fact, $(\bar{a})^k(\bar{a})^{r-k} = \bar{1}$. Also verify that the entries in G must be distinct. Now consider the sets $\bar{b}G \equiv \left\{\bar{b}(\bar{a})^k : k = 0, \cdots, r-1\right\}$

where $\overline{b} \in \{\overline{1}, \overline{2}, \cdots, \overline{p-1}\}$. Show that two of these sets are either the same or disjoint and that they all consist of r elements. Explain why it follows that $p - 1 = lr$ for some positive integer l equal to the number of these distinct sets. Then explain why $(\overline{a})^{p-1} = (\overline{a})^{lr} = \overline{1}$.

1.10 Fundamental theorem of algebra*

The fundamental theorem of algebra is a major result in mathematics and is very important in linear algebra. It was first proved completely by Argand in 1806 although Gauss essentially proved it in 1797.

Theorem 1.3. *Every non constant polynomial $p(z)$ with complex coefficients has a root. That is $p(z) = 0$ for some $z \in \mathbb{C}$.*

In this short section, we give an outline of a proof of this major theorem. Then, for those who know some analysis, we give a rigorous proof. The first plausibility argument is based on DeMoivre's theorem, Problems 10–13 on Page 21. It is not a complete proof but will give you a good reason for believing this theorem which is better than simply accepting the decrees of the establishment. It follows [1]. Rigorous proofs are in [16] or [6]. However, this theorem also comes very easily from Liouville's theorem in complex analysis. See any book on functions of a complex variable.

Dividing by the leading coefficient, there is no loss of generality in assuming that the polynomial is of the form

$$p(z) = z^n + a_{n-1}z^{n-1} + \cdots + a_1 z + a_0$$

If $a_0 = 0$, there is nothing to prove because $p(0) = 0$. Therefore, assume $a_0 \neq 0$. From the above problems any complex number z can be written as $|z|(\cos\theta + i\sin\theta)$. Thus, by DeMoivre's theorem,

$$z^n = |z|^n(\cos(n\theta) + i\sin(n\theta))$$

It follows that z^n is some point on the circle of radius $|z|^n$

Denote by C_r the circle of radius r in the complex plane which is centered at 0. Then if r is sufficiently large and $|z| = r$, the term z^n is far larger than the rest of the polynomial. It is on the circle of radius $|z|^n$ while the other terms are on circles of fixed multiples of $|z|^k$ for $k \leq n-1$. Thus, for r large enough, $A_r = \{p(z) : z \in C_r\}$ describes a closed curve which misses the inside of some circle having 0 as its center. It will not be as simple as implied in the following picture, but it is a closed curve as stated. This follows from the fact that $\cos 2\pi = \cos 0$ and $\sin 2\pi = \sin 0$. Thus when θ, the angle identifying a point on C_r, goes from 0 to 2π, the corresponding point on A_r returns to its starting point.

Now shrink r. Eventually, for r small enough, the non constant terms are negligible and so A_r is a curve which is contained in some circle centered at a_0 which has 0 on the outside.

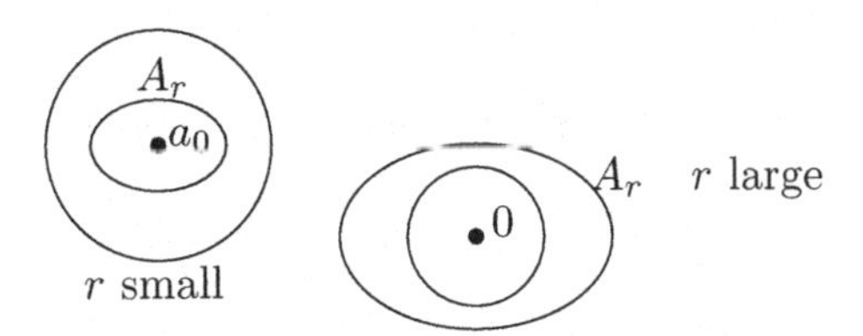

Thus it is reasonable to believe that for some r during this shrinking process, the set A_r must hit 0. It follows that $p(z) = 0$ for some z. This is not a proof because it depends too much on pictures.

Now we give a rigorous proof for those who have studied analysis.

Proof. Suppose the nonconstant polynomial $p(z) = a_0 + a_1 z + \cdots + a_n z^n$, $a_n \neq 0$, has no zero in $\mathbb{C}$. Since $\lim_{|z| \to \infty} |p(z)| = \infty$, there is a z_0 with

$$|p(z_0)| = \min_{z \in \mathbb{C}} |p(z)| > 0$$

Replacing $p(z)$ by

$$\frac{p(z + z_0)}{p(z_0)},$$

we can take $z_0 = 0, p(z_0) = p(0) = 1$. Thus we can assume our nonconstant polynomial is of the form

$$p(z) = 1 + a_1 z + \cdots + a_n z^n$$

and the minimum value of $|p(z)|$ occurs when $z = 0$ and equals 1.

Now if a_k is the first nonzero element in $\{a_1, \cdots, a_n\}$,

$$\begin{aligned} p(z) &= 1 + a_k z^k + \cdots + a_n z^n \\ &= 1 + a_k z^k + r(z) \end{aligned}$$

There is a $\delta > 0$ such that $|a_k| \delta^k < 1$ and

$$|r(z)| \leq \frac{|a_k z^k|}{2} \text{ if } |z| \leq \delta$$

Choose

$$z = \delta (\cos \theta + i \sin \theta)$$

where

$$a_k (\cos k\theta + i \sin k\theta) = -|a_k|$$

Simply let $\cos \theta + i \sin \theta$ be a kth root of the complex number $-|a_k| / a_k$ as in Problem 15 on Page 21. Then

$$|p(z)| \leq 1 - |a_k| \delta^k + |r(z)| \leq 1 - |a_k| \delta^k + \frac{|a_k|}{2} \delta^k < 1.$$

This is a contradiction, and the theorem is proved. □

Chapter 2

Matrices

Chapter summary

This chapter describes how to do row operations to solve linear systems of equations and gives some examples of physical situations in which this is interesting. After this, it presents the rules for matrix operations including addition, multiplication, and the transpose.

2.1 Systems of equations

Sometimes it is necessary to solve systems of equations. For example the problem could be to find x and y such that

$$x + y = 7 \text{ and } 2x - y = 8. \tag{2.1}$$

The set of ordered pairs, (x, y) which solve both equations is called the solution set. For example, you can see that $(5, 2) = (x, y)$ is a solution to the above system. To solve this, note that the solution set does not change if any equation is replaced by a non zero multiple of itself. It also does not change if one equation is replaced by itself added to a multiple of the other equation. For example, x and y solve the above system if and only if x and y solve the system

$$x + y = 7, \overbrace{2x - y + (-2)(x + y) = 8 + (-2)(7)}^{-3y=-6}. \tag{2.2}$$

The second equation was replaced by -2 times the first equation added to the second. Thus the solution is $y = 2$, from $-3y = -6$ and now, knowing $y = 2$, it follows from the other equation that $x + 2 = 7$, and so $x = 5$.

Why exactly does the replacement of one equation with a multiple of another added to it not change the solution set? The two equations of (2.1) are of the form

$$E_1 = f_1, E_2 = f_2 \tag{2.3}$$

where E_1 and E_2 are expressions involving the variables. The claim is that if a is a number, then (2.3) has the same solution set as

$$E_1 = f_1, \; E_2 + aE_1 = f_2 + af_1. \tag{2.4}$$

Why is this?

If (x, y) solves (2.3), then it solves the first equation in (2.4). Also, it satisfies $aE_1 = af_1$ and so, since it also solves $E_2 = f_2$, it must solve the second equation in (2.4). If (x, y) solves (2.4), then it solves the first equation of (2.3). Also $aE_1 = af_1$ and it is given that the second equation of (2.4) is satisfied. Therefore, $E_2 = f_2$, and it follows that (x, y) is a solution of the second equation in (2.3). This shows that the solutions to (2.3) and (2.4) are exactly the same, which means they have the same solution set. Of course the same reasoning applies with no change if there are many more variables than two and many more equations than two. It is still the case that when one equation is replaced with a multiple of another one added to itself, the solution set of the whole system does not change.

Two other operations which do not change the solution set are multiplying an equation by a nonzero number or listing the equations in a different order. None of these observations change with equations which have coefficients in any field. Therefore, it will always be assumed the equations have coefficients which are in F, a field, although the majority of the examples will involve coefficients in $\mathbb{R}$.

Here is another example.

Example 2.1. Find the solutions to the system

$$\begin{aligned} x + 3y + 6z &= 25 \\ 2x + 7y + 14z &= 58 \\ 2y + 5z &= 19 \end{aligned} \tag{2.5}$$

To solve this system, replace the second equation by (-2) times the first equation added to the second. This yields the system

$$\begin{aligned} x + 3y + 6z &= 25 \\ y + 2z &= 8 \\ 2y + 5z &= 19 \end{aligned} \tag{2.6}$$

Now take (-2) times the second and add to the third. More precisely, replace the third equation with (-2) times the second added to the third. This yields the system

$$\begin{aligned} x + 3y + 6z &= 25 \\ y + 2z &= 8 \\ z &= 3 \end{aligned} \tag{2.7}$$

At this point, you can determine the solution. This system has the same solution as the original system and in the above, $z = 3$. Then using this in the second equation, it follows that $y + 6 = 8$, and so $y = 2$. Now using this in the top equation yields $x + 6 + 18 = 25$, and so $x = 1$.

It is foolish to write the variables every time you do these operations. It is easier to write the system (2.5) as the following "**augmented matrix**"

$$\begin{pmatrix} 1 & 3 & 6 & 25 \\ 2 & 7 & 14 & 58 \\ 0 & 2 & 5 & 19 \end{pmatrix}.$$

It has exactly the same information as the original system, but here it is understood that there is an x column, $\begin{pmatrix} 1 \\ 2 \\ 0 \end{pmatrix}$, a y column, $\begin{pmatrix} 3 \\ 7 \\ 2 \end{pmatrix}$, and a z column, $\begin{pmatrix} 6 \\ 14 \\ 5 \end{pmatrix}$. The rows correspond to the equations in the system. Thus the top row in the augmented matrix corresponds to the equation,

$$x + 3y + 6z = 25.$$

Now when you replace an equation with a multiple of another equation added to itself, you are just taking a row of this augmented matrix and replacing it with a multiple of another row added to it. Thus the first step in solving (2.5) would be to take (-2) times the first row of the augmented matrix above and add it to the second row,

$$\begin{pmatrix} 1 & 3 & 6 & 25 \\ 0 & 1 & 2 & 8 \\ 0 & 2 & 5 & 19 \end{pmatrix}.$$

Note how this corresponds to (2.6). Next take (-2) times the second row and add to the third,

$$\begin{pmatrix} 1 & 3 & 6 & 25 \\ 0 & 1 & 2 & 8 \\ 0 & 0 & 1 & 3 \end{pmatrix}$$

which is the same as (2.7). You get the idea, we hope. Write the system as an augmented matrix and follow the procedure of either switching rows, multiplying a row by a non zero number, or replacing a row by a multiple of another row added to it. Each of these operations leaves the solution set unchanged. These operations are called **row operations**.

Definition 2.1. The row operations consist of the following

(1) Switch two rows.
(2) Multiply a row by a nonzero number.
(3) Replace a row by the same row added to a multiple of another row.

Example 2.2. Give the complete solution to the system of equations, $5x + 10y - 7z = -2$, $2x + 4y - 3z = -1$, and $3x + 6y + 5z = 9$.

The augmented matrix for this system is

$$\begin{pmatrix} 2 & 4 & -3 & -1 \\ 5 & 10 & -7 & -2 \\ 3 & 6 & 5 & 9 \end{pmatrix}.$$

Now here is a sequence of row operations leading to a simpler system for which the solution is easily seen. Multiply the second row by 2, the first row by 5, and then take (-1) times the first row and add to the second. Then multiply the first row by 1/5. This yields

$$\begin{pmatrix} 2 & 4 & -3 & -1 \\ 0 & 0 & 1 & 1 \\ 3 & 6 & 5 & 9 \end{pmatrix}.$$

Now, combining some row operations, take (-3) times the first row and add this to 2 times the last row and replace the last row with this. This yields.

$$\begin{pmatrix} 2 & 4 & -3 & -1 \\ 0 & 0 & 1 & 1 \\ 0 & 0 & 1 & 21 \end{pmatrix}.$$

Putting in the variables, the last two rows say that $z = 1$ and $z = 21$. This is impossible, so the last system of equations determined by the above augmented matrix has no solution. However, it has the same solution set as the first system of equations. This shows that there is no solution to the three given equations. When this happens, the system is called **inconsistent.**

It should not be surprising that something like this can take place. It can even happen for one equation in one variable. Consider for example, $x = x + 1$. There is clearly no solution to this.

Example 2.3. Give the complete solution to the system of equations, $3x - y - 5z = 9$, $y - 10z = 0$, and $-2x + y = -6$.

Then the following sequence of row operations yields the solution.

$$\begin{pmatrix} 3 & -1 & -5 & 9 \\ 0 & 1 & -10 & 0 \\ -2 & 1 & 0 & -6 \end{pmatrix} \overset{2\times\text{top } +3\times\text{bottom}}{\rightarrow} \begin{pmatrix} 3 & -1 & -5 & 9 \\ 0 & 1 & -10 & 0 \\ 0 & 1 & -10 & 0 \end{pmatrix}$$

$$\rightarrow \begin{pmatrix} 3 & -1 & -5 & 9 \\ 0 & 1 & -10 & 0 \\ 0 & 0 & 0 & 0 \end{pmatrix} \rightarrow \begin{pmatrix} 1 & 0 & -5 & 3 \\ 0 & 1 & -10 & 0 \\ 0 & 0 & 0 & 0 \end{pmatrix}.$$

This says $y = 10z$ and $x = 3 + 5z$. Apparently z can equal any number. Therefore, the solution set of this system is $x = 3 + 5t, y = 10t$, and $z = t$ where t is completely arbitrary. The system has an infinite set of solutions and this is a good description of the solutions.

Definition 2.2. Suppose that a system has a solution with a particular variable (say z) arbitrary, that is $z = t$ where t is arbitrary. Then the variable z is called a **free variable.**

The phenomenon of an infinite solution set occurs in equations having only one variable also. For example, consider the equation $x = x$. It doesn't matter what x equals. Recall that

$$\sum_{j=1}^{n} a_j = a_1 + a_2 + \cdots + a_n.$$

Definition 2.3. A system of linear equations is a list of equations,

$$\sum_{j=1}^{n} a_{ij}x_j = f_i, \ i = 1, 2, 3, \cdots, m$$

where a_{ij}, f_i are in F, and it is desired to find $(x_1, \cdots, x_n)$ solving each of the equations listed. The entry a_{ij} is in the ith row and j^{th} column.

As illustrated above, such a system of linear equations may have a unique solution, no solution, or infinitely many solutions. It turns out that these are the only three cases which can occur for linear systems. Furthermore, you do exactly the same things to solve any linear system. You write the augmented matrix and do row operations until you get a simpler system in which the solution is easy to find. All is based on the observation that the row operations do not change the solution set. You can have more equations than variables or fewer equations than variables. It doesn't matter. You always set up the augmented matrix and so on.

2.1.1 *Balancing chemical reactions*

Consider the chemical reaction

$$SnO_2 + H_2 \to Sn + H_2O$$

Here the elements involved are tin Sn oxygen O and Hydrogen H. Some chemical reaction happens and you end up with some tin and some water. The question is, how much do you start with and how much do you end up with.

The balance of mass requires that you have the same number of oxygen, tin, and hydrogen on both sides of the reaction. However, this does not happen in the above. For example, there are two oxygen atoms on the left and only one on the right. The problem is to find numbers x, y, z, w such that

$$xSnO_2 + yH_2 \to zSn + wH_2O$$

and both sides have the same number of atoms of the various substances. You can do this in a systematic way by setting up a system of equations which will require that this take place. Thus you need

$$\begin{array}{rl} Sn: & x = z \\ O: & 2x = w \\ H: & 2y = 2w \end{array}$$

The augmented matrix for this system of equations is then

$$\begin{pmatrix} 1 & 0 & -1 & 0 & 0 \\ 2 & 0 & 0 & -1 & 0 \\ 0 & 2 & 0 & -2 & 0 \end{pmatrix}$$

Row reducing this yields

$$\begin{pmatrix} 1 & 0 & 0 & -\frac{1}{2} & 0 \\ 0 & 1 & 0 & -1 & 0 \\ 0 & 0 & 1 & -\frac{1}{2} & 0 \end{pmatrix}$$

Thus you could let $w = 2$ and this would yield $x = 1, y = 2$, and $z = 1$. Hence, the description of the reaction which has the same numbers of atoms on both sides would be

$$SnO_2 + 2H_2 \to Sn + 2H_2O$$

You see that this preserves the total number of atoms and so the chemical equation is balanced. Here is another example

Example 2.4. Potassium is denoted by K, oxygen by O, phosphorus by P and hydrogen by H. The reaction is

$$KOH + H_3PO_4 \to K_3PO_4 + H_2O$$

balance this equation.

You need to have

$$xKOH + yH_3PO_4 \to zK_3PO_4 + wH_2O$$

Equations which preserve the total number of atoms of each element on both sides of the equation are

$$\begin{array}{rr} K: & x = 3z \\ O: & x + 4y = 4z + w \\ H: & x + 3y = 2w \\ P: & y = z \end{array}$$

The augmented matrix for this system is

$$\begin{pmatrix} 1 & 0 & -3 & 0 & 0 \\ 1 & 4 & -4 & -1 & 0 \\ 1 & 3 & 0 & -2 & 0 \\ 0 & 1 & -1 & 0 & 0 \end{pmatrix}$$

Then the row reduced echelon form is

$$\begin{pmatrix} 1 & 0 & 0 & -1 & 0 \\ 0 & 1 & 0 & -\frac{1}{3} & 0 \\ 0 & 0 & 1 & -\frac{1}{3} & 0 \\ 0 & 0 & 0 & 0 & 0 \end{pmatrix}$$

You could let $w = 3$ and this yields $x = 3, y = 1, z = 1$. Then the balanced equation is

$$3KOH + 1H_3PO_4 \to 1K_3PO_4 + 3H_2O$$

Note that this results in the same number of atoms on both sides.

Of course these numbers you are finding would typically be the number of moles of the molecules on each side. Thus three moles of KOH added to one mole of H_3PO_4 yields one mole of K_3PO_4 and three moles of H_2O, water.

Note that in this example, you have a row of zeros. This means that some of the information in computing the appropriate numbers was redundant. If this can happen with a single reaction, think how much more it could happen if you were dealing with hundreds of reactions. This aspect of the problem can be understood later in terms of the rank of a matrix.

For an introduction to the chemical considerations mentioned here, there is a nice site on the web
http://chemistry.about.com/od/chemicalreactions/a/reactiontypes.htm
where there is a sample test and examples of chemical reactions. For names of the various elements symbolized by the various letters, you can go to the site http://chemistry.about.com/od/elementfacts/a/elementlist.htm. Of course these things are in standard chemistry books, but if you have not seen much chemistry, these sites give a nice introduction to these concepts.

2.1.2 *Dimensionless variables**

This section shows how solving systems of equations can be used to determine appropriate dimensionless variables. It is only an introduction to this topic. We got this example from [5]. This considers a specific example of a simple airplane wing shown below. We assume for simplicity that it is just a flat plane at an angle to the wind which is blowing against it with speed V as shown.

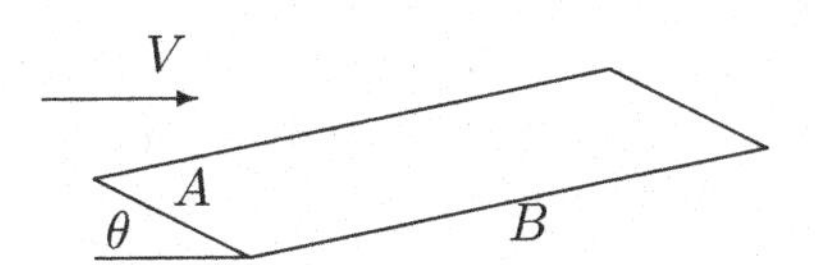

The angle is called the angle of incidence, B is the span of the wing and A is called the chord. Denote by l the lift. Then this should depend on various quantities like θ, V, B, A and so forth. Here is a table which indicates various quantities on which it is reasonable to expect l to depend.

Variable	Symbol	Units
chord	A	m
span	B	m
angle incidence	θ	$m^0 kg^0 \sec^0$
speed of wind	V	$m \sec^{-1}$
speed of sound	V_0	$m \sec^{-1}$
density of air	ρ	kgm^{-3}
viscosity	μ	$kg \sec^{-1} m^{-1}$
lift	l	$kg \sec^{-2} m$

Here m denotes meters, sec refers to seconds and kg refers to kilograms. All of these are likely familiar except for μ. One can simply decree that these are the dimensions of something called viscosity but it might be better to consider this a little more.

Viscosity is a measure of how much internal friction is experienced when the fluid moves. It is roughly a measure of how "sticky" the fluid is. Consider a piece of area parallel to the direction of motion of the fluid. To say that the viscosity is large is to say that the tangential force applied to this area must be large in order to achieve a given change in speed of the fluid in a direction normal to the tangential force. Thus

$$\mu\,(\text{area})\,(\text{velocity gradient}) = \text{tangential force}.$$

Hence

$$(\text{units on } \mu)\, m^2 \left(\frac{m}{\sec m}\right) = kg \sec^{-2} m$$

Thus the units on μ are

$$kg \sec^{-1} m^{-1}$$

as claimed above.

Then one would think that you would want

$$l = f\,(A, B, \theta, V, V_0, \rho, \mu)$$

However, this is very cumbersome because it depends on seven variables. Also, it doesn't make very good sense. It is likely that a change in the units such as going from meters to feet would result in an incorrect value for l. The way to get around this problem is to look for l as a function of dimensionless variables multiplied by something which has units of force. It is helpful because first of all, you will likely have fewer independent variables and secondly, you could expect the formula to hold independent of the way of specifying length, mass and so forth. One looks for

$$l = f\,(g_1, \cdots, g_k)\, \rho V^2 AB$$

where the units on $\rho V^2 AB$ are

$$\frac{kg}{m^3}\left(\frac{m}{\sec}\right)^2 m^2 = \frac{kg \times m}{\sec^2}$$

which are the units of force. Each of these g_i is of the form

$$A^{x_1} B^{x_2} \theta^{x_3} V^{x_4} V_0^{x_5} \rho^{x_6} \mu^{x_7} \tag{2.8}$$

and each g_i is independent of the diminsions. That is, this expression must not depend on meters, kilograms, seconds, etc. Thus, placing in the units for each of these quantities, one needs

$$m^{x_1} m^{x_2} \left(m^{x_4} \sec^{-x_4}\right) \left(m^{x_5} \sec^{-x_5}\right) \left(kg m^{-3}\right)^{x_6} \left(kg \sec^{-1} m^{-1}\right)^{x_7} = m^0 kg^0 \sec^0$$

Notice that there are no units on θ because it is just the radian measure of an angle. Hence its dimensions consist of length divided by length, thus it is dimensionless. Then this leads to the following equations for the x_i.

$$\begin{aligned} m: &\quad x_1 + x_2 + x_4 + x_5 - 3x_6 - x_7 = 0 \\ \sec: &\quad -x_4 - x_5 - x_7 = 0 \\ kg: &\quad x_6 + x_7 = 0 \end{aligned}$$

Then the augmented matrix for this matrix is

$$\begin{pmatrix} 1 & 1 & 0 & 1 & 1 & -3 & -1 & 0 \\ 0 & 0 & 0 & 1 & 1 & 0 & 1 & 0 \\ 0 & 0 & 0 & 0 & 0 & 1 & 1 & 0 \end{pmatrix}$$

The row reduced echelon form is then

$$\begin{pmatrix} 1 & 1 & 0 & 0 & 0 & 0 & 1 & 0 \\ 0 & 0 & 0 & 1 & 1 & 0 & 1 & 0 \\ 0 & 0 & 0 & 0 & 0 & 1 & 1 & 0 \end{pmatrix}$$

and so the solutions are of the form

$$x_1 = -x_2 - x_7,\ x_3 = x_3,\ x_4 = -x_5 - x_7,\ x_6 = -x_7$$

Thus the free variables are x_2, x_3, x_5, x_7. By assigning values to these, we can obtain dimensionless variables by placing the values obtained for the x_i in the formula (2.8). For example, let $x_2 = 1$ and all the rest of the free variables are 0. This yields

$$x_1 = -1, x_2 = 1, x_3 = 0, x_4 = 0, x_5 = 0, x_6 = 0, x_7 = 0.$$

The dimensionless variable is then $A^{-1}B^1$. This is the ratio between the span and the chord. It is called the aspect ratio, denoted as AR. Next let $x_3 = 1$ and all others equal zero. This gives for a dimensionless quantity the angle θ. Next let $x_5 = 1$ and all others equal zero. This gives

$$x_1 = 0, x_2 = 0, x_3 = 0, x_4 = -1, x_5 = 1, x_6 = 0, x_7 = 0.$$

Then the dimensionless variable is $V^{-1}V_0^1$. However, it is written as V/V_0. This is called the Mach number $\mathcal{M}$. Finally, let $x_7 = 1$ and all the other free variables equal 0. Then

$$x_1 = -1, x_2 = 0, x_3 = 0, x_4 = -1, x_5 = 0, x_6 = -1, x_7 = 1$$

then the dimensionless variable which results from this is $A^{-1}V^{-1}\rho^{-1}\mu$. It is customary to write it as $\text{Re} = (AV\rho)/\mu$. This one is called the Reynolds number. It is the one which involves viscosity. Thus we would look for

$$l = f(\text{Re}, AR, \theta, \mathcal{M})\, kg \times m/\sec^2$$

This is quite interesting because it is easy to vary Re by simply adjusting the velocity or A but it is hard to vary things like μ or ρ. Note that all the quantities are easy to adjust. Now this could be used, along with wind tunnel experiments to get a formula for the lift which would be reasonable. Obviously, you could consider more variables and more complicated situations in the same way.

2.2 Matrix operations and algebra

2.2.1 *Addition and scalar multiplication of matrices*

You have now solved systems of equations by writing them in terms of an augmented matrix and then doing row operations on this augmented matrix. It turns out that such rectangular arrays of numbers are important from many other different points of view. As before, the entries of a matrix will be elements of a field F and are called **scalars**.

A **matrix** is a rectangular array of scalars. If we have several, we use the term **matrices**. For example, here is a matrix.

$$\begin{pmatrix} 1 & 2 & 3 & 4 \\ 5 & 2 & 8 & 7 \\ 6 & -9 & 1 & 2 \end{pmatrix}.$$

The **size** or **dimension** of a matrix is defined as $m \times n$ where m is the number of rows and n is the number of columns. The above matrix is a 3×4 matrix because there are three rows and four columns. The first row is $(1\ 2\ 3\ 4)$, the second row is $(5\ 2\ 8\ 7)$ and so forth. The first column is $\begin{pmatrix} 1 \\ 5 \\ 6 \end{pmatrix}$. When specifying the size of a matrix, you always list the number of rows before the number of columns. Also, you can remember the columns are like columns in a Greek temple. They stand upright while the rows just lay there like rows made by a tractor in a plowed field. Elements of the matrix are identified according to position in the matrix. For example, 8 is in position $2, 3$ because it is in the second row and the third column. You might remember that you always list the rows before the columns by using the phrase **Row**man **C**atholic. The symbol (a_{ij}) refers to a matrix. The entry in the ith row and the jth column of this matrix is denoted by a_{ij}. Using this notation on the above matrix, $a_{23} = 8, a_{32} = -9, a_{12} = 2$, etc.

We shall often need to discuss the tuples which occur as rows and columns of a given matrix A. When you see

$$\begin{pmatrix} \mathbf{a}_1 & \cdots & \mathbf{a}_p \end{pmatrix},$$

this equation tells you that A has p columns and that column j is written as $\mathbf{a}_j$. Similarly, if you see

$$A = \begin{pmatrix} \mathbf{r}_1 \\ \vdots \\ \mathbf{r}_q \end{pmatrix},$$

this equation reveals that A has q rows and that row i is written $\mathbf{r}_i$.

For example, if

$$A = \begin{pmatrix} 1 & 2 \\ 3 & 2 \\ 1 & -2 \end{pmatrix},$$

we could write

$$A = \begin{pmatrix} \mathbf{a}_1 & \mathbf{a}_2 \end{pmatrix},\ \mathbf{a}_1 = \begin{pmatrix} 1 \\ 3 \\ 1 \end{pmatrix},\ \mathbf{a}_2 = \begin{pmatrix} 2 \\ 2 \\ -2 \end{pmatrix}$$

and

$$A = \begin{pmatrix} \mathbf{r}_1 \\ \mathbf{r}_2 \\ \mathbf{r}_3 \end{pmatrix}$$

where $\mathbf{r}_1 = \begin{pmatrix} 1 & 2 \end{pmatrix}, \mathbf{r}_2 = \begin{pmatrix} 3 & 2 \end{pmatrix}$, and $\mathbf{r}_3 = \begin{pmatrix} 1 & -2 \end{pmatrix}$.

There are various operations which are done on matrices. Matrices can be added, multiplied by a scalar, and multiplied by other matrices. To illustrate scalar multiplication, consider the following example in which a matrix is being multiplied by the scalar, 3.

$$3 \begin{pmatrix} 1 & 2 & 3 & 4 \\ 5 & 2 & 8 & 7 \\ 6 & -9 & 1 & 2 \end{pmatrix} = \begin{pmatrix} 3 & 6 & 9 & 12 \\ 15 & 6 & 24 & 21 \\ 18 & -27 & 3 & 6 \end{pmatrix}.$$

The new matrix is obtained by multiplying every entry of the original matrix by the given scalar. If A is an $m \times n$ matrix, $-A$ is defined to equal $(-1) A$.

Two matrices must be the same size to be added. The sum of two matrices is a matrix which is obtained by adding the corresponding entries. Thus

$$\begin{pmatrix} 1 & 2 \\ 3 & 4 \\ 5 & 2 \end{pmatrix} + \begin{pmatrix} -1 & 4 \\ 2 & 8 \\ 6 & -4 \end{pmatrix} = \begin{pmatrix} 0 & 6 \\ 5 & 12 \\ 11 & -2 \end{pmatrix}.$$

Two matrices are equal exactly when they are the same size and the corresponding entries are identical. Thus

$$\begin{pmatrix} 0 & 0 \\ 0 & 0 \\ 0 & 0 \end{pmatrix} \neq \begin{pmatrix} 0 & 0 \\ 0 & 0 \end{pmatrix}$$

because they are different sizes. As noted above, you write (c_{ij}) for the matrix C whose ij^{th} entry is c_{ij}. In doing arithmetic with matrices you must define what happens in terms of the c_{ij}, sometimes called the **entries** of the matrix or the **components** of the matrix.

The above discussion, stated for general matrices, is given in the following definition.

Definition 2.4. (Scalar Multiplication) If $A = (a_{ij})$ and k is a scalar, then $kA = (ka_{ij})$.

Definition 2.5. (Addition) Let $A = (a_{ij})$ and $B = (b_{ij})$ be two $m \times n$ matrices. Then $A + B = C$, where

$$C = (c_{ij})$$

for $c_{ij} = a_{ij} + b_{ij}$.

To save on notation, we will often abuse notation and use A_{ij} to refer to the ij^{th} entry of the matrix A instead of using lower case a_{ij}.

Definition 2.6. (The zero matrix) The $m \times n$ zero matrix is the $m \times n$ matrix having every entry equal to zero. It is denoted by 0.

Example 2.5. The 2×3 zero matrix is $\begin{pmatrix} 0 & 0 & 0 \\ 0 & 0 & 0 \end{pmatrix}$.

Note that there are 2×3 zero matrices, 3×4 zero matrices and so on.

Definition 2.7. (Equality of matrices) Let A and B be two matrices. Then $A = B$ means that the two matrices are of the same size and for $A = (a_{ij})$ and $B = (b_{ij})$, $a_{ij} = b_{ij}$ for all $1 \leq i \leq m$ and $1 \leq j \leq n$.

The following properties of matrices can be easily verified. You should do so. They are called the vector space axioms.

- Commutative law for addition,
$$A + B = B + A, \tag{2.9}$$
- Associative law for addition,
$$(A + B) + C = A + (B + C), \tag{2.10}$$
- Existence of an additive identity,
$$A + 0 = A, \tag{2.11}$$
- Existence of an additive inverse,
$$A + (-A) = 0. \tag{2.12}$$

Also for scalars α, β, the following additional properties hold.

- Distributive law over matrix addition,
$$\alpha (A + B) = \alpha A + \alpha B, \tag{2.13}$$
- Distributive law over scalar addition,
$$(\alpha + \beta) A = \alpha A + \beta A, \tag{2.14}$$
- Associative law for scalar multiplication,
$$\alpha (\beta A) = \alpha\beta (A), \tag{2.15}$$
- Rule for multiplication by 1,
$$1A = A. \tag{2.16}$$

2.2.2 *Multiplication of a vector by a matrix*

Definition 2.8. Matrices which are $n \times 1$ or $1 \times n$ are called **vectors** and are often denoted by a bold letter. Thus the $n \times 1$ matrix

$$\mathbf{x} = \begin{pmatrix} x_1 \\ \vdots \\ x_n \end{pmatrix}$$

is also called a **column vector.** The $1 \times n$ matrix

$$\begin{pmatrix} x_1 & \cdots & x_n \end{pmatrix}$$

is called a **row vector**.

Although the following description of matrix multiplication may seem strange, it is in fact the most important and useful of the matrix operations. To begin with, consider the case where a matrix is multiplied by a column vector. We will illustrate the general definition by first considering a special case.

$$\begin{pmatrix} 1 & 2 & 3 \\ 4 & 5 & 6 \end{pmatrix} \begin{pmatrix} 7 \\ 8 \\ 9 \end{pmatrix} = 7 \begin{pmatrix} 1 \\ 4 \end{pmatrix} + 8 \begin{pmatrix} 2 \\ 5 \end{pmatrix} + 9 \begin{pmatrix} 3 \\ 6 \end{pmatrix}.$$

In general, here is the definition of how to multiply an $(m \times n)$ matrix times a $(n \times 1)$ matrix.

Definition 2.9. Let A be an $m \times n$ matrix of the form

$$A = \begin{pmatrix} \mathbf{a}_1 & \cdots & \mathbf{a}_n \end{pmatrix}$$

where each $\mathbf{a}_k$ is a vector in F^m. Let

$$\mathbf{v} = \begin{pmatrix} v_1 \\ \vdots \\ v_n \end{pmatrix}.$$

Then $A\mathbf{v}$ is defined as the following **linear combination**.

$$A\mathbf{v} = v_1\mathbf{a}_1 + v_2\mathbf{a}_2 + \cdots + v_n\mathbf{a}_n \tag{2.17}$$

We refer to this as a linear combination of the columns because it is a sum of scalars times columns. Note that the jth column of A is

$$\mathbf{a}_j = \begin{pmatrix} A_{1j} \\ A_{2j} \\ \vdots \\ A_{mj} \end{pmatrix},$$

so (2.17) reduces to

$$v_1 \begin{pmatrix} A_{11} \\ A_{21} \\ \vdots \\ A_{m1} \end{pmatrix} + v_2 \begin{pmatrix} A_{12} \\ A_{22} \\ \vdots \\ A_{m2} \end{pmatrix} + \cdots + v_n \begin{pmatrix} A_{1n} \\ A_{2n} \\ \vdots \\ A_{mn} \end{pmatrix} = \begin{pmatrix} \sum_{j=1}^n A_{1j} v_j \\ \sum_{j=1}^n A_{2j} v_j \\ \vdots \\ \sum_{j=1}^n A_{mj} v_j \end{pmatrix}.$$

Note also that multiplication by an $m \times n$ matrix takes as input an $n \times 1$ matrix, and produces an $m \times 1$ matrix.

Here is another example.

Example 2.6. Compute

$$\begin{pmatrix} 1 & 2 & 1 & 3 \\ 0 & 2 & 1 & -2 \\ 2 & 1 & 4 & 1 \end{pmatrix} \begin{pmatrix} 1 \\ 2 \\ 0 \\ 1 \end{pmatrix}.$$

First of all, this is of the form $(3 \times 4)(4 \times 1)$, and so the result should be a (3×1). Note that the inside numbers cancel. The product of these two matrices equals

$$1 \begin{pmatrix} 1 \\ 0 \\ 2 \end{pmatrix} + 2 \begin{pmatrix} 2 \\ 2 \\ 1 \end{pmatrix} + 0 \begin{pmatrix} 1 \\ 1 \\ 4 \end{pmatrix} + 1 \begin{pmatrix} 3 \\ -2 \\ 1 \end{pmatrix} = \begin{pmatrix} 8 \\ 2 \\ 5 \end{pmatrix}.$$

Another thing to notice is the relation between matrix multiplication of a vector and systems of equations.

Example 2.7. Here is a vector

$$\begin{pmatrix} x + 3y + z \\ 2x - y + z \\ 3x + y - z \end{pmatrix}$$

Write this in the form

$$A \begin{pmatrix} x \\ y \\ z \end{pmatrix}$$

for A a matrix.

Note that the given vector can be written as

$$x \begin{pmatrix} 1 \\ 2 \\ 3 \end{pmatrix} + y \begin{pmatrix} 3 \\ -1 \\ 1 \end{pmatrix} + z \begin{pmatrix} 1 \\ 1 \\ -1 \end{pmatrix}.$$

From the above discussion, this equals

$$\begin{pmatrix} 1 & 3 & 1 \\ 2 & -1 & 1 \\ 3 & 1 & -1 \end{pmatrix} \begin{pmatrix} x \\ y \\ z \end{pmatrix}$$

Thus the system of equations

$$\begin{aligned} x + 3y + z &= 3 \\ 2x - y + z &= 2 \\ 3x + y - z &= -1 \end{aligned}$$

could be written in the form

$$\begin{pmatrix} 1 & 3 & 1 \\ 2 & -1 & 1 \\ 3 & 1 & -1 \end{pmatrix} \begin{pmatrix} x \\ y \\ z \end{pmatrix} = \begin{pmatrix} 3 \\ 2 \\ -1 \end{pmatrix}$$

2.2.3 *Multiplication of two matrices*

The next task is to multiply an $m \times n$ matrix times an $n \times p$ matrix. Before doing so, the following may be helpful.

For A and B matrices, in order to form the product AB, the number of columns of A must equal the number of rows of B.

$$(m \times \overbrace{n)\ (n}^{\text{These numbers must match!}} \times p) \quad = m \times p.$$

Note that the two outside numbers give the size of the product. Remember that to multiply,

The middle numbers must be the same.

Definition 2.10. When the number of columns of A equals the number of rows of B the two matrices are said to be **conformable** and the product AB is obtained as follows. Let A be an $m \times n$ matrix and let B be an $n \times p$ matrix. Then B is of the form

$$B = \begin{pmatrix} \mathbf{b}_1 & \cdots & \mathbf{b}_p \end{pmatrix}$$

where $\mathbf{b}_k$ is an $n \times 1$ matrix or column vector. Then the $m \times p$ matrix AB is defined as follows:

$$AB \equiv \begin{pmatrix} A\mathbf{b}_1 & \cdots & A\mathbf{b}_p \end{pmatrix} \tag{2.18}$$

where $A\mathbf{b}_k$ is an $m \times 1$ matrix or column vector which gives the kth column of AB.

Example 2.8. Multiply the following.

$$\begin{pmatrix} 1 & 2 & 1 \\ 0 & 2 & 1 \end{pmatrix} \begin{pmatrix} 1 & 2 & 0 \\ 0 & 3 & 1 \\ -2 & 1 & 1 \end{pmatrix}.$$

The first thing you need to check before doing anything else is whether it is possible to do the multiplication. The first matrix is a 2×3 and the second matrix is a 3×3. Therefore, is it possible to multiply these matrices. According to the above discussion it should be a 2×3 matrix of the form

$$\left(\overbrace{\begin{pmatrix} 1 & 2 & 1 \\ 0 & 2 & 1 \end{pmatrix} \begin{pmatrix} 1 \\ 0 \\ -2 \end{pmatrix}}^{\text{First column}}, \overbrace{\begin{pmatrix} 1 & 2 & 1 \\ 0 & 2 & 1 \end{pmatrix} \begin{pmatrix} 2 \\ 3 \\ 1 \end{pmatrix}}^{\text{Second column}}, \overbrace{\begin{pmatrix} 1 & 2 & 1 \\ 0 & 2 & 1 \end{pmatrix} \begin{pmatrix} 0 \\ 1 \\ 1 \end{pmatrix}}^{\text{Third column}} \right).$$

You know how to multiply a matrix times a vector, and so you do so to obtain each of the three columns. Thus

$$\begin{pmatrix} 1 & 2 & 1 \\ 0 & 2 & 1 \end{pmatrix} \begin{pmatrix} 1 & 2 & 0 \\ 0 & 3 & 1 \\ -2 & 1 & 1 \end{pmatrix} = \begin{pmatrix} -1 & 9 & 3 \\ -2 & 7 & 3 \end{pmatrix}.$$

Example 2.9. Multiply the following.

$$\begin{pmatrix} 1 & 2 & 0 \\ 0 & 3 & 1 \\ -2 & 1 & 1 \end{pmatrix} \begin{pmatrix} 1 & 2 & 1 \\ 0 & 2 & 1 \end{pmatrix}.$$

First check if it is possible. This is of the form $(3 \times 3)(2 \times 3)$. The inside numbers do not match, and so you can't do this multiplication. This means that anything you write will be absolute nonsense because it is impossible to multiply these matrices in this order. Aren't they the same two matrices considered in the previous example? Yes they are. It is just that here they are in a different order. This shows something you must always remember about matrix multiplication.

Order Matters!

Matrix Multiplication Is Not Commutative!

This is very different than multiplication of numbers!

2.2.4 *The ij^{th} entry of a product*

It is important to describe matrix multiplication in terms of entries of the matrices. What is the ij^{th} entry of AB? It would be the ith entry of the jth column of AB. Thus it would be the ith entry of $A\mathbf{b}_j$. Now

$$\mathbf{b}_j = \begin{pmatrix} B_{1j} \\ \vdots \\ B_{nj} \end{pmatrix}$$

and from the above definition, the ith entry is

$$\sum_{k=1}^{n} A_{ik} B_{kj}. \tag{2.19}$$

In terms of pictures of the matrix, you are doing

$$\begin{pmatrix} A_{11} & A_{12} & \cdots & A_{1n} \\ A_{21} & A_{22} & \cdots & A_{2n} \\ \vdots & \vdots & & \vdots \\ A_{m1} & A_{m2} & \cdots & A_{mn} \end{pmatrix} \begin{pmatrix} B_{11} & B_{12} & \cdots & B_{1p} \\ B_{21} & B_{22} & \cdots & B_{2p} \\ \vdots & \vdots & & \vdots \\ B_{n1} & B_{n2} & \cdots & B_{np} \end{pmatrix}.$$

Then as explained above, the jth column is of the form

$$\begin{pmatrix} A_{11} & A_{12} & \cdots & A_{1n} \\ A_{21} & A_{22} & \cdots & A_{2n} \\ \vdots & \vdots & & \vdots \\ A_{m1} & A_{m2} & \cdots & A_{mn} \end{pmatrix} \begin{pmatrix} B_{1j} \\ B_{2j} \\ \vdots \\ B_{nj} \end{pmatrix}$$

which is a $m \times 1$ matrix or column vector which equals

$$\begin{pmatrix} A_{11} \\ A_{21} \\ \vdots \\ A_{m1} \end{pmatrix} B_{1j} + \begin{pmatrix} A_{12} \\ A_{22} \\ \vdots \\ A_{m2} \end{pmatrix} B_{2j} + \cdots + \begin{pmatrix} A_{1n} \\ A_{2n} \\ \vdots \\ A_{mn} \end{pmatrix} B_{nj}.$$

The second entry of this $m \times 1$ matrix is

$$A_{21}B_{1j} + A_{22}B_{2j} + \cdots + A_{2n}B_{nj} = \sum_{k=1}^{n} A_{2k} B_{kj}.$$

Similarly, the ith entry of this $m \times 1$ matrix is

$$A_{i1}B_{1j} + A_{i2}B_{2j} + \cdots + A_{in}B_{nj} = \sum_{k=1}^{n} A_{ik} B_{kj}.$$

This shows the following definition for matrix multiplication in terms of the ij^{th} entries of the product coincides with Definition 2.10.

Definition 2.11. Let $A = (A_{ij})$ be an $m \times n$ matrix and let $B = (B_{ij})$ be an $n \times p$ matrix. Then AB is an $m \times p$ matrix and

$$(AB)_{ij} = \sum_{k=1}^{n} A_{ik} B_{kj}. \tag{2.20}$$

Another way to think of this is as follows. To get the ij^{th} entry of the product AB, you take the dot product of the ith row of A with the jth column of B.

Although matrix multiplication is not commutative, there are some properties of matrix multiplication which will appear familiar.

Proposition 2.1. If all multiplications and additions make sense, the following hold for matrices A, B, C and a, b scalars.

$$A(aB + bC) = a(AB) + b(AC). \tag{2.21}$$

$$(B + C)A = BA + CA. \tag{2.22}$$

$$A(BC) = (AB)C. \tag{2.23}$$

Proof. Using Definition 2.11,

$$\begin{aligned}(A(aB+bC))_{ij} &= \sum_k A_{ik}(aB+bC)_{kj} = \sum_k A_{ik}(aB_{kj}+bC_{kj}) \\ &= a\sum_k A_{ik}B_{kj} + b\sum_k A_{ik}C_{kj} = a(AB)_{ij} + b(AC)_{ij} \\ &= (a(AB)+b(AC))_{ij}.\end{aligned}$$

Thus $A(B+C) = AB + AC$ as claimed. Formula (2.22) is entirely similar.

Formula (2.23) is the associative law of multiplication. Using Definition 2.11,

$$(A(BC))_{ij} = \sum_k A_{ik}(BC)_{kj} = \sum_k A_{ik}\sum_l B_{kl}C_{lj} = \sum_l (AB)_{il} C_{lj} = ((AB)C)_{ij}.$$

This proves (2.23). □

Example 2.10. Multiply if possible $\begin{pmatrix} 1 & 2 \\ 3 & 1 \\ 2 & 6 \end{pmatrix}\begin{pmatrix} 2 & 3 & 1 \\ 7 & 6 & 2 \end{pmatrix}$.

First check to see if this is possible. It is of the form $(3 \times 2)(2 \times 3)$ and since the inside numbers match, the two matrices are conformable and it is possible to do the multiplication. The result should be a 3×3 matrix. The answer is of the form

$$\left(\begin{pmatrix} 1 & 2 \\ 3 & 1 \\ 2 & 6 \end{pmatrix}\begin{pmatrix} 2 \\ 7 \end{pmatrix}, \begin{pmatrix} 1 & 2 \\ 3 & 1 \\ 2 & 6 \end{pmatrix}\begin{pmatrix} 3 \\ 6 \end{pmatrix}, \begin{pmatrix} 1 & 2 \\ 3 & 1 \\ 2 & 6 \end{pmatrix}\begin{pmatrix} 1 \\ 2 \end{pmatrix}\right)$$

where the commas separate the columns in the resulting product. Thus the above product equals

$$\begin{pmatrix} 16 & 15 & 5 \\ 13 & 15 & 5 \\ 46 & 42 & 14 \end{pmatrix},$$

a 3×3 matrix as desired. In terms of the ij^{th} entries and the above definition, the entry in the third row and second column of the product should equal

$$\sum_j a_{3k}b_{k2} = a_{31}b_{12} + a_{32}b_{22} = 2 \times 3 + 6 \times 6 = 42.$$

You should try a few more such examples to verify that the above definition in terms of the ij^{th} entries works for other entries.

Example 2.11. Find if possible $\begin{pmatrix} 1 & 2 \\ 3 & 1 \\ 2 & 6 \end{pmatrix} \begin{pmatrix} 2 & 3 & 1 \\ 7 & 6 & 2 \\ 0 & 0 & 0 \end{pmatrix}$.

This is not possible because it is of the form $(3 \times 2)(3 \times 3)$ and the middle numbers don't match. In other words the two matrices are not conformable in the indicated order.

Example 2.12. Multiply if possible $\begin{pmatrix} 2 & 3 & 1 \\ 7 & 6 & 2 \\ 0 & 0 & 0 \end{pmatrix} \begin{pmatrix} 1 & 2 \\ 3 & 1 \\ 2 & 6 \end{pmatrix}$.

This is possible because in this case it is of the form $(3 \times 3)(3 \times 2)$ and the middle numbers do match, so the matrices are conformable. When the multiplication is done it equals

$$\begin{pmatrix} 13 & 13 \\ 29 & 32 \\ 0 & 0 \end{pmatrix}.$$

Check this and be sure you come up with the same answer.

Example 2.13. Multiply if possible $\begin{pmatrix} 1 \\ 2 \\ 1 \end{pmatrix} \begin{pmatrix} 1 & 2 & 1 & 0 \end{pmatrix}$.

In this case you are trying to do $(3 \times 1)(1 \times 4)$. The inside numbers match, so you can do it. Verify that

$$\begin{pmatrix} 1 \\ 2 \\ 1 \end{pmatrix} \begin{pmatrix} 1 & 2 & 1 & 0 \end{pmatrix} = \begin{pmatrix} 1 & 2 & 1 & 0 \\ 2 & 4 & 2 & 0 \\ 1 & 2 & 1 & 0 \end{pmatrix}.$$

Definition 2.12. Let I be the matrix

$$I = \begin{pmatrix} 1 & 0 & \cdots & 0 \\ 0 & \ddots & & \vdots \\ \vdots & & \ddots & 0 \\ 0 & 0 & \cdots & 1 \end{pmatrix}.$$

That is $I_{ij} = 1$ if $i = j$ and $I_{ij} = 0$ if $i \neq j$. We call I the $n \times n$ identity matrix. The symbol δ_{ij} is often used for the ij^{th} entry of the identity matrix. It is called the Kronecker delta symbol. Thus δ_{ij} equals 1 if $i = j$ and 0 if $i \neq j$.

The importance of this matrix is that if A is an $m \times n$ matrix and I is the $m \times m$ identity matrix, then

$$(IA)_{ij} = \sum_{k=1}^{m} I_{ik} A_{kj} = A_{ij}$$

because I_{ik} in the sum is equal to zero unless $k = i$ when it is equal to 1. Similarly, if I is the $n \times n$ identity matrix,

$$(AI)_{ij} = \sum_{k} A_{ik} I_{kj} = A_{ij}.$$

Thus the identity matrix acts like 1 in the sense that when it is multiplied on either side of a matrix A then the result of the multiplication yields A.

Example 2.14. Let I be the 3×3 identity matrix. Then

$$\begin{pmatrix} 5 & 1 & 2 \\ 3 & 4 & 3 \\ 2 & -5 & 4 \end{pmatrix} \begin{pmatrix} 1 & 0 & 0 \\ 0 & 1 & 0 \\ 0 & 0 & 1 \end{pmatrix} = \begin{pmatrix} 5 & 1 & 2 \\ 3 & 4 & 3 \\ 2 & -5 & 4 \end{pmatrix}.$$

You should verify that if the identity matrix is placed on the left, the result is the same.

2.2.5 *The transpose*

Another important operation on matrices is that of taking the **transpose**. The following example shows what is meant by this operation, denoted by placing a t as an exponent on the matrix.

$$\begin{pmatrix} 1 & 4 \\ 3 & 1 \\ 2 & 6 \end{pmatrix}^t = \begin{pmatrix} 1 & 3 & 2 \\ 4 & 1 & 6 \end{pmatrix}.$$

What happened? The first column became the first row and the second column became the second row. Thus the 3×2 matrix became a 2×3 matrix. The number 3 was in the second row and the first column and it ended up in the first row and second column. Here is the definition.

Definition 2.13. Let A be an $m \times n$ matrix. Then A^t denotes the $n \times m$ matrix which is defined as follows.

$$\left(A^t\right)_{ij} = A_{ji}.$$

Example 2.15.

$$\begin{pmatrix} 1 & 2 & -6 \\ 3 & 5 & 4 \end{pmatrix}^t = \begin{pmatrix} 1 & 3 \\ 2 & 5 \\ -6 & 4 \end{pmatrix}.$$

The transpose of a matrix has the following important properties.

Lemma 2.1. *Let A be an $m \times n$ matrix and let B be a $n \times p$ matrix. Then*

$$(AB)^t = B^t A^t \tag{2.24}$$

and if α and β are scalars,

$$(\alpha A + \beta B)^t = \alpha A^t + \beta B^t. \tag{2.25}$$

Proof. From the definition,

$$\left((AB)^t\right)_{ij} = (AB)_{ji} = \sum_k A_{jk} B_{ki} = \sum_k \left(B^t\right)_{ik} \left(A^t\right)_{kj} = \left(B^t A^t\right)_{ij}.$$

The proof of Formula (2.25) is left as an exercise and this proves the lemma.

Definition 2.14. An $n \times n$ matrix A is said to be **symmetric** if $A = A^t$. It is said to be **skew symmetric** if $A = -A^t$.

Example 2.16. Let

$$A = \begin{pmatrix} 2 & 1 & 3 \\ 1 & 5 & -3 \\ 3 & -3 & 7 \end{pmatrix}.$$

Then A is symmetric.

Example 2.17. Let

$$A = \begin{pmatrix} 0 & 1 & 3 \\ -1 & 0 & 2 \\ -3 & -2 & 0 \end{pmatrix}.$$

Then A is skew symmetric.

2.3 Block multiplication of matrices

Consider the following problem

$$\begin{pmatrix} A & B \\ C & D \end{pmatrix} \begin{pmatrix} E & F \\ G & H \end{pmatrix}.$$

You know how to do this. You get

$$\begin{pmatrix} AE + BG & AF + BH \\ CE + DG & CF + DH \end{pmatrix}.$$

Now what if instead of numbers, the entries, A, B, C, D, E, F, G are matrices of a size such that the multiplications and additions needed in the above formula all make sense. Would the formula be true in this case?

Suppose A is a matrix of the form

$$A = \begin{pmatrix} A_{11} & \cdots & A_{1m} \\ \vdots & \ddots & \vdots \\ A_{r1} & \cdots & A_{rm} \end{pmatrix} \tag{2.26}$$

where A_{ij} is a $s_i \times p_j$ matrix where s_i is constant for $j = 1, \cdots, m$ for each $i = 1, \cdots, r$. Such a matrix is called a **block matrix,** also a **partitioned matrix**. How do you get the block A_{ij}? Here is how for A an $m \times n$ matrix:

$$\overbrace{\begin{pmatrix} \mathbf{0} & I_{s_i \times s_i} & \mathbf{0} \end{pmatrix}}^{s_i \times m} A \overbrace{\begin{pmatrix} \mathbf{0} \\ I_{p_j \times p_j} \\ \mathbf{0} \end{pmatrix}}^{n \times p_j}. \tag{2.27}$$

In the block column matrix on the right, you need to have $c_j - 1$ rows of zeros above the small $p_j \times p_j$ identity matrix where the columns of A involved in A_{ij} are $c_j, \cdots, c_j + p_j - 1$ and in the block row matrix on the left, you need to have $r_i - 1$ columns of zeros to the left of the $s_i \times s_i$ identity matrix where the rows of A involved in A_{ij} are $r_i, \cdots, r_i + s_i$. An important observation to make is that the matrix on the right specifies columns to use in the block and the one on the left specifies the rows. Thus the block A_{ij}, in this case, is a matrix of size $s_i \times p_j$. There is no overlap between the blocks of A. Thus the identity $n \times n$ identity matrix corresponding to multiplication on the right of A is of the form

$$\begin{pmatrix} I_{p_1 \times p_1} & & 0 \\ & \ddots & \\ 0 & & I_{p_m \times p_m} \end{pmatrix},$$

where these little identity matrices don't overlap. A similar conclusion follows from consideration of the matrices $I_{s_i \times s_i}$. Note that in (2.27), the matrix on the right is a block column matrix for the above block diagonal matrix, and the matrix on the left in (2.27) is a block row matrix taken from a similar block diagonal matrix consisting of the $I_{s_i \times s_i}$.

Next consider the question of multiplication of two block matrices. Let B be a block matrix of the form

$$\begin{pmatrix} B_{11} & \cdots & B_{1p} \\ \vdots & \ddots & \vdots \\ B_{r1} & \cdots & B_{rp} \end{pmatrix} \tag{2.28}$$

and A is a block matrix of the form

$$\begin{pmatrix} A_{11} & \cdots & A_{1m} \\ \vdots & \ddots & \vdots \\ A_{p1} & \cdots & A_{pm} \end{pmatrix} \tag{2.29}$$

such that for all i, j, it makes sense to multiply $B_{is}A_{sj}$ for all $s \in \{1, \cdots, p\}$. (That is the two matrices B_{is} and A_{sj} are conformable.) and that for fixed ij, it follows that $B_{is}A_{sj}$ is the same size for each s so that it makes sense to write $\sum_s B_{is}A_{sj}$.

The following theorem says essentially that when you take the product of two matrices, you can partition both matrices, formally multiply the blocks to get another block matrix and this one will be BA partitioned. Before presenting this theorem, here is a simple lemma which is really a special case of the theorem.

Lemma 2.2. *Consider the following product.*

$$\begin{pmatrix} \mathbf{0} \\ I \\ \mathbf{0} \end{pmatrix} \begin{pmatrix} \mathbf{0} & I & \mathbf{0} \end{pmatrix}$$

where the first is $n \times r$ and the second is $r \times n$. The small identity matrix I is an $r \times r$ matrix and there are l zero rows above I and l zero columns to the left of I in the right matrix. Then the product of these matrices is a block matrix of the form

$$\begin{pmatrix} \mathbf{0} & \mathbf{0} & \mathbf{0} \\ \mathbf{0} & I & \mathbf{0} \\ \mathbf{0} & \mathbf{0} & \mathbf{0} \end{pmatrix}.$$

Proof. From the definition of matrix multiplication, the product is

$$\left(\begin{pmatrix} \mathbf{0} \\ I \\ \mathbf{0} \end{pmatrix} \mathbf{0} \cdots \begin{pmatrix} \mathbf{0} \\ I \\ \mathbf{0} \end{pmatrix} \mathbf{e}_1 \cdots \begin{pmatrix} \mathbf{0} \\ I \\ \mathbf{0} \end{pmatrix} \mathbf{e}_r \cdots \begin{pmatrix} \mathbf{0} \\ I \\ \mathbf{0} \end{pmatrix} \mathbf{0} \right)$$

which yields the claimed result. In the formula $\mathbf{e}_j$ refers to the column vector of length r which has a 1 in the jth position. This proves the lemma. □

Theorem 2.1. *Let B be a $q \times p$ block matrix as in (2.28) and let A be a $p \times n$ block matrix as in (2.29) such that B_{is} is conformable with A_{sj} and each product, $B_{is}A_{sj}$ for $s = 1, \cdots, p$ is of the same size, so that they can be added. Then BA can be obtained as a block matrix such that the ij^{th} block is of the form*

$$\sum_s B_{is}A_{sj}. \tag{2.30}$$

Proof. From (2.27)

$$B_{is}A_{sj} = \begin{pmatrix} \mathbf{0} & I_{r_i \times r_i} & \mathbf{0} \end{pmatrix} B \begin{pmatrix} \mathbf{0} \\ I_{p_s \times p_s} \\ \mathbf{0} \end{pmatrix} \begin{pmatrix} \mathbf{0} & I_{p_s \times p_s} & \mathbf{0} \end{pmatrix} A \begin{pmatrix} \mathbf{0} \\ I_{q_j \times q_j} \\ \mathbf{0} \end{pmatrix}$$

where here it is assumed B_{is} is $r_i \times p_s$ and A_{sj} is $p_s \times q_j$. The product involves the sth block in the ith row of blocks for B and the sth block in the jth column of A. Thus there are the same number of rows above the $I_{p_s \times p_s}$ as there are columns to the left of $I_{p_s \times p_s}$ in those two inside matrices. Then from Lemma 2.2

$$\begin{pmatrix} \mathbf{0} \\ I_{p_s \times p_s} \\ \mathbf{0} \end{pmatrix} \begin{pmatrix} \mathbf{0} & I_{p_s \times p_s} & \mathbf{0} \end{pmatrix} = \begin{pmatrix} \mathbf{0} & \mathbf{0} & \mathbf{0} \\ \mathbf{0} & I_{p_s \times p_s} & \mathbf{0} \\ \mathbf{0} & \mathbf{0} & \mathbf{0} \end{pmatrix}.$$

Since the blocks of small identity matrices do not overlap,

$$\sum_s \begin{pmatrix} \mathbf{0} & \mathbf{0} & \mathbf{0} \\ \mathbf{0} & I_{p_s \times p_s} & \mathbf{0} \\ \mathbf{0} & \mathbf{0} & \mathbf{0} \end{pmatrix} = \begin{pmatrix} I_{p_1 \times p_1} & & 0 \\ & \ddots & \\ 0 & & I_{p_p \times p_p} \end{pmatrix} = I,$$

and so,

$$\sum_s B_{is} A_{sj} = \sum_s \begin{pmatrix} \mathbf{0} & I_{r_i \times r_i} & \mathbf{0} \end{pmatrix} B \begin{pmatrix} \mathbf{0} \\ I_{p_s \times p_s} \\ \mathbf{0} \end{pmatrix} \begin{pmatrix} \mathbf{0} & I_{p_s \times p_s} & \mathbf{0} \end{pmatrix} A \begin{pmatrix} \mathbf{0} \\ I_{q_j \times q_j} \\ \mathbf{0} \end{pmatrix}$$

$$= \begin{pmatrix} \mathbf{0} & I_{r_i \times r_i} & \mathbf{0} \end{pmatrix} B \sum_s \begin{pmatrix} \mathbf{0} \\ I_{p_s \times p_s} \\ \mathbf{0} \end{pmatrix} \begin{pmatrix} \mathbf{0} & I_{p_s \times p_s} & \mathbf{0} \end{pmatrix} A \begin{pmatrix} \mathbf{0} \\ I_{q_j \times q_j} \\ \mathbf{0} \end{pmatrix}$$

$$= \begin{pmatrix} \mathbf{0} & I_{r_i \times r_i} & \mathbf{0} \end{pmatrix} BIA \begin{pmatrix} \mathbf{0} \\ I_{q_j \times q_j} \\ \mathbf{0} \end{pmatrix} = \begin{pmatrix} \mathbf{0} & I_{r_i \times r_i} & \mathbf{0} \end{pmatrix} BA \begin{pmatrix} \mathbf{0} \\ I_{q_j \times q_j} \\ \mathbf{0} \end{pmatrix}$$

which equals the ij^{th} block of BA. Hence the ij^{th} block of BA equals the formal multiplication according to matrix multiplication,

$$\sum_s B_{is} A_{sj}.$$

This proves the theorem. □

Example 2.18. Multiply the following pair of partitioned matrices using the above theorem by multiplying the blocks as described above and then in the conventional manner.

$$\left(\begin{array}{cc|c} 1 & 2 & 3 \\ \hline -1 & 2 & 3 \\ 3 & -2 & 1 \end{array}\right) \left(\begin{array}{c|cc} 1 & -1 & 2 \\ 2 & 3 & 0 \\ \hline -2 & 2 & 1 \end{array}\right)$$

Doing it in terms of the blocks, this yields, after the indicated multiplications of the blocks,

$$\left(\begin{array}{c|c} 5 + (-6) & \begin{pmatrix} 5 & 2 \end{pmatrix} + 3 \begin{pmatrix} 2 & 1 \end{pmatrix} \\ \hline \begin{pmatrix} 3 \\ -1 \end{pmatrix} + \begin{pmatrix} 3 \\ 1 \end{pmatrix} (-2) & \begin{pmatrix} 7 & -2 \\ -9 & 6 \end{pmatrix} + \begin{pmatrix} 6 & 3 \\ 2 & 1 \end{pmatrix} \end{array}\right)$$

This is

$$\left(\begin{array}{c|c} -1 & \begin{pmatrix} 11 & 5 \end{pmatrix} \\ \hline \begin{pmatrix} -3 \\ -3 \end{pmatrix} & \begin{pmatrix} 13 & 1 \\ -7 & 7 \end{pmatrix} \end{array}\right)$$

Multiplying it out the usual way, you have

$$\begin{pmatrix} 1 & 2 & 3 \\ -1 & 2 & 3 \\ 3 & -2 & 1 \end{pmatrix} \begin{pmatrix} 1 & -1 & 2 \\ 2 & 3 & 0 \\ -2 & 2 & 1 \end{pmatrix} = \begin{pmatrix} -1 & 11 & 5 \\ -3 & 13 & 1 \\ -3 & -7 & 7 \end{pmatrix}$$

you see this is the same thing without the partition lines.

2.4 Exercises

(1) Find the general solution of the system whose augmented matrix is

$$\begin{pmatrix} 1 & 2 & 0 & 2 \\ 1 & 3 & 4 & 2 \\ 1 & 0 & 2 & 1 \end{pmatrix}.$$

(2) Find the general solution of the system whose augmented matrix is

$$\begin{pmatrix} 1 & 2 & 0 & 2 \\ 2 & 0 & 1 & 1 \\ 3 & 2 & 1 & 3 \end{pmatrix}.$$

(3) Find the general solution of the system whose augmented matrix is

$$\begin{pmatrix} 1 & 1 & 0 & 1 \\ 1 & 0 & 4 & 2 \end{pmatrix}.$$

(4) Solve the system

$$\begin{aligned} x + 2y + z - w &= 2 \\ x - y + z + w &= 1 \\ 2x + y - z &= 1 \\ 4x + 2y + z &= 5 \end{aligned}$$

(5) Solve the system

$$\begin{aligned} x + 2y + z - w &= 2 \\ x - y + z + w &= 0 \\ 2x + y - z &= 1 \\ 4x + 2y + z &= 3 \end{aligned}$$

(6) Find the general solution of the system whose augmented matrix is

$$\begin{pmatrix} 1 & 0 & 2 & 1 & 1 & 2 \\ 0 & 1 & 0 & 1 & 2 & 1 \\ 1 & 2 & 0 & 0 & 1 & 3 \\ 1 & 0 & 1 & 0 & 2 & 2 \end{pmatrix}.$$

(7) Find the general solution of the system whose augmented matrix is

$$\begin{pmatrix} 1 & 0 & 2 & 1 & 1 & 2 \\ 0 & 1 & 0 & 1 & 2 & 1 \\ 0 & 2 & 0 & 0 & 1 & 3 \\ 1 & -1 & 2 & 2 & 2 & 0 \end{pmatrix}.$$

(8) Give the complete solution to the system of equations, $7x + 14y + 15z = 22$, $2x + 4y + 3z = 5$, and $3x + 6y + 10z = 13$.

(9) Give the complete solution to the system of equations, $3x - y + 4z = 6$, $y + 8z = 0$, and $-2x + y = -4$.

(10) Give the complete solution to the system of equations, $9x - 2y + 4z = -17$, $13x - 3y + 6z = -25$, and $-2x - z = 3$.
(11) Give the complete solution to the system of equations, $8x+2y+3z = -3, 8x+3y+3z = -1$, and $4x + y + 3z = -9$.
(12) Give the complete solution to the system of equations, $-8x + 2y + 5z = 18, -8x + 3y + 5z = 13$, and $-4x + y + 5z = 19$.
(13) Give the complete solution to the system of equations, $3x - y - 2z = 3$, $y - 4z = 0$, and $-2x + y = -2$.
(14) Give the complete solution to the system of equations, $-9x+15y = 66, -11x+18y = 79$,$-x + y = 4$, and $z = 3$.
(15) Consider the system $-5x + 2y - z = 0$ and $-5x - 2y - z = 0$. Both equations equal zero and so $-5x + 2y - z = -5x - 2y - z$ which is equivalent to $y = 0$. Thus x and z can equal anything. But when $x = 1$, $z = -4$, and $y = 0$ are plugged in to the equations, it does not work. Why?
(16) Here are some chemical reactions. Balance them.

    (a) $KNO_3 + H_2CO_3 \to K_2CO_3 + HNO_3$
    (b) $AgI + Na_2S \to Ag_2S + NaI$
    (c) $Ba_3N_2 + H_2O \to Ba\left(OH\right)_2 + NH_3$
    (d) $CaCl_2 + Na_3PO_4 \to Ca_3\left(PO_4\right)_2 + NaCl$

(17) In the section on dimensionless variables 33 it was observed that $\rho V^2 AB$ has the units of force. Describe a systematic way to obtain such combinations of the variables which will yield something which has the units of force.
(18) The steady state temperature, u in a plate solves Laplace's equation, $\Delta u = 0$. One way to approximate the solution which is often used is to divide the plate into a square mesh and require the temperature at each node to equal the average of the temperature at the four adjacent nodes. This procedure is justified by the mean value property of harmonic functions. In the following picture, the numbers represent the observed temperature at the indicated nodes. Your task is to find the temperature at the interior nodes, indicated by x, y, z, and w. One of the equations is $z = \frac{1}{4}\left(10 + 0 + w + x\right)$.

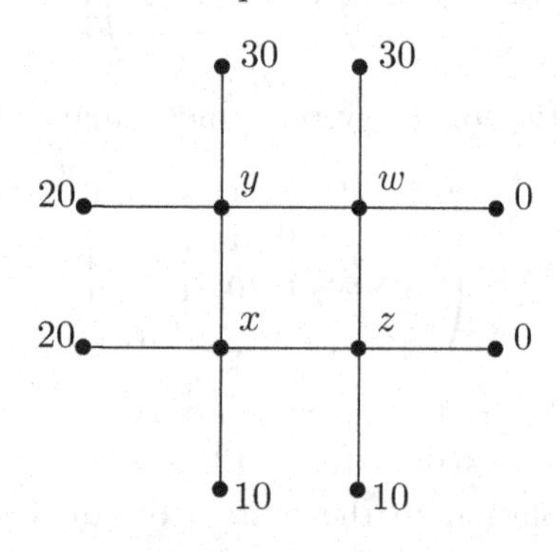

(19) Consider the following diagram of four circuits.

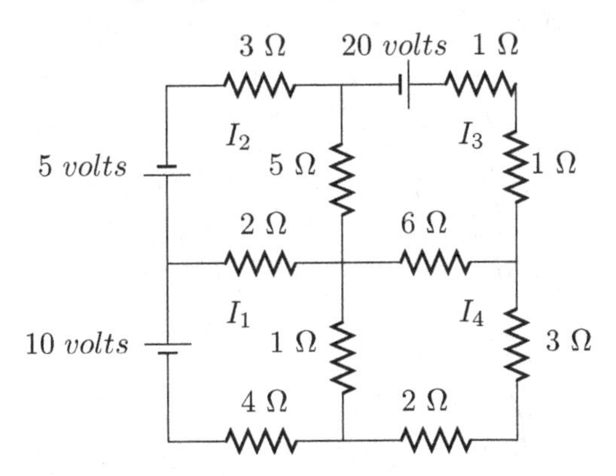

Those jagged places denote resistors and the numbers next to them give their resistance in ohms, written as Ω. The breaks in the lines having one short line and one long line denote a voltage source which causes the current to flow in the direction which goes from the longer of the two lines toward the shorter along the unbroken part of the circuit. The current in amps in the four circuits is denoted by I_1, I_2, I_3, I_4 and it is understood that the motion is in the counter clockwise direction. If I_k ends up being negative, then it just means the current flows in the clockwise direction. Then Kirchhoff's law states that

The sum of the resistance times the amps in the counter clockwise direction around a loop equals the sum of the voltage sources in the same direction around the loop.

In the above diagram, the top left circuit should give the equation

$$2I_2 - 2I_1 + 5I_2 - 5I_3 + 3I_2 = 5$$

For the circuit on the lower left, you should have

$$4I_1 + I_1 - I_4 + 2I_1 - 2I_2 = -10$$

Write equations for each of the other two circuits and then give a solution to the resulting system of equations. You might use a computer algebra system to find the solution. It might be more convenient than doing it by hand.

(20) Consider the following diagram of three circuits.

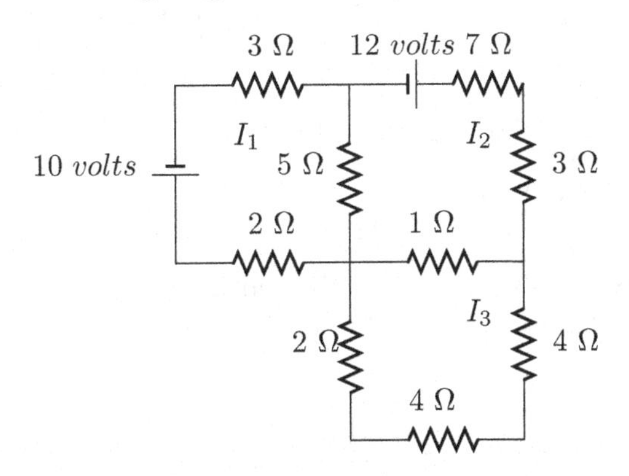

Find I_1, I_2, I_3. Use the conventions of the preceding problem.

(21) Here are some matrices:

$$A = \begin{pmatrix} 1 & 2 & 3 \\ 2 & 1 & 7 \end{pmatrix}, B = \begin{pmatrix} 3 & -1 & 2 \\ -3 & 2 & 1 \end{pmatrix},$$
$$C = \begin{pmatrix} 1 & 2 \\ 3 & 1 \end{pmatrix}, D = \begin{pmatrix} -1 & 2 \\ 2 & -3 \end{pmatrix}, E = \begin{pmatrix} 2 \\ 3 \end{pmatrix}.$$

Find if possible $-3A, 3B - A, AC, CB, AE, EA$. If it is not possible, explain why.

(22) Here are some matrices:

$$A = \begin{pmatrix} 1 & 2 \\ 3 & 2 \\ 1 & -1 \end{pmatrix}, B = \begin{pmatrix} 2 & -5 & 2 \\ -3 & 2 & 1 \end{pmatrix},$$
$$C = \begin{pmatrix} 1 & 2 \\ 5 & 0 \end{pmatrix}, D = \begin{pmatrix} -1 & 1 \\ 4 & -3 \end{pmatrix}, E = \begin{pmatrix} 1 \\ 3 \end{pmatrix}.$$

Find if possible $-3A, 3B - A, AC, CA, AE, EA, BE, DE$. If it is not possible, explain why.

(23) Here are some matrices:

$$A = \begin{pmatrix} 1 & 2 \\ 3 & 2 \\ 1 & -1 \end{pmatrix}, B = \begin{pmatrix} 2 & -5 & 2 \\ -3 & 2 & 1 \end{pmatrix},$$
$$C = \begin{pmatrix} 1 & 2 \\ 5 & 0 \end{pmatrix}, D = \begin{pmatrix} -1 & 1 \\ 4 & -3 \end{pmatrix}, E = \begin{pmatrix} 1 \\ 3 \end{pmatrix}.$$

Find if possible $-3A^t, 3B - A^t, AC, CA, AE, E^tB, BE, DE, EE^t, E^tE$. If it is not possible, explain why.

(24) Here are some matrices:

$$A = \begin{pmatrix} 1 & 2 \\ 3 & 2 \\ 1 & -1 \end{pmatrix}, B = \begin{pmatrix} 2 & -5 & 2 \\ -3 & 2 & 1 \end{pmatrix},$$
$$C = \begin{pmatrix} 1 & 2 \\ 5 & 0 \end{pmatrix}, D = \begin{pmatrix} -1 \\ 4 \end{pmatrix}, E = \begin{pmatrix} 1 \\ 3 \end{pmatrix}.$$

Find the following if possible, and explain why it is not possible if this is the case. $AD, DA, D^tB, D^tBE, E^tD, DE^t$.

(25) Let $A = \begin{pmatrix} 1 & 1 \\ -2 & -1 \\ 1 & 2 \end{pmatrix}$, $B = \begin{pmatrix} 1 & -1 & -2 \\ 2 & 1 & -2 \end{pmatrix}$, and $C = \begin{pmatrix} 1 & 1 & -3 \\ -1 & 2 & 0 \\ -3 & -1 & 0 \end{pmatrix}$.

Find if possible: AB, BA, AC, CA, CB, and BC.

(26) Suppose A and B are square matrices of the same size. Which of the following are correct?

(a) $(A-B)^2 = A^2 - 2AB + B^2$
(b) $(AB)^2 = A^2B^2$
(c) $(A+B)^2 = A^2 + 2AB + B^2$
(d) $(A+B)^2 = A^2 + AB + BA + B^2$
(e) $A^2B^2 = A(AB)B$
(f) $(A+B)^3 = A^3 + 3A^2B + 3AB^2 + B^3$
(g) $(A+B)(A-B) = A^2 - B^2$

(27) Let $A = \begin{pmatrix} -1 & -1 \\ 3 & 3 \end{pmatrix}$. Find $\boxed{\underline{\textbf{all}}}$ 2×2 matrices B such that $AB = 0$.

(28) Let $\mathbf{x} = (-1, -1, 1)$ and $\mathbf{y} = (0, 1, 2)$. Find $\mathbf{x}^t\mathbf{y}$ and $\mathbf{x}\mathbf{y}^t$ if possible.

(29) Let $A = \begin{pmatrix} 1 & 2 \\ 3 & 4 \end{pmatrix}$, $B = \begin{pmatrix} 1 & 2 \\ 3 & k \end{pmatrix}$. Is it possible to choose k such that $AB = BA$? If so, what should k equal?

(30) Let $A = \begin{pmatrix} 1 & 2 \\ 3 & 4 \end{pmatrix}$, $B = \begin{pmatrix} 1 & 2 \\ 1 & k \end{pmatrix}$. Is it possible to choose k such that $AB = BA$? If so, what should k equal?

(31) In (2.9) - (2.16) describe $-A$ and 0.

(32) Let A be an $n \times n$ matrix. Show that A equals the sum of a symmetric and a skew symmetric matrix. (M is skew symmetric if $M = -M^t$. M is symmetric if $M^t = M$.) Next show that there is only one way to do this. **Hint:** Show that $\frac{1}{2}(A^t + A)$ is symmetric and then consider using this as one of the matrices.

(33) Show that every skew symmetric matrix has all zeros down the main diagonal. The **main diagonal** consists of every entry of the matrix which is of the form a_{ii}. It runs from the upper left down to the lower right.

(34) Suppose M is a 3×3 skew symmetric matrix. Show that there exists a vector $\boldsymbol{\Omega}$ such that for all $\mathbf{u} \in \mathbb{R}^3$

$$M\mathbf{u} = \boldsymbol{\Omega} \times \mathbf{u}$$

Hint: Explain why, since M is skew symmetric, it is of the form

$$M = \begin{pmatrix} 0 & -\omega_3 & \omega_2 \\ \omega_3 & 0 & -\omega_1 \\ -\omega_2 & \omega_1 & 0 \end{pmatrix}$$

where the ω_i are numbers. Then consider $\omega_1\mathbf{i} + \omega_2\mathbf{j} + \omega_3\mathbf{k}$.

(35) Using only the properties (2.9) - (2.16), show that $-A$ is unique.

(36) Using only the properties (2.9) - (2.16), show that 0 is unique.

(37) Using only the properties (2.9) - (2.16), show that $0A = 0$. Here the 0 on the left is the scalar 0 and the 0 on the right is the zero for $m \times n$ matrices.

(38) Using only the properties (2.9) - (2.16) and previous problems, show that $(-1)A = -A$.

(39) Prove (2.25).

(40) Prove that $I_mA = A$ where A is an $m \times n$ matrix.

(41) Give an example of matrices A, B, C such that $B \neq C$, $A \neq 0$, and yet $AB = AC$.

(42) Give an example of matrices A, B such that neither A nor B equals zero and yet $AB = 0$.

(43) Give another example of two square matrices A and B such that $AB \neq BA$.

(44) Write $\begin{pmatrix} x_1 - x_2 + 2x_3 \\ 2x_3 + x_1 \\ 3x_3 \\ 3x_4 + 3x_2 + x_1 \end{pmatrix}$ in the form $A \begin{pmatrix} x_1 \\ x_2 \\ x_3 \\ x_4 \end{pmatrix}$ where A is an appropriate matrix.

(45) Write $\begin{pmatrix} x_1 + 3x_2 + 2x_3 \\ 2x_3 + x_1 \\ 6x_3 \\ x_4 + 3x_2 + x_1 \end{pmatrix}$ in the form $A \begin{pmatrix} x_1 \\ x_2 \\ x_3 \\ x_4 \end{pmatrix}$ where A is an appropriate matrix.

(46) Write $\begin{pmatrix} x_1 + x_2 + x_3 \\ 2x_3 + x_1 + x_2 \\ x_3 - x_1 \\ 3x_4 + x_1 \end{pmatrix}$ in the form $A \begin{pmatrix} x_1 \\ x_2 \\ x_3 \\ x_4 \end{pmatrix}$ where A is an appropriate matrix.

(47) Let A be a real $m \times n$ matrix and let $\mathbf{x} \in \mathbb{R}^n$ and $\mathbf{y} \in \mathbb{R}^m$. Show that $\langle A\mathbf{x}, \mathbf{y} \rangle_{\mathbb{R}^m} = \langle \mathbf{x}, A^t\mathbf{y} \rangle_{\mathbb{R}^n}$ where $\langle \cdot, \cdot \rangle_{\mathbb{R}^k}$ denotes the dot product in $\mathbb{R}^k$. In the notation above, $A\mathbf{x} \cdot \mathbf{y} = \mathbf{x} \cdot A^t\mathbf{y}$. Use the definition of matrix multiplication to do this.

(48) Use the result of Problem 47 to verify directly that $(AB)^t = B^t A^t$ without making any reference to subscripts.

(49) Multiply the following two block matrices by using the blocks and then do the multiplication in the usual way and verify that you get the same thing either way.

$$\left(\begin{array}{cc|c} 0 & 2 & 2 \\ \hline -1 & 2 & -2 \\ \hline -1 & 1 & 1 \end{array}\right) \left(\begin{array}{c|cc} 0 & 2 & 3 \\ 2 & 3 & 1 \\ \hline 1 & 1 & 3 \end{array}\right)$$

(50) Let A be an $n \times n$ matrix and suppose the following is obtained.

$$A\mathbf{u}_1 = \lambda \mathbf{u}_1$$

The matrix $P \equiv \begin{pmatrix} \mathbf{u}_1 & \cdots & \mathbf{u}_n \end{pmatrix}$ having $\mathbf{u}_k$ as the kth column, is invertible. Then explain why

$$P^{-1}AP = \begin{pmatrix} \lambda & * \\ \mathbf{0} & B \end{pmatrix}$$

where B is an $(n-1) \times (n-1)$ matrix and $*$ is a $1 \times (n-1)$ matrix. This is an important part of the proof of Schur's theorem which will be presented much later in the book. This proof will use simple block multiplication.

Chapter 3

Row operations

Chapter summary

This chapter is centered on elementary matrices and their relation to row operations. It presents the row reduced echelon form of a matrix and its uniqueness. This is used to define the rank of a matrix. The row reduced echelon form is also used to give a method for finding the inverse of a matrix. Also the LU and PLU factorizations are presented along with a convenient method for computing them.

3.1 Elementary matrices

The elementary matrices result from doing a row operation to the identity matrix.

As before, everything will apply to matrices having coefficients in an arbitrary field of scalars, although we will mainly feature the real numbers in the examples.

Definition 3.1. The row operations consist of the following

(1) Switch two rows.
(2) Multiply a row by a nonzero number.
(3) Replace a row by the same row added to a multiple of another row.

We refer to these as the row operations of type 1,2, and 3 respectively.

The elementary matrices are given in the following definition.

Definition 3.2. The elementary matrices consist of those matrices which result by applying a row operation to an identity matrix. Those which involve switching rows of the identity are called permutation matrices. More generally, a permutation matrix is a matrix which comes by permuting the rows of the identity matrix, not just switching two rows.

As an example of why these elementary matrices are interesting, consider the following.

$$\begin{pmatrix} 0 & 1 & 0 \\ 1 & 0 & 0 \\ 0 & 0 & 1 \end{pmatrix} \begin{pmatrix} a & b & c & d \\ x & y & z & w \\ f & g & h & i \end{pmatrix} = \begin{pmatrix} x & y & z & w \\ a & b & c & d \\ f & g & h & i \end{pmatrix}.$$

A 3×4 matrix was multiplied on the left by an elementary matrix which was obtained from row operation 1 applied to the identity matrix. This resulted in applying the operation 1 to the given matrix. This is what happens in general.

Now consider what these elementary matrices look like. First P_{ij}, which involves switching row i and row j of the identity where $i < j$. We write

$$I = \begin{pmatrix} \mathbf{r}_1 \\ \vdots \\ \mathbf{r}_i \\ \vdots \\ \mathbf{r}_j \\ \vdots \\ \mathbf{r}_n \end{pmatrix}$$

where

$$\mathbf{r}_j = (0 \cdots 1 \cdots 0)$$

with the 1 in the jth position from the left.

This matrix P^{ij} is of the form

$$\begin{pmatrix} \mathbf{r}_1 \\ \vdots \\ \mathbf{r}_j \\ \vdots \\ \mathbf{r}_i \\ \vdots \\ \mathbf{r}_n \end{pmatrix}.$$

Now consider what this does to a column vector.

$$\begin{pmatrix} \mathbf{r}_1 \\ \vdots \\ \mathbf{r}_j \\ \vdots \\ \mathbf{r}_i \\ \vdots \\ \mathbf{r}_n \end{pmatrix} \begin{pmatrix} v_1 \\ \vdots \\ v_i \\ \vdots \\ v_j \\ \vdots \\ v_n \end{pmatrix} = \begin{pmatrix} v_1 \\ \vdots \\ v_j \\ \vdots \\ v_i \\ \vdots \\ v_n \end{pmatrix}.$$

Now we try multiplication of a matrix on the left by this elementary matrix P^{ij}. Consider

$$P^{ij}\begin{pmatrix} a_{11} & a_{12} & \cdots & \cdots & \cdots & \cdots & a_{1p} \\ \vdots & \vdots & & & & & \vdots \\ a_{i1} & a_{i2} & \cdots & \cdots & \cdots & \cdots & a_{ip} \\ \vdots & \vdots & & & & & \vdots \\ a_{j1} & a_{j2} & \cdots & \cdots & \cdots & \cdots & a_{jp} \\ \vdots & \vdots & & & & & \vdots \\ a_{n1} & a_{n2} & \cdots & \cdots & \cdots & \cdots & a_{np} \end{pmatrix}.$$

From the way you multiply matrices this is a matrix which has the indicated columns.

$$\left(P^{ij}\begin{pmatrix} a_{11} \\ \vdots \\ a_{i1} \\ \vdots \\ a_{j1} \\ \vdots \\ a_{n1} \end{pmatrix}, P^{ij}\begin{pmatrix} a_{12} \\ \vdots \\ a_{i2} \\ \vdots \\ a_{j2} \\ \vdots \\ a_{n2} \end{pmatrix}, \cdots, P^{ij}\begin{pmatrix} a_{1p} \\ \vdots \\ a_{ip} \\ \vdots \\ a_{jp} \\ \vdots \\ a_{np} \end{pmatrix} \right)$$

$$= \left(\begin{pmatrix} a_{11} \\ \vdots \\ a_{j1} \\ \vdots \\ a_{i1} \\ \vdots \\ a_{n1} \end{pmatrix}, \begin{pmatrix} a_{12} \\ \vdots \\ a_{j2} \\ \vdots \\ a_{i2} \\ \vdots \\ a_{n2} \end{pmatrix}, \cdots, \begin{pmatrix} a_{1p} \\ \vdots \\ a_{jp} \\ \vdots \\ a_{ip} \\ \vdots \\ a_{np} \end{pmatrix} \right)$$

$$= \begin{pmatrix} a_{11} & a_{12} & \cdots & \cdots & \cdots & \cdots & a_{1p} \\ \vdots & \vdots & & & & & \vdots \\ a_{j1} & a_{j2} & \cdots & \cdots & \cdots & \cdots & a_{jp} \\ \vdots & \vdots & & & & & \vdots \\ a_{i1} & a_{i2} & \cdots & \cdots & \cdots & \cdots & a_{ip} \\ \vdots & \vdots & & & & & \vdots \\ a_{n1} & a_{n2} & \cdots & \cdots & \cdots & \cdots & a_{np} \end{pmatrix}.$$

This has established the following lemma.

Lemma 3.1. *Let P^{ij} denote the elementary matrix which involves switching the ith and the jth rows of I. Then if P^{ij}, A are conformable, we have*

$$P^{ij}A = B$$

where B is obtained from A by switching the ith and the jth rows.

Next consider the row operation which involves multiplying the ith row by a nonzero constant, c. We write

$$I = \begin{pmatrix} \mathbf{r}_1 \\ \mathbf{r}_2 \\ \vdots \\ \mathbf{r}_n \end{pmatrix}$$

where

$$\mathbf{r}_j = (0 \cdots 1 \cdots 0)$$

with the 1 in the jth position from the left. The elementary matrix which results from applying this operation to the ith row of the identity matrix is of the form

$$E(c,i) = \begin{pmatrix} \mathbf{r}_1 \\ \vdots \\ c\mathbf{r}_i \\ \vdots \\ \mathbf{r}_n \end{pmatrix}.$$

Now consider what this does to a column vector.

$$\begin{pmatrix} \mathbf{r}_1 \\ \vdots \\ c\mathbf{r}_i \\ \vdots \\ \mathbf{r}_n \end{pmatrix} \begin{pmatrix} v_1 \\ \vdots \\ v_i \\ \vdots \\ v_n \end{pmatrix} = \begin{pmatrix} v_1 \\ \vdots \\ cv_i \\ \vdots \\ v_n \end{pmatrix}.$$

Denote by $E(c,i)$ this elementary matrix which multiplies the ith row of the identity by the nonzero constant, c. Then from what was just discussed and the way matrices are multiplied,

$$E(c,i) \begin{pmatrix} a_{11} & a_{12} & \cdots & a_{1p} \\ \vdots & \vdots & & \vdots \\ a_{i1} & a_{i2} & \cdots & a_{ip} \\ \vdots & \vdots & & \vdots \\ a_{n1} & a_{n2} & \cdots & a_{np} \end{pmatrix}$$

equals a matrix having the columns indicated below.

$$= \begin{pmatrix} a_{11} & a_{12} & \cdots & a_{1p} \\ \vdots & \vdots & & \vdots \\ ca_{i1} & ca_{i2} & \cdots & ca_{ip} \\ \vdots & \vdots & & \vdots \\ a_{n1} & a_{n2} & \cdots & a_{np} \end{pmatrix}.$$

This proves the following lemma.

Lemma 3.2. *Let $E(c,i)$ denote the elementary matrix corresponding to the row operation in which the ith row is multiplied by the nonzero constant c. Thus $E(c,i)$ involves multiplying the ith row of the identity matrix by c. Then*

$$E(c,i)\,A = B$$

where B is obtained from A by multiplying the ith row of A by c.

Finally consider the third of these row operations. Letting $\mathbf{r}_j$ be the jth row of the identity matrix, denote by $E(c \times i + j)$ the elementary matrix obtained from the identity matrix by replacing $\mathbf{r}_j$ with $\mathbf{r}_j + c\mathbf{r}_i$. In case $i < j$ this will be of the form

$$\begin{pmatrix} \mathbf{r}_1 \\ \vdots \\ \mathbf{r}_i \\ \vdots \\ c\mathbf{r}_i + \mathbf{r}_j \\ \vdots \\ \mathbf{r}_n \end{pmatrix}.$$

Now consider what this does to a column vector.

$$\begin{pmatrix} \mathbf{r}_1 \\ \vdots \\ \mathbf{r}_i \\ \vdots \\ c\mathbf{r}_i + \mathbf{r}_j \\ \vdots \\ \mathbf{r}_n \end{pmatrix} \begin{pmatrix} v_1 \\ \vdots \\ v_i \\ \vdots \\ v_j \\ \vdots \\ v_n \end{pmatrix} = \begin{pmatrix} v_1 \\ \vdots \\ v_i \\ \vdots \\ cv_i + v_j \\ \vdots \\ v_n \end{pmatrix}.$$

Now from this and the way matrices are multiplied,

$$E(c \times i + j) \begin{pmatrix} a_{11} & a_{12} & \cdots & \cdots & \cdots & \cdots & a_{1p} \\ \vdots & \vdots & & & & & \vdots \\ a_{i1} & a_{i2} & \cdots & \cdots & \cdots & \cdots & a_{ip} \\ \vdots & \vdots & & & & & \vdots \\ a_{j2} & a_{j2} & \cdots & \cdots & \cdots & \cdots & a_{jp} \\ \vdots & \vdots & & & & & \vdots \\ a_{n1} & a_{n2} & \cdots & \cdots & \cdots & \cdots & a_{np} \end{pmatrix}$$

equals a matrix of the following form having the indicated columns.

$$\left(E(c\times i+j)\begin{pmatrix} a_{11} \\ \vdots \\ a_{i1} \\ \vdots \\ a_{j2} \\ \vdots \\ a_{n1} \end{pmatrix}, E(c\times i+j)\begin{pmatrix} a_{12} \\ \vdots \\ a_{i2} \\ \vdots \\ a_{j2} \\ \vdots \\ a_{n2} \end{pmatrix}, \cdots E(c\times i+j)\begin{pmatrix} a_{1p} \\ \vdots \\ a_{ip} \\ \vdots \\ a_{jp} \\ \vdots \\ a_{np} \end{pmatrix}\right)$$

$$= \begin{pmatrix} a_{11} & a_{12} & \cdots & a_{1p} \\ \vdots & \vdots & & \vdots \\ a_{i1} & a_{i2} & \cdots & a_{ip} \\ \vdots & \vdots & & \vdots \\ a_{j2}+ca_{i1} & a_{j2}+ca_{i2} & \cdots & a_{jp}+ca_{ip} \\ \vdots & \vdots & & \vdots \\ a_{n1} & a_{n2} & \cdots & a_{np} \end{pmatrix}.$$

The case where $i > j$ is similar. This proves the following lemma in which, as above, the ith row of the identity is $\mathbf{r}_i$.

Lemma 3.3. *Let $E(c\times i+j)$ denote the elementary matrix obtained from I by replacing the jth row of the identity $\mathbf{r}_j$ with $c\mathbf{r}_i+\mathbf{r}_j$. Letting the kth row of A be $\mathbf{a}_k$,*

$$E(c\times i+j)A = B$$

where B has the same rows as A except the jth row of B is $c\mathbf{a}_i+\mathbf{a}_j$.

The above lemmas are summarized in the following theorem.

Theorem 3.1. *To perform any of the three row operations on a matrix A it suffices to do the row operation on the identity matrix, obtaining an elementary matrix E, and then take the product, EA. In addition to this, the following identities hold for the elementary matrices described above.*

$$E(c\times i+j)E(-c\times i+j) = E(-c\times i+j)E(c\times i+j) = I. \tag{3.1}$$

$$E(c,i)E\left(c^{-1},i\right) = E\left(c^{-1},i\right)E(c,i) = I. \tag{3.2}$$

$$P^{ij}P^{ij} = I. \tag{3.3}$$

Proof. Consider (3.1). Starting with I and taking $-c$ times the ith row added to the jth yields $E(-c \times i + j)$ which differs from I only in the jth row. Now multiplying on the left by $E(c \times i + j)$ takes c times the ith row and adds to the jth thus restoring the jth row to its original state. Thus $E(c \times i + j) E(-c \times i + j) = I$. Similarly $E(-c \times i + j) E(c \times i + j) = I$. The reasoning is similar for (3.2) and (3.3).

Each of these elementary matrices has a significant geometric significance. The following picture shows the effect of doing $E\left(\frac{1}{2} \times 3 + 1\right)$ on a box. You will see that it shears the box in one direction. Of course there would be corresponding shears in the other directions also. Note that this does not change the volume. You should think about the geometric effect of the other elementary matrices on a box.

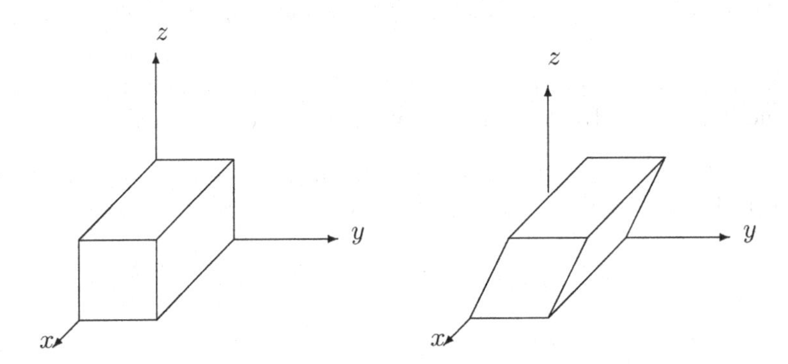

Definition 3.3. For an $n \times n$ matrix A, an $n \times n$ matrix B which has the property that $AB = BA = I$ is denoted by A^{-1}. Such a matrix is called an **inverse.** When A has an inverse, it is called **invertible.**

The following lemma says that if a matrix acts like an inverse, then it is **the** inverse. Also, the product of invertible matrices is invertible.

Lemma 3.4. *If B, C are both inverses of A, then $B = C$. That is, there exists at most one inverse of a matrix. If $A_1, \cdots, A_m$ are each invertible $m \times m$ matrices, then the product $A_1 A_2 \cdots A_m$ is also invertible and*

$$(A_1 A_2 \cdots A_m)^{-1} = A_m^{-1} A_{m-1}^{-1} \cdots A_1^{-1}.$$

Proof. From the definition and associative law of matrix multiplication,

$$B = BI = B(AC) = (BA)C = IC = C.$$

This proves the uniqueness of the inverse.

Next suppose A, B are invertible. Then

$$AB\left(B^{-1}A^{-1}\right) = A\left(BB^{-1}\right)A^{-1} = AIA^{-1} = AA^{-1} = I$$

and also

$$\left(B^{-1}A^{-1}\right)AB = B^{-1}\left(A^{-1}A\right)B = B^{-1}IB = B^{-1}B = I.$$

It follows from Definition 3.3 that AB has an inverse and it is $B^{-1}A^{-1}$. Thus the case of $m = 1, 2$ in the claim of the lemma is true. Suppose this claim is true for k. Then

$$A_1A_2 \cdots A_kA_{k+1} = (A_1A_2 \cdots A_k)\, A_{k+1}.$$

By induction, the two matrices $(A_1A_2 \cdots A_k)\,,\ A_{k+1}$ are both invertible and

$$(A_1A_2 \cdots A_k)^{-1} = A_k^{-1} \cdots A_2^{-1}A_1^{-1}.$$

By the case of the product of two invertible matrices shown above,

$$\begin{aligned}((A_1A_2 \cdots A_k)\, A_{k+1})^{-1} &= A_{k+1}^{-1}\,(A_1A_2 \cdots A_k)^{-1} \\ &= A_{k+1}^{-1}A_k^{-1} \cdots A_2^{-1}A_1^{-1}.\end{aligned}$$

This proves the lemma. □

We will discuss methods for finding the inverse later. For now, observe that Theorem 3.1 says that elementary matrices are invertible and that the inverse of such a matrix is also an elementary matrix. The major conclusion of the above Lemma and Theorem is the following lemma about linear relationships.

Definition 3.4. Let $\mathbf{v}_1, \cdots, \mathbf{v}_k, \mathbf{u}$ be vectors. Then $\mathbf{u}$ is said to be a **linear combination** of the vectors $\{\mathbf{v}_1, \cdots, \mathbf{v}_k\}$ if there exist scalars $c_1, \cdots, c_k$ such that

$$\mathbf{u} = \sum_{i=1}^{k} c_i \mathbf{v}_i.$$

We also say that when the above holds for some scalars $c_1, \cdots, c_k$, there exists a **linear relationship** between the vector $\mathbf{u}$ and the vectors $\{\mathbf{v}_1, \cdots, \mathbf{v}_k\}$.

We will discuss this more later, but the following picture illustrates the geometric significance of the vectors which have a linear relationship with two vectors $\mathbf{u}, \mathbf{v}$ pointing in different directions.

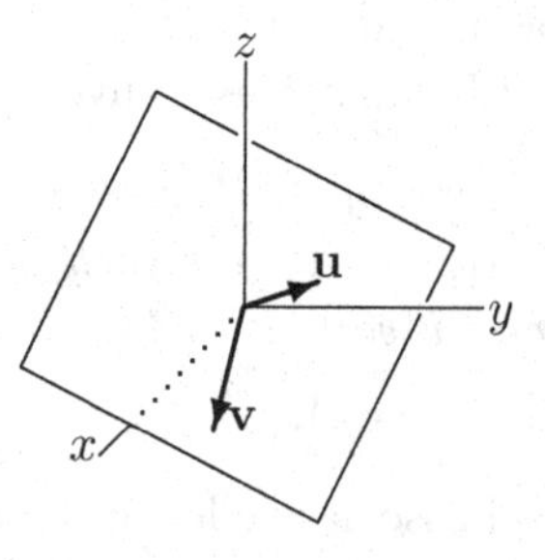

The following lemma states that linear relationships between columns in a matrix are preserved by row operations. This simple lemma is the main result in understanding all the major questions related to the row reduced echelon form as well as many other topics.

Lemma 3.5. *Let A and B be two $m \times n$ matrices and suppose B results from a row operation applied to A. Then the kth column of B is a linear combination of the*

$i_1, \cdots, i_r$ columns of B if and only if the kth column of A is a linear combination of the $i_1, \cdots, i_r$ columns of A. Furthermore, the scalars in the linear combinations are the same. (The linear relationship between the kth column of A and the $i_1, \cdots, i_r$ columns of A is the same as the linear relationship between the kth column of B and the $i_1, \cdots, i_r$ columns of B.)

Proof. Let A be the following matrix in which the $\mathbf{a}_k$ are the columns

$$\begin{pmatrix} \mathbf{a}_1 & \mathbf{a}_2 & \cdots & \mathbf{a}_n \end{pmatrix}$$

and let B be the following matrix in which the columns are given by the $\mathbf{b}_k$

$$\begin{pmatrix} \mathbf{b}_1 & \mathbf{b}_2 & \cdots & \mathbf{b}_n \end{pmatrix}.$$

Then by Theorem 3.1 on Page 62, $\mathbf{b}_k = E\mathbf{a}_k$ where E is an elementary matrix. Suppose then that one of the columns of A is a linear combination of some other columns of A. Say

$$\mathbf{a}_k = c_1\mathbf{a}_{i_1} + \cdots + c_r\mathbf{a}_{i_r}.$$

Then multiplying by E,

$$\mathbf{b}_k = E\mathbf{a}_k = c_1E\mathbf{a}_{i_1} + \cdots + c_rE\mathbf{a}_{i_r} = c_1\mathbf{b}_{i_1} + \cdots + c_r\mathbf{b}_{i_r}.$$

This proves the lemma. □

Example 3.1. Find linear relationships between the columns of the matrix

$$A = \begin{pmatrix} 1 & 3 & 11 & 10 & 36 \\ 1 & 2 & 8 & 9 & 23 \\ 1 & 1 & 5 & 8 & 10 \end{pmatrix}.$$

It is not clear what the relationships are, so we do row operations to this matrix. Lemma 3.5 says that all the linear relationships between columns are preserved, so the idea is to do row operations until a matrix results which has the property that the linear relationships are obvious. First take -1 times the top row and add to the two bottom rows. This yields

$$\begin{pmatrix} 1 & 3 & 11 & 10 & 36 \\ 0 & -1 & -3 & -1 & -13 \\ 0 & -2 & -6 & -2 & -26 \end{pmatrix}$$

Next take -2 times the middle row and add to the bottom row followed by multiplying the middle row by -1 :

$$\begin{pmatrix} 1 & 3 & 11 & 10 & 36 \\ 0 & 1 & 3 & 1 & 13 \\ 0 & 0 & 0 & 0 & 0 \end{pmatrix}.$$

Next take -3 times the middle row added to the top:

$$\begin{pmatrix} 1 & 0 & 2 & 7 & -3 \\ 0 & 1 & 3 & 1 & 13 \\ 0 & 0 & 0 & 0 & 0 \end{pmatrix}. \tag{3.4}$$

At this point it is clear that the last column is -3 times the first column added to 13 times the second. By Lemma 3.5, the same is true of the corresponding columns in the original matrix A. As a check,

$$-3\begin{pmatrix} 1 \\ 1 \\ 1 \end{pmatrix} + 13\begin{pmatrix} 3 \\ 2 \\ 1 \end{pmatrix} = \begin{pmatrix} 36 \\ 23 \\ 10 \end{pmatrix}.$$

You should notice that other linear relationships are also easily seen from (3.4). For example the fourth column is 7 times the first added to the second. This is obvious from (3.4) and Lemma 3.5 says the same relationship holds for A.

This is really just an extension of the technique for finding solutions to a linear system of equations. In solving a system of equations earlier, row operations were used to exhibit the last column of an augmented matrix as a linear combination of the preceding columns. The **row reduced echelon form** just extends this by making obvious the linear relationships between **every** column, not just the last, and those columns preceding it. The matrix in (3.4) is in row reduced echelon form. The row reduced echelon form is the topic of the next section.

3.2 The row reduced echelon form of a matrix

When you do row operations on a matrix, there is an ultimate conclusion. It is called the **row reduced echelon form**. We show here that every matrix has such a row reduced echelon form and that this row reduced echelon form is unique. The significance is that it becomes possible to use the definite article in referring to **the** row reduced echelon form. Hence important conclusions about the original matrix may be logically deduced from an examination of its unique row reduced echelon form. First we need the following definition.

Definition 3.5. Define special column vectors $\mathbf{e}_i$ as follows.

$$\mathbf{e}_i = \begin{pmatrix} 0 & \cdots & 1 & \cdots & 0 \end{pmatrix}^t.$$

Recall that ^{t} says to take the transpose. Thus $\mathbf{e}_i$ is the column vector which has all zero entries except for a 1 in the ith position down from the top.

Now here is the description of the row reduced echelon form.

Definition 3.6. An $m \times n$ matrix is said to be in **row reduced echelon form** if, in viewing successive columns from left to right, the first nonzero column encountered is $\mathbf{e}_1$ and if, in viewing the columns of the matrix from left to right, you have encountered $\mathbf{e}_1, \mathbf{e}_2, \cdots, \mathbf{e}_k$, the next column is either $\mathbf{e}_{k+1}$ or this next column is a linear combination of the vectors, $\mathbf{e}_1, \mathbf{e}_2, \cdots, \mathbf{e}_k$.

Example 3.2. The following matrices are in row reduced echelon form.

$$\begin{pmatrix} 1 & 0 & 4 & 0 \\ 0 & 1 & 3 & 0 \\ 0 & 0 & 0 & 1 \end{pmatrix}, \begin{pmatrix} 0 & 1 & 0 & 0 & 7 \\ 0 & 0 & 0 & 1 & 3 \\ 0 & 0 & 0 & 0 & 0 \\ 0 & 0 & 0 & 0 & 0 \end{pmatrix}, \begin{pmatrix} 0 & 1 & 0 & 3 \\ 0 & 0 & 1 & -5 \\ 0 & 0 & 0 & 0 \end{pmatrix}.$$

The above definition emphasizes the linear relationships between columns. However, it is sometimes easier to use a different description of the row reduced echelon form, although it is equivalent to the above. This description also employs some standard terminology.

Definition 3.7. The first nonzero entry (element) of a nonzero row (reading from left to right) is the **leading entry** of the row. A matrix is row reduced, (or, in **row reduced echelon form**) if

(i) all zero rows are below nonzero rows,
(ii) all leading entries are 1,
(iii) the leading entry of row j is to the right of the leading entry of row $j-1$, for all possible j,
(iv) if a column contains a leading entry, its other entries are 0.

The $n \times n$ matrix

$$I = \begin{pmatrix} 1 & 0 & \cdots & 0 \\ 0 & 1 & \cdots & 0 \\ \vdots & & \ddots & \vdots \\ 0 & 0 & \cdots & 1 \end{pmatrix} \quad \text{(the \textbf{identity matrix})}$$

is row reduced. So too are,

$$A = \begin{pmatrix} 0 & 1 \\ 0 & 0 \end{pmatrix}, B = \begin{pmatrix} 1 & 0 & 0 \\ 0 & 1 & 3 \\ 0 & 0 & 0 \end{pmatrix}, C = \begin{pmatrix} 1 & 2 & 0 & 0 \\ 0 & 0 & 1 & 0 \\ 0 & 0 & 0 & 1 \end{pmatrix}.$$

A column of a row reduced matrix is called a **pivot column** if it contains a leading entry. The pivot columns of C (above) are the first, third and fourth columns. Clearly the pivot columns in the identity matrix are $\begin{pmatrix} 1 \\ 0 \\ \vdots \\ 0 \end{pmatrix}, \begin{pmatrix} 0 \\ 1 \\ \vdots \\ 0 \end{pmatrix}$, etc.

If row operations are done to a matrix A to obtain one in row reduced echelon form, and the ith row of this row reduced matrix is a pivot column, then we also refer to the ith column of A as a pivot column.

Definition 3.8. Given a matrix A, row reduction produces one and only one row reduced matrix B with $A \sim B$. See Corollary 3.1. We call B **the** row reduced echelon form of A.

Theorem 3.2. *Let A be an $m \times n$ matrix. Then A has a row reduced echelon form determined by a simple process.*

Proof. Viewing the columns of A from left to right, take the first nonzero column. Pick a nonzero entry in this column and switch the row containing this entry with the top row of A. Now divide this new top row by the value of this nonzero entry to get a 1 in this position and then use row operations to make all entries below this equal to zero. Thus the first nonzero column is now $\mathbf{e}_1$. Denote the resulting matrix by A_1. Consider the sub-matrix of A_1 to the right of this column and below the first row. Do exactly the same thing for this sub-matrix that was done for A. This time the $\mathbf{e}_1$ will refer to F^{m-1}. Use the first 1 obtained by the above process which is in the top row of this sub-matrix and row operations, to produce a zero in place of every entry above it and below it. Call the resulting matrix A_2. Thus A_2 satisfies the conditions of the above definition up to the column just encountered. Continue this way till every column has been dealt with and the result must be in row reduced echelon form. $\square$

Example 3.3. Row reduce the following matrix to row reduced echelon form.

$$\begin{pmatrix} 2 & 1 & 7 & 7 \\ 0 & 2 & 6 & 5 \\ 1 & 1 & 5 & 5 \end{pmatrix}$$

For convenience, switch the top row and the bottom row[1]:

$$\begin{pmatrix} 1 & 1 & 5 & 5 \\ 0 & 2 & 6 & 5 \\ 2 & 1 & 7 & 7 \end{pmatrix}$$

Next take -2 times the top row and add to the bottom and then multiply the middle row by $1/2$:

$$\begin{pmatrix} 1 & 1 & 5 & 5 \\ 0 & 1 & 3 & 5/2 \\ 0 & -1 & -3 & -3 \end{pmatrix}$$

Next add the middle row to the bottom row and then take -1 times the middle row and add to the top:

$$\begin{pmatrix} 1 & 0 & 2 & 5/2 \\ 0 & 1 & 3 & 3 \\ 0 & 0 & 0 & -1/2 \end{pmatrix}$$

Next take 6 times the bottom row and add to the middle row and then take 5 times the bottom row and add to the top row. Finally, divide the bottom row by $-1/2$:

$$\begin{pmatrix} 1 & 0 & 2 & 0 \\ 0 & 1 & 3 & 0 \\ 0 & 0 & 0 & 1 \end{pmatrix}$$

[1]Numerical implementations would not make this switch because $2 > 1$

This matrix is in row reduced echelon form. Note how the third column is a linear combination of the preceding columns because it is not equal to $\mathbf{e}_3$.

As mentioned earlier, here is some terminology about pivot columns.

Definition 3.9. The first **pivot column** of A is the first nonzero column of A which becomes $\mathbf{e}_1$ in the row reduced echelon form. The next pivot column is the first column after this which becomes $\mathbf{e}_2$ in the row reduced echelon form. The third is the next column which becomes $\mathbf{e}_3$ in the row reduced echelon form and so forth.

The algorithm just described for obtaining a row reduced echelon form shows that these columns are well defined, but we will deal with this issue more carefully in Corollary 3.1 where we show that every matrix corresponds to exactly one row reduced echelon form.

Example 3.4. Determine the pivot columns for the matrix

$$A = \begin{pmatrix} 2 & 1 & 3 & 6 & 2 \\ 1 & 7 & 8 & 4 & 0 \\ 1 & 3 & 4 & -2 & 2 \end{pmatrix} \tag{3.5}$$

As described above, the kth pivot column of A is the column in which $\mathbf{e}_k$ appears in the row reduced echelon form for the first time in reading from left to right. A row reduced echelon form for A is

$$\begin{pmatrix} 1 & 0 & 1 & 0 & \frac{64}{35} \\ 0 & 1 & 1 & 0 & -\frac{4}{35} \\ 0 & 0 & 0 & 1 & -\frac{9}{35} \end{pmatrix}$$

It follows that columns 1,2, and 4 in A are pivot columns.

Note that from Lemma 3.5 the last column of A has a linear relationship to the first four columns. Namely

$$\begin{pmatrix} 2 \\ 0 \\ 2 \end{pmatrix} = \frac{64}{35}\begin{pmatrix} 2 \\ 1 \\ 1 \end{pmatrix} + \left(-\frac{4}{35}\right)\begin{pmatrix} 1 \\ 7 \\ 3 \end{pmatrix} + \left(-\frac{9}{35}\right)\begin{pmatrix} 6 \\ 4 \\ -2 \end{pmatrix}$$

This linear relationship is revealed by the row reduced echelon form but it was not apparent from the original matrix.

Definition 3.10. Two matrices A, B are said to be **row equivalent** if B can be obtained from A by a sequence of row operations. When A is row equivalent to B, we write $A \sim B$.

Proposition 3.1. In the notation of Definition 3.10. $A \sim A$. If $A \sim B$, then $B \sim A$. If $A \sim B$ and $B \sim C$, then $A \sim C$.

Proof. That $A \sim A$ is obvious. Consider the second claim. By Theorem 3.1, there exist elementary matrices $E_1, E_2, \cdots, E_m$ such that

$$B = E_1 E_2 \cdots E_m A.$$

It follows from Lemma 3.4 that $(E_1E_2\cdots E_m)^{-1}$ exists and equals the product of the inverses of these matrices in the reverse order. Thus

$$E_m^{-1}E_{m-1}^{-1}\cdots E_1^{-1}B = (E_1E_2\cdots E_m)^{-1}B$$

$$= (E_1E_2\cdots E_m)^{-1}(E_1E_2\cdots E_m)A = A.$$

By Theorem 3.1, each E_k^{-1} is an elementary matrix. By Theorem 3.1 again, the above shows that A results from a sequence of row operations applied to B. The last claim is left for an exercise. See Exercise 7 on Page 87. This proves the proposition. □

There are three choices for row operations at each step in Theorem 3.2. A natural question is whether the same row reduced echelon matrix always results in the end from following any sequence of row operations.

We have already made use of the following observation in finding a linear relationship between the columns of the matrix A in (3.5), but here it is stated more formally.

$$\begin{pmatrix} x_1 \\ \vdots \\ x_n \end{pmatrix} = x_1\mathbf{e}_1 + \cdots + x_n\mathbf{e}_n,$$

so to say two column vectors are equal, is to say the column vectors are the same linear combination of the special vectors $\mathbf{e}_j$.

Corollary 3.1. *The row reduced echelon form is unique. That is if B, C are two matrices in row reduced echelon form and both are obtained from A by a sequence of row operations, then $B = C$.*

Proof. Suppose B and C are both row reduced echelon forms for the matrix A. It follows that B and C have zero columns in the same positions because row operations do not affect zero columns. By Proposition 3.1, B and C are row equivalent. In reading from left to right in B, suppose $\mathbf{e}_1, \cdots, \mathbf{e}_r$ occur first in positions $i_1, \cdots, i_r$ respectively. Then from the description of the row reduced echelon form, each of these columns of B, in positions $i_1, \cdots, i_r$, is not a linear combination of the preceding columns. Since C is row equivalent to B, it follows from Lemma 3.5, that each column of C in positions $i_1, \cdots, i_r$ is not a linear combination of the preceding columns of C. By the description of the row reduced echelon form, $\mathbf{e}_1, \cdots, \mathbf{e}_r$ occur first in C, in positions $i_1, \cdots, i_r$ respectively. Therefore, both B and C have the sequence $\mathbf{e}_1, \mathbf{e}_2, \cdots, \mathbf{e}_r$ occurring first (reading from left to right) in the positions, $i_1, i_2, \cdots, i_r$. Since these matrices are row equivalent, it follows from Lemma 3.5, that the columns between the i_k and i_{k+1} position in the two matrices are linear combinations involving the same scalars, of the columns in the $i_1, \cdots, i_k$ position. Similarly, the columns after the i_r position are linear combinations of the columns in the $i_1, \cdots, i_r$ positions involving the same scalars in both matrices. This is equivalent to the assertion that each of these columns is identical in B and C. □

Now with the above corollary, here is a very fundamental observation. The number of nonzero rows in the row reduced echelon form is the same as the number of pivot columns. Namely, this number is r in both cases where $\mathbf{e}_1, \cdots, \mathbf{e}_r$ are the pivot columns in the row reduced echelon form. This number r is called the **rank** of the matrix. This is discussed more later, but first here are some other applications.

Consider a matrix which looks like this: (More columns than rows.)

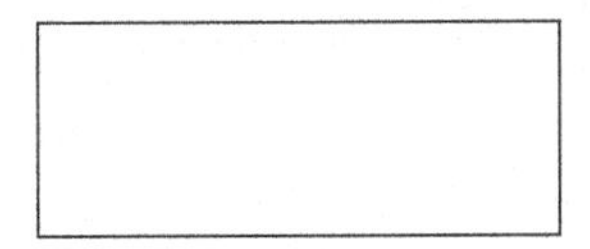

Corollary 3.2. *Suppose A is an $m \times n$ matrix and that $m < n$. That is, the number of rows is less than the number of columns. Then one of the columns of A is a linear combination of the preceding columns of A. Also, there exists $\mathbf{x} \in F^n$ such that $\mathbf{x} \neq \mathbf{0}$ and $A\mathbf{x} = \mathbf{0}$.*

Proof. Since $m < n$, not all the columns of A can be pivot columns. In reading from left to right, pick the first one which is not a pivot column. Then from the description of the row reduced echelon form, this column is a linear combination of the preceding columns. Say

$$\mathbf{a}_j = x_1\mathbf{a}_1 + \cdots + x_{j-1}\mathbf{a}_{j-1}.$$

Therefore, from the way we multiply a matrix times a vector,

$$A \begin{pmatrix} x_1 \\ \vdots \\ x_{j-1} \\ -1 \\ 0 \\ \vdots \\ 0 \end{pmatrix} = (\mathbf{a}_1 \cdots \mathbf{a}_{j-1}\mathbf{a}_j \cdots \mathbf{a}_n) \begin{pmatrix} x_1 \\ \vdots \\ x_{j-1} \\ -1 \\ 0 \\ \vdots \\ 0 \end{pmatrix} = \mathbf{0}. \quad \square$$

Example 3.5. Find the row reduced echelon form of the matrix

$$\begin{pmatrix} 0 & 0 & 2 & 3 \\ 0 & 2 & 0 & 1 \\ 0 & 1 & 1 & 5 \end{pmatrix}.$$

The first nonzero column is the second in the matrix. We switch the third and first rows to obtain

$$\begin{pmatrix} 0 & 1 & 1 & 5 \\ 0 & 2 & 0 & 1 \\ 0 & 0 & 2 & 3 \end{pmatrix} \overset{\text{-2}\times\text{top+second}}{\rightarrow} \begin{pmatrix} 0 & 1 & 1 & 5 \\ 0 & 0 & -2 & -9 \\ 0 & 0 & 2 & 3 \end{pmatrix}$$

$$\overset{\text{second+bottom}}{\to} \begin{pmatrix} 0 & 1 & 1 & 5 \\ 0 & 0 & -2 & -9 \\ 0 & 0 & 0 & -6 \end{pmatrix} \to \begin{pmatrix} 0 & 1 & 1 & 5 \\ 0 & 0 & -2 & -9 \\ 0 & 0 & 0 & 1 \end{pmatrix}$$

Next use the bottom row to obtain zeros in the last column above the 1 and divide the second row by -2

$$\begin{pmatrix} 0 & 1 & 1 & 0 \\ 0 & 0 & 1 & 0 \\ 0 & 0 & 0 & 1 \end{pmatrix} \overset{\text{-1}\times\text{second+top}}{\to} \begin{pmatrix} 0 & 1 & 0 & 0 \\ 0 & 0 & 1 & 0 \\ 0 & 0 & 0 & 1 \end{pmatrix}$$

This is in row reduced echelon form.

Example 3.6. Find the row reduced echelon form for the matrix

$$\begin{pmatrix} 1 & 2 & 0 & 2 \\ -1 & 3 & 4 & 3 \\ 0 & 5 & 4 & 5 \end{pmatrix}$$

You should verify that the row reduced echelon form is

$$\begin{pmatrix} 1 & 0 & -\frac{8}{5} & 0 \\ 0 & 1 & \frac{4}{5} & 1 \\ 0 & 0 & 0 & 0 \end{pmatrix}.$$

3.3 Finding the inverse of a matrix

We have already discussed the idea of an inverse of a matrix in Definition 3.3. Recall that the inverse of an $n \times n$ matrix A is a matrix B such that

$$AB = BA = I$$

where I is the identity matrix discussed in Theorem 3.1, it was shown that an elementary matrix is invertible and that its inverse is also an elementary matrix. We also showed in Lemma 3.4 that the product of invertible matrices is invertible and that the inverse of this product is the product of the inverses in the reverse order. In this section, we consider the problem of finding an inverse for a given $n \times n$ matrix.

Unlike ordinary multiplication of numbers, it can happen that $A \neq 0$ but A may fail to have an inverse. This is illustrated in the following example.

Example 3.7. Let $A = \begin{pmatrix} 1 & 1 \\ 1 & 1 \end{pmatrix}$. Does A have an inverse?

One might think A would have an inverse because it does not equal zero. However,

$$\begin{pmatrix} 1 & 1 \\ 1 & 1 \end{pmatrix} \begin{pmatrix} -1 \\ 1 \end{pmatrix} = \begin{pmatrix} 0 \\ 0 \end{pmatrix}$$

and if A^{-1} existed, this could not happen because you could write

$$\begin{pmatrix} 0 \\ 0 \end{pmatrix} = A^{-1}\begin{pmatrix} 0 \\ 0 \end{pmatrix} = A^{-1}\left(A\begin{pmatrix} -1 \\ 1 \end{pmatrix}\right) =$$

$$= (A^{-1}A)\begin{pmatrix} -1 \\ 1 \end{pmatrix} = I\begin{pmatrix} -1 \\ 1 \end{pmatrix} = \begin{pmatrix} -1 \\ 1 \end{pmatrix},$$

a contradiction. Here we have used the associative law of matrix multiplication found in Proposition 2.1. Thus the answer is that A does not have an inverse.

Example 3.8. Let $A = \begin{pmatrix} 1 & 1 \\ 1 & 2 \end{pmatrix}$. Show that $\begin{pmatrix} 2 & -1 \\ -1 & 1 \end{pmatrix}$ is the inverse of A.

To check this, multiply

$$\begin{pmatrix} 1 & 1 \\ 1 & 2 \end{pmatrix}\begin{pmatrix} 2 & -1 \\ -1 & 1 \end{pmatrix} = \begin{pmatrix} 1 & 0 \\ 0 & 1 \end{pmatrix},$$

and

$$\begin{pmatrix} 2 & -1 \\ -1 & 1 \end{pmatrix}\begin{pmatrix} 1 & 1 \\ 1 & 2 \end{pmatrix} = \begin{pmatrix} 1 & 0 \\ 0 & 1 \end{pmatrix},$$

showing that this matrix is indeed the inverse of A.

In the last example, how would you find A^{-1}? You wish to find a matrix $\begin{pmatrix} x & z \\ y & w \end{pmatrix}$ such that

$$\begin{pmatrix} 1 & 1 \\ 1 & 2 \end{pmatrix}\begin{pmatrix} x & z \\ y & w \end{pmatrix} = \begin{pmatrix} 1 & 0 \\ 0 & 1 \end{pmatrix}.$$

This requires the solution of the systems of equations,

$$x + y = 1, x + 2y = 0$$

and

$$z + w = 0, z + 2w = 1.$$

Writing the augmented matrix for these two systems gives

$$\left(\begin{array}{cc|c} 1 & 1 & 1 \\ 1 & 2 & 0 \end{array}\right) \tag{3.6}$$

for the first system and

$$\left(\begin{array}{cc|c} 1 & 1 & 0 \\ 1 & 2 & 1 \end{array}\right) \tag{3.7}$$

for the second. Let's solve the first system. Take (-1) times the first row and add to the second to get

$$\left(\begin{array}{cc|c} 1 & 1 & 1 \\ 0 & 1 & -1 \end{array}\right)$$

Now take (-1) times the second row and add to the first to get

$$\left(\begin{array}{cc|c} 1 & 0 & 2 \\ 0 & 1 & -1 \end{array}\right).$$

Putting in the variables, this says $x = 2$ and $y = -1$.

Now solve the second system, (3.7) to find z and w. Take (-1) times the first row and add to the second to get

$$\left(\begin{array}{cc|c} 1 & 1 & 0 \\ 0 & 1 & 1 \end{array}\right).$$

Now take (-1) times the second row and add to the first to get

$$\left(\begin{array}{cc|c} 1 & 0 & -1 \\ 0 & 1 & 1 \end{array}\right).$$

Putting in the variables, this says $z = -1$ and $w = 1$. Therefore, the inverse is

$$\begin{pmatrix} 2 & -1 \\ -1 & 1 \end{pmatrix}.$$

Didn't the above seem rather repetitive? Exactly the same row operations were used in both systems. In each case, the end result was something of the form $(I|\mathbf{v})$ where I is the identity and $\mathbf{v}$ gave a column of the inverse. In the above $\begin{pmatrix} x \\ y \end{pmatrix}$, the first column of the inverse was obtained first and then the second column $\begin{pmatrix} z \\ w \end{pmatrix}$.

To simplify this procedure, you could have written

$$\left(\begin{array}{cc|cc} 1 & 1 & 1 & 0 \\ 1 & 2 & 0 & 1 \end{array}\right)$$

and row reduced till you obtained

$$\left(\begin{array}{cc|cc} 1 & 0 & 2 & -1 \\ 0 & 1 & -1 & 1 \end{array}\right).$$

Then you could have read off the inverse as the 2×2 matrix on the right side. You should be able to see that it is valid by adapting the argument used in the simple case above.

This is the reason for the following simple procedure for finding the inverse of a matrix. This procedure is called the **Gauss-Jordan procedure**.

Procedure 3.3. Suppose A is an $n \times n$ matrix. To find A^{-1} if it exists, form the augmented $n \times 2n$ matrix

$$(A|I)$$

and then if possible, do row operations until you obtain an $n \times 2n$ matrix of the form

$$(I|B)\,. \tag{3.8}$$

When this has been done, $B = A^{-1}$. If it is impossible to row reduce to a matrix of the form $(I|B)$, then A has no inverse.

The procedure just described along with the preceding explanation shows that this procedure actually yields a **right inverse**. This is a matrix B such that $AB = I$. We will show in Theorem 3.4 that this right inverse is really **the** inverse. This is a stronger result than that of Lemma 3.4 about the uniqueness of the inverse. For now, here are some examples.

Example 3.9. Let $A = \begin{pmatrix} 1 & 2 & 2 \\ 1 & 0 & 2 \\ 3 & 1 & -1 \end{pmatrix}$. Find A^{-1} if it exists.

Set up the augmented matrix $(A|I)$:

$$\left(\begin{array}{ccc|ccc} 1 & 2 & 2 & 1 & 0 & 0 \\ 1 & 0 & 2 & 0 & 1 & 0 \\ 3 & 1 & -1 & 0 & 0 & 1 \end{array}\right)$$

Next take (-1) times the first row and add to the second followed by (-3) times the first row added to the last. This yields

$$\left(\begin{array}{ccc|ccc} 1 & 2 & 2 & 1 & 0 & 0 \\ 0 & -2 & 0 & -1 & 1 & 0 \\ 0 & -5 & -7 & -3 & 0 & 1 \end{array}\right).$$

Then take 5 times the second row and add to -2 times the last row.

$$\left(\begin{array}{ccc|ccc} 1 & 2 & 2 & 1 & 0 & 0 \\ 0 & -10 & 0 & -5 & 5 & 0 \\ 0 & 0 & 14 & 1 & 5 & -2 \end{array}\right)$$

Next take the last row and add to (-7) times the top row. This yields

$$\left(\begin{array}{ccc|ccc} -7 & -14 & 0 & -6 & 5 & -2 \\ 0 & -10 & 0 & -5 & 5 & 0 \\ 0 & 0 & 14 & 1 & 5 & -2 \end{array}\right).$$

Now take $(-7/5)$ times the second row and add to the top.

$$\left(\begin{array}{ccc|ccc} -7 & 0 & 0 & 1 & -2 & -2 \\ 0 & -10 & 0 & -5 & 5 & 0 \\ 0 & 0 & 14 & 1 & 5 & -2 \end{array}\right).$$

Finally divide the top row by -7, the second row by -10 and the bottom row by 14, which yields

$$\left(\begin{array}{ccc|ccc} 1 & 0 & 0 & -\frac{1}{7} & \frac{2}{7} & \frac{2}{7} \\ 0 & 1 & 0 & \frac{1}{2} & -\frac{1}{2} & 0 \\ 0 & 0 & 1 & \frac{1}{14} & \frac{5}{14} & -\frac{1}{7} \end{array}\right).$$

Therefore, the inverse is

$$\begin{pmatrix} -\frac{1}{7} & \frac{2}{7} & \frac{2}{7} \\ \frac{1}{2} & -\frac{1}{2} & 0 \\ \frac{1}{14} & \frac{5}{14} & -\frac{1}{7} \end{pmatrix}.$$

Example 3.10. Let $A = \begin{pmatrix} 1 & 2 & 2 \\ 1 & 0 & 2 \\ 2 & 2 & 4 \end{pmatrix}$. Find A^{-1} if it exists.

Write the augmented matrix $(A|I)$:

$$\left(\begin{array}{ccc|ccc} 1 & 2 & 2 & 1 & 0 & 0 \\ 1 & 0 & 2 & 0 & 1 & 0 \\ 2 & 2 & 4 & 0 & 0 & 1 \end{array}\right)$$

and proceed to do row operations attempting to obtain $\left(I|A^{-1}\right)$. Take (-1) times the top row and add to the second. Then take (-2) times the top row and add to the bottom:

$$\left(\begin{array}{ccc|ccc} 1 & 2 & 2 & 1 & 0 & 0 \\ 0 & -2 & 0 & -1 & 1 & 0 \\ 0 & -2 & 0 & -2 & 0 & 1 \end{array}\right)$$

Next add (-1) times the second row to the bottom row:

$$\left(\begin{array}{ccc|ccc} 1 & 2 & 2 & 1 & 0 & 0 \\ 0 & -2 & 0 & -1 & 1 & 0 \\ 0 & 0 & 0 & -1 & -1 & 1 \end{array}\right)$$

At this point, you can see that there will be no inverse. To see this, consider the matrix

$$\left(\begin{array}{ccc|c} 1 & 2 & 2 & 0 \\ 0 & -2 & 0 & 0 \\ 0 & 0 & 0 & 1 \end{array}\right)$$

which is an augmented matrix for a system of equations which has no solution. In other words, the last column is not a linear combination of the first three. The inverses of the row operations used to obtain

$$\begin{pmatrix} 1 & 2 & 2 \\ 0 & -2 & 0 \\ 0 & 0 & 0 \end{pmatrix}$$

from A can be performed in reverse order to obtain an augmented matrix $(A|\mathbf{b})$ and by Lemma 3.5, $\mathbf{b}$ is not a linear combination of the columns of A. Thus there is no solution $\mathbf{x}$ to the system

$$A\mathbf{x} = \mathbf{b}$$

This is impossible if A has an inverse because you could simply multiply on both sides by A^{-1} resulting in

$$\mathbf{x} = A^{-1}(A\mathbf{x}) = \left(A^{-1}A\right)\mathbf{x} = I\mathbf{x} = A^{-1}\mathbf{b}$$

Similar considerations are valid in general. A row of zeros in the row reduced echelon form for A indicates that A^{-1} does not exist.

Example 3.11. Let $A = \begin{pmatrix} 1 & 0 & 1 \\ 1 & -1 & 1 \\ 1 & 1 & -1 \end{pmatrix}$. Find A^{-1} if it exists.

Form the augmented matrix

$$\left(\begin{array}{ccc|ccc} 1 & 0 & 1 & 1 & 0 & 0 \\ 1 & -1 & 1 & 0 & 1 & 0 \\ 1 & 1 & -1 & 0 & 0 & 1 \end{array}\right).$$

Now do row operations until the $n \times n$ matrix on the left becomes the identity matrix. This yields after some computations,

$$\left(\begin{array}{ccc|ccc} 1 & 0 & 0 & 0 & \frac{1}{2} & \frac{1}{2} \\ 0 & 1 & 0 & 1 & -1 & 0 \\ 0 & 0 & 1 & 1 & -\frac{1}{2} & -\frac{1}{2} \end{array}\right)$$

and so the inverse of A is the matrix on the right,

$$\begin{pmatrix} 0 & 1/2 & 1/2 \\ 1 & -1 & 0 \\ 1 & -1/2 & -1/2 \end{pmatrix}.$$

Checking the answer is easy. Just multiply the matrices and see if your answer works.

$$\begin{pmatrix} 1 & 0 & 1 \\ 1 & -1 & 1 \\ 1 & 1 & -1 \end{pmatrix}\begin{pmatrix} 0 & 1/2 & 1/2 \\ 1 & -1 & 0 \\ 1 & -1/2 & -1/2 \end{pmatrix} = \begin{pmatrix} 1 & 0 & 0 \\ 0 & 1 & 0 \\ 0 & 0 & 1 \end{pmatrix}.$$

Always check your answer because if you are like some of us, you will usually have made a mistake.

As mentioned earlier, what you have really found in the above algorithm is a **right inverse.** Is this right inverse matrix, which we have called the inverse, really **the** inverse, the matrix which when multiplied on both sides gives the identity?

Theorem 3.4. *Suppose A, B are $n \times n$ matrices and $AB = I$. Then it follows that $BA = I$ also, and so $B = A^{-1}$. For $n \times n$ matrices, the left inverse, right inverse and inverse are all the same thing.*

Proof. If $AB = I$ for A, B $n \times n$ matrices, is $BA = I$? If $AB = I$, there exists a unique solution $\mathbf{x}$ to the equation

$$B\mathbf{x} = \mathbf{y}$$

for any choice of $\mathbf{y}$. In fact,

$$\mathbf{x} = A(B\mathbf{x}) = A\mathbf{y}.$$

This means the row reduced echelon form of B must be I. Thus every column is a pivot column. Otherwise, there exists a free variable and the solution, if it exists, would not be unique, contrary to what was just shown must happen if $AB = I$. It follows that a right inverse B^{-1} for B exists. The above procedure yields

$$\begin{pmatrix} B & I \end{pmatrix} \to \begin{pmatrix} I & B^{-1} \end{pmatrix}.$$

Now multiply both sides of the equation $AB = I$ on the right by B^{-1}. Then

$$A = A\left(BB^{-1}\right) = (AB)B^{-1} = B^{-1}.$$

Thus A is the right inverse of B, and so $BA = I$. This shows that if $AB = I$, then $BA = I$ also. Exchanging roles of A and B, we see that if $BA = I$, then $AB = I$. This proves the theorem. □

This has shown that in the context of $n \times n$ matrices, right inverses, left inverses and inverses are all the same and this matrix is called A^{-1}.

The following corollary is also of interest.

Corollary 3.3. *An $n \times n$ matrix A has an inverse if and only if the row reduced echelon form of A is I.*

Proof. First suppose the row reduced echelon form of A is I. Then Procedure 3.3 yields a right inverse for A. By Theorem 3.4 this is **the** inverse. Next suppose A has an inverse. Then there exists a unique solution $\mathbf{x}$ to the equation

$$A\mathbf{x} = \mathbf{y}$$

given by $\mathbf{x} = A^{-1}\mathbf{y}$. It follows that in the augmented matrix $(A|\mathbf{0})$ there are no free variables, and so every column to the left of — is a pivot column. Therefore, the row reduced echelon form of A is I. □

Example 3.12. In this example, it is shown how to use the inverse of a matrix to find the solution to a system of equations. Consider the following system of equations. Use the inverse of a suitable matrix to give the solutions to this system.

$$\begin{aligned} x + z &= 1 \\ x - y + z &= 3 \\ x + y - z &= 2 \end{aligned}$$

The system of equations can be written in terms of matrices as

$$\begin{pmatrix} 1 & 0 & 1 \\ 1 & -1 & 1 \\ 1 & 1 & -1 \end{pmatrix} \begin{pmatrix} x \\ y \\ z \end{pmatrix} = \begin{pmatrix} 1 \\ 3 \\ 2 \end{pmatrix}. \tag{3.9}$$

More simply, this is of the form $A\mathbf{x} = \mathbf{b}$. Suppose you find the inverse of the matrix A^{-1}. Then you could multiply both sides of this equation by A^{-1} to obtain

$$\mathbf{x} = \left(A^{-1}A\right)\mathbf{x} = A^{-1}(A\mathbf{x}) = A^{-1}\mathbf{b}.$$

This gives the solution as $\mathbf{x} = A^{-1}\mathbf{b}$. Note that once you have found the inverse, you can easily get the solution for different right hand sides without any effort. The solution is always $A^{-1}\mathbf{b}$. In the given example, the inverse of the matrix is

$$\begin{pmatrix} 0 & \frac{1}{2} & \frac{1}{2} \\ 1 & -1 & 0 \\ 1 & -\frac{1}{2} & -\frac{1}{2} \end{pmatrix}$$

This was shown in Example 3.11. Therefore, from what was just explained, the solution to the given system is

$$\begin{pmatrix} x \\ y \\ z \end{pmatrix} = \begin{pmatrix} 0 & \frac{1}{2} & \frac{1}{2} \\ 1 & -1 & 0 \\ 1 & -\frac{1}{2} & -\frac{1}{2} \end{pmatrix} \begin{pmatrix} 1 \\ 3 \\ 2 \end{pmatrix} = \begin{pmatrix} \frac{5}{2} \\ -2 \\ -\frac{3}{2} \end{pmatrix}.$$

What if the right side of (3.9) had been

$$\begin{pmatrix} 0 \\ 1 \\ 3 \end{pmatrix}?$$

What would be the solution to

$$\begin{pmatrix} 1 & 0 & 1 \\ 1 & -1 & 1 \\ 1 & 1 & -1 \end{pmatrix} \begin{pmatrix} x \\ y \\ z \end{pmatrix} = \begin{pmatrix} 0 \\ 1 \\ 3 \end{pmatrix}?$$

By the above discussion, this solution is

$$\begin{pmatrix} x \\ y \\ z \end{pmatrix} = \begin{pmatrix} 0 & \frac{1}{2} & \frac{1}{2} \\ 1 & -1 & 0 \\ 1 & -\frac{1}{2} & -\frac{1}{2} \end{pmatrix} \begin{pmatrix} 0 \\ 1 \\ 3 \end{pmatrix} = \begin{pmatrix} 2 \\ -1 \\ -2 \end{pmatrix}.$$

This illustrates why, once you have found the inverse of a given matrix, you can use this inverse to solve many different systems easily.

3.4 The rank of a matrix

With the existence and uniqueness of the row reduced echelon form, it is a natural step to define the rank of a matrix.

Definition 3.11. Let A be an $m \times n$ matrix. Then the **rank** of A is defined to be the number of pivot columns. From the description of the row reduced echelon form, this is also equal to the number of nonzero rows in the row reduced echelon form. The **nullity** of A is the number of non pivot columns. This is the same as the number of free variables in the augmented matrix $(A|\mathbf{0})$.

The rank is important because of the following proposition.

Proposition 3.2. Let A be an $m \times n$ matrix which has rank r. Then there exists a set of r columns of A such that every other column is a linear combination of these columns. Furthermore, none of these columns is a linear combination of the other $r-1$ columns in the set. The rank of A is no larger than the minimum of m and n. Also the rank added to the nullity equals n.

Proof. Since the rank of A is r it follows that A has exactly r pivot columns. Thus, in the row reduced echelon form, every column is a linear combination of these pivot columns and none of the pivot columns is a linear combination of the others pivot columns. By Lemma 3.5 the same is true of the columns in the original matrix A. There are at most $\min(m, n)$ pivot columns (nonzero rows). Therefore, the rank of A is no larger than $\min(m, n)$ as claimed. Since every column is either a pivot column or isn't a pivot column, this shows that the rank added to nullity equals n. $\square$

This will be discussed more later when we define the concept of a **vector space** and a **basis.** We will see that the pivot columns form a basis for a vector space called $\mathrm{Col}(A)$, the **column space** of A.

3.5 The *LU* and *PLU* factorization

For A an $m \times n$ matrix, the LU factorization is of the form $A = LU$ where L is an $m \times m$ lower triangular matrix which has all ones down the main diagonal and U is an "upper triangular" $m \times n$ matrix, one with the property that $U_{ij} = 0$ if $i > j$. It turns out that not all matrices have an LU factorization, but there are many very important examples of matrices which do. However, the existence of matrices which do not have an LU factorization shows that the method lacks generality. Consider the following example.

Example 3.13. Can you write $\begin{pmatrix} 0 & 1 \\ 1 & 0 \end{pmatrix}$ in the form LU as just described?

To do so you would need

$$\begin{pmatrix} 1 & 0 \\ x & 1 \end{pmatrix} \begin{pmatrix} a & b \\ 0 & c \end{pmatrix} = \begin{pmatrix} a & b \\ xa & xb+c \end{pmatrix} = \begin{pmatrix} 0 & 1 \\ 1 & 0 \end{pmatrix}.$$

Therefore, $b = 1$ and $a = 0$. Also, from the bottom rows, $xa = 1$ which can't happen and have $a = 0$. Therefore, you can't write this matrix in the form LU. It has no LU factorization. This is what we mean above by saying the method lacks generality.

The problem with this example is that in order to row reduce the matrix to row reduced echelon form, you must switch some rows. It turns out that a matrix has an LU factorization if and only if it is possible to row reduce the given matrix to upper triangular form using only row operations of type 3. Recall that this row operation involves replacing a given row by a multiple of another row added to the given row.

Lemma 3.6. *Let L be a lower (upper) triangular matrix $m \times m$ which has ones down the main diagonal. Then L^{-1} also is a lower (upper) triangular matrix which has ones down the main diagonal. In the case that L is of the form*

$$L = \begin{pmatrix} 1 & & & \\ a_1 & 1 & & \\ \vdots & & \ddots & \\ a_n & & & 1 \end{pmatrix} \tag{3.10}$$

where all entries are zero except for the left column and main diagonal, it is also the case that L^{-1} is obtained from L by simply multiplying each entry below the main diagonal in L with -1. The same is true if the single nonzero column is in another position.

Proof. Consider the usual setup for finding the inverse $\begin{pmatrix} L & I \end{pmatrix}$. Then each row operation done to L to reduce to row reduced echelon form results in changing only the entries in I below the main diagonal. In the special case of L given in (3.10) or the single nonzero column is in another position, the matrix described in the lemma is of the form

$$\begin{pmatrix} 1 & & & \\ -a_1 & 1 & & \\ \vdots & & \ddots & \\ -a_n & & & 1 \end{pmatrix} = \begin{pmatrix} 1 & & & \\ -a_1 & 1 & & \\ \vdots & & \ddots & \\ 0 & & & 1 \end{pmatrix} \begin{pmatrix} 1 & & & \\ 0 & 1 & & \\ -a_2 & & \ddots & \\ \vdots & & & 1 \end{pmatrix}$$

$$\cdots \begin{pmatrix} 1 & & & \\ 0 & 1 & & \\ \vdots & & \ddots & \\ -a_n & & & 1 \end{pmatrix}$$

Thus this matrix is the product of elementary matrices which accomplish the row operations necessary to zero out the left column of L beneath the 1 in the upper left corner. If the non-zero column were elsewhere, the same reasoning applies. □

Now let A be an $m \times n$ matrix, say

$$A = \begin{pmatrix} a_{11} & a_{12} & \cdots & a_{1n} \\ a_{21} & a_{22} & \cdots & a_{2n} \\ \vdots & \vdots & & \vdots \\ a_{m1} & a_{m2} & \cdots & a_{mn} \end{pmatrix}$$

and assume A can be row reduced to an upper triangular form using only row operation 3. Thus, in particular, $a_{11} \neq 0$. Multiply on the left by $E_1 =$

$$\begin{pmatrix} 1 & 0 & \cdots & 0 \\ -\frac{a_{21}}{a_{11}} & 1 & \cdots & 0 \\ \vdots & \vdots & \ddots & \vdots \\ -\frac{a_{m1}}{a_{11}} & 0 & \cdots & 1 \end{pmatrix}$$

This is the product of elementary matrices which make modifications in the first column only. It is equivalent to taking $-a_{21}/a_{11}$ times the first row and adding to the second. Then taking $-a_{21}/a_{11}$ times the first row and adding to the second and so forth. Thus the result is of the form

$$E_1 A = \begin{pmatrix} a_{11} & a_{12} & \cdots & a'_{1n} \\ 0 & a'_{22} & \cdots & a'_{2n} \\ \vdots & \vdots & & \vdots \\ 0 & a'_{m2} & \cdots & a'_{mn} \end{pmatrix}$$

By assumption, $a'_{22} \neq 0$ and so it is possible to use this entry to zero out all the entries below it in the matrix on the right by multiplication by a matrix of the form $E_2 = \begin{pmatrix} 1 & \mathbf{0} \\ \mathbf{0} & E \end{pmatrix}$ where E is an $(m-1) \times (m-1)$ matrix of the form

$$\begin{pmatrix} 1 & 0 & \cdots & 0 \\ -\frac{a'_{32}}{a'_{22}} & 1 & \cdots & 0 \\ \vdots & \vdots & \ddots & \vdots \\ -\frac{a'_{m2}}{a'_{22}} & 0 & \cdots & 1 \end{pmatrix}$$

Continuing this way, zeroing out the entries below the diagonal entries, finally leads to

$$E_{m-1} E_{n-2} \cdots E_1 A = U$$

where U is upper triangular. Each E_j has all ones down the main diagonal and is lower triangular. Now multiply both sides by the inverses of the E_j. This yields

$$A = E_1^{-1} E_2^{-1} \cdots E_{m-1}^{-1} U$$

By Lemma 3.6, this implies that the product of those E_j^{-1} is a lower triangular matrix having all ones down the main diagonal.

The above discussion and lemma also gives a convenient way to compute an LU factorization. The expressions $-a_{21}/a_{11}, -a_{31}/a_{11}, \cdots - a_{m1}/a_{11}$ denoted respectively by $m_{21}, \cdots, m_{m1}$ to save notation which were obtained in building E_1 are called **multipliers.** . Then according to the lemma, to find E_1^{-1} you simply write

$$\begin{pmatrix} 1 & 0 & \cdots & 0 \\ -m_{21} & 1 & \cdots & 0 \\ \vdots & \vdots & \ddots & \vdots \\ -m_{m1} & 0 & \cdots & 1 \end{pmatrix}$$

Similar considerations apply to the other E_j^{-1}. Thus L is of the form

$$\begin{pmatrix} 1 & 0 & \cdots & 0 & 0 \\ -m_{21} & 1 & \cdots & 0 & 0 \\ \vdots & \vdots & \ddots & \vdots & \vdots \\ -m_{(m-1)1} & 0 & \cdots & 1 & 0 \\ -m_{m1} & 0 & \cdots & 0 & 1 \end{pmatrix} \begin{pmatrix} 1 & 0 & \cdots & 0 & 0 \\ 0 & 1 & \cdots & 0 & 0 \\ \vdots & -m_{32} & \ddots & \vdots & \vdots \\ 0 & \vdots & \cdots & 1 & 0 \\ 0 & -m_{m2} & \cdots & 0 & 1 \end{pmatrix} \cdots$$

$$\begin{pmatrix} 1 & 0 & \cdots & 0 & 0 \\ 0 & 1 & \cdots & 0 & 0 \\ \vdots & 0 & \ddots & \vdots & \vdots \\ 0 & \vdots & \cdots & 1 & 0 \\ 0 & 0 & \cdots & -m_{mm-1} & 1 \end{pmatrix}$$

It follows from Theorem 3.1 about the effect of multiplying on the left by an elementary matrix that the above product is of the form

$$\begin{pmatrix} 1 & 0 & \cdots & 0 & 0 \\ -m_{21} & 1 & \cdots & 0 & 0 \\ \vdots & -m_{32} & \ddots & \vdots & \vdots \\ -m_{(m-1)1} & \vdots & \cdots & 1 & 0 \\ -m_{m1} & -m_{m2} & \cdots & -m_{mm-1} & 1 \end{pmatrix}$$

In words, you simply insert, into the corresponding position in the identity matrix, -1 times the multiplier which was used to zero out an entry in that position below the main diagonal in A, while retaining the main diagonal which consists entirely of ones. This is L. The following example shows that beginning with the expression $A = IA$, one can obtain LU by updating the identity matrix while doing the row operations on the original matrix.

Example 3.14. Find an LU factorization for $A = \begin{pmatrix} 1 & 2 & 1 & 2 & 1 \\ 2 & 0 & 2 & 1 & 1 \\ 2 & 3 & 1 & 3 & 2 \\ 1 & 0 & 1 & 1 & 2 \end{pmatrix}$.

Write $A =$

$$\begin{pmatrix} 1 & 0 & 0 & 0 \\ 0 & 1 & 0 & 0 \\ 0 & 0 & 1 & 0 \\ 0 & 0 & 0 & 1 \end{pmatrix} \begin{pmatrix} 1 & 2 & 1 & 2 & 1 \\ 2 & 0 & 2 & 1 & 1 \\ 2 & 3 & 1 & 3 & 2 \\ 1 & 0 & 1 & 1 & 2 \end{pmatrix}$$

First multiply the first row by (-1) and then add to the last row in the second matrix above. Next take (-2) times the first and add to the second and then (-2)

times the first and add to the third.

$$\begin{pmatrix} 1 & 0 & 0 & 0 \\ 2 & 1 & 0 & 0 \\ 2 & 0 & 1 & 0 \\ 1 & 0 & 0 & 1 \end{pmatrix} \begin{pmatrix} 1 & 2 & 1 & 2 & 1 \\ 0 & -4 & 0 & -3 & -1 \\ 0 & -1 & -1 & -1 & 0 \\ 0 & -2 & 0 & -1 & 1 \end{pmatrix}.$$

This finishes the first column of L and the first column of U. Now take $-(1/4)$ times the second row in the matrix on the right and add to the third followed by $-(1/2)$ times the second added to the last.

$$\begin{pmatrix} 1 & 0 & 0 & 0 \\ 2 & 1 & 0 & 0 \\ 2 & 1/4 & 1 & 0 \\ 1 & 1/2 & 0 & 1 \end{pmatrix} \begin{pmatrix} 1 & 2 & 1 & 2 & 1 \\ 0 & -4 & 0 & -3 & -1 \\ 0 & 0 & -1 & -1/4 & 1/4 \\ 0 & 0 & 0 & 1/2 & 3/2 \end{pmatrix}$$

This finishes the second column of L as well as the second column of U. Since the matrix on the right is upper triangular, stop. The LU factorization has now been obtained. It is also worth noting that, from the above discussion, the product of the two matrices in the iteration will always equal the original matrix. This technique is called **Dolittle's method**.

So what happens in general? In general, it may be necessary to switch rows in order to row reduce A to an upper triangular form. If this happens, then in the above process you would encounter a permutation matrix. Therefore, you could first multiply A on the left by this permutation matrix so that the problem will no longer occur when you follow the above procedure. Thus, it is **always possible** to obtain a permutation matrix P such that

$$L^{-1}PA = U$$

where U is upper triangular and L^{-1} is lower triangular having all ones down the diagonal. It follows that A can always be written in the form

$$A = PLU.$$

Computer algebra systems are set up to compute a PLU factorization for a given matrix. For example, in Maple 15, you enter the matrix, then right click on it and select "solvers and forms". Then choose "LU decomposition" followed by "Gaussian elimination" and finally "Gaussian elimination (P, L, U)".

One other thing which is interesting to observe is that LU factorizations are not necessarily unique. See Problem 44 in the following exercises. Thus they don't always exist and they are not even unique when they do exist. How could something so mathematically inferior be of any interest?

One reason that this is of interest is that it takes about half the number of operations to produce an LU factorization as it does to find the row reduced echelon form, and once an LU factorization has been obtained, it becomes easy for a computer to find the solution to a system of equations. This is illustrated in the following example. However, it must be admitted that if you are doing it by hand, you will have an easier time if you do not use an *LU* factorization. Thus this technique is of use because it makes computers happy.

The existence of an *LU* factorization is also a hypothesis in certain theorems involving the QR algorithm for finding eigenvalues. This latter algorithm is of tremendous significance and is discussed briefly in a later exercise. In addition, the *LU* factorization leads to a nice factorization called the Cholesky factorization for positive definite symmetric matrices.

Example 3.15. Suppose you want to find the solutions to

$$\begin{pmatrix} 1 & 2 & 3 & 2 \\ 4 & 3 & 1 & 1 \\ 1 & 2 & 3 & 0 \end{pmatrix} \begin{pmatrix} x \\ y \\ z \\ w \end{pmatrix} = \begin{pmatrix} 1 \\ 2 \\ 3 \end{pmatrix}.$$

Of course one way is to write the augmented matrix and grind away. However, this involves more row operations than the computation of the *LU* factorization and it turns out that the *LU* factorization can give the solution quickly. Here is how. The following is an *LU* factorization for the matrix.

$$\begin{pmatrix} 1 & 2 & 3 & 2 \\ 4 & 3 & 1 & 1 \\ 1 & 2 & 3 & 0 \end{pmatrix} = \begin{pmatrix} 1 & 0 & 0 \\ 4 & 1 & 0 \\ 1 & 0 & 1 \end{pmatrix} \begin{pmatrix} 1 & 2 & 3 & 2 \\ 0 & -5 & -11 & -7 \\ 0 & 0 & 0 & -2 \end{pmatrix}.$$

Let $U\mathbf{x} = \mathbf{y}$ and consider $L\mathbf{y} = \mathbf{b}$ where in this case, $\mathbf{b} = (1, 2, 3)^t$. Thus

$$\begin{pmatrix} 1 & 0 & 0 \\ 4 & 1 & 0 \\ 1 & 0 & 1 \end{pmatrix} \begin{pmatrix} y_1 \\ y_2 \\ y_3 \end{pmatrix} = \begin{pmatrix} 1 \\ 2 \\ 3 \end{pmatrix}$$

which yields very quickly that $\mathbf{y} = \begin{pmatrix} 1 \\ -2 \\ 2 \end{pmatrix}$. Now you can find $\mathbf{x}$ by solving $U\mathbf{x} = \mathbf{y}$.

Thus in this case,

$$\begin{pmatrix} 1 & 2 & 3 & 2 \\ 0 & -5 & -11 & -7 \\ 0 & 0 & 0 & -2 \end{pmatrix} \begin{pmatrix} x \\ y \\ z \\ w \end{pmatrix} = \begin{pmatrix} 1 \\ -2 \\ 2 \end{pmatrix}$$

which yields

$$\mathbf{x} = \begin{pmatrix} -\frac{3}{5} + \frac{7}{5}t \\ \frac{9}{5} - \frac{11}{5}t \\ t \\ -1 \end{pmatrix}, t \in \mathbb{R}.$$

3.6 Exercises

(1) Let $\{\mathbf{u}_1, \cdots, \mathbf{u}_n\}$ be vectors in $\mathbb{R}^n$. The parallelepiped determined by these vectors $P(\mathbf{u}_1, \cdots, \mathbf{u}_n)$ is defined as

$$P(\mathbf{u}_1, \cdots, \mathbf{u}_n) \equiv \left\{ \sum_{k=1}^{n} t_k \mathbf{u}_k : t_k \in (0,1) \text{ for all } k \right\}.$$

Now let A be an $n \times n$ matrix. Show that

$$\{A\mathbf{x} : \mathbf{x} \in P(\mathbf{u}_1, \cdots, \mathbf{u}_n)\}$$

is also a parallelepiped.

(2) In the context of Problem 1, draw $P(\mathbf{e}_1, \mathbf{e}_2)$ where $\mathbf{e}_1, \mathbf{e}_2$ are the standard basis vectors for $\mathbb{R}^2$. Thus $\mathbf{e}_1 = (1,0)^t, \mathbf{e}_2 = (0,1)^t$. Now suppose

$$E = \begin{pmatrix} 1 & 1 \\ 0 & 1 \end{pmatrix}$$

where E is the elementary matrix which takes the second row of I and adds it to the first. Draw

$$\{E\mathbf{x} : \mathbf{x} \in P(\mathbf{e}_1, \mathbf{e}_2)\}.$$

In other words, draw the result left multiplying the vectors in $P(\mathbf{e}_1, \mathbf{e}_2)$ by E. We must regard the vectors in $P(\mathbf{e}_1, \mathbf{e}_2)$ as column vectors, so the multiplication on the left by the matrix E makes sense. Next draw the results of applying the other elementary matrices to $P(\mathbf{e}_1, \mathbf{e}_2)$ in the same sense.

(3) In the context of Problem 1, either draw or describe the result of applying elementary matrices to $P(\mathbf{e}_1, \mathbf{e}_2, \mathbf{e}_3)$. This is in the chapter, but try to go through the details yourself and also discuss the elementary operation not considered.

(4) Determine which matrices are in row reduced echelon form.

(a) $\begin{pmatrix} 1 & 2 & 0 \\ 0 & 1 & 7 \end{pmatrix}$

(b) $\begin{pmatrix} 1 & 0 & 0 & 0 \\ 0 & 0 & 1 & 2 \\ 0 & 0 & 0 & 0 \end{pmatrix}$

(c) $\begin{pmatrix} 1 & 1 & 0 & 0 & 0 & 5 \\ 0 & 0 & 1 & 2 & 0 & 4 \\ 0 & 0 & 0 & 0 & 1 & 3 \end{pmatrix}$

(5) Row reduce the following matrices to obtain the row reduced echelon form. List the pivot columns in the original matrix.

(a) $\begin{pmatrix} 1 & 2 & 0 & 3 \\ 2 & 1 & 2 & 2 \\ 1 & 1 & 0 & 3 \end{pmatrix}$

(b) $\begin{pmatrix} 1 & 2 & 3 \\ 2 & 1 & -2 \\ 3 & 0 & 0 \\ 3 & 2 & 1 \end{pmatrix}$

(c) $\begin{pmatrix} 1 & 2 & 1 & 3 \\ -3 & 2 & 1 & 0 \\ 3 & 2 & 1 & 1 \end{pmatrix}$

(6) Find the rank of the following matrices.

(a) $\begin{pmatrix} 1 & 2 & 0 \\ 3 & 2 & 1 \\ 2 & 1 & 0 \\ 0 & 2 & 1 \end{pmatrix}$

(b) $\begin{pmatrix} 1 & 0 & 0 \\ 4 & 1 & 1 \\ 2 & 1 & 0 \\ 0 & 2 & 0 \end{pmatrix}$

(c) $\begin{pmatrix} 0 & 1 & 0 & 2 & 1 & 2 & 2 \\ 0 & 3 & 2 & 12 & 1 & 6 & 8 \\ 0 & 1 & 1 & 5 & 0 & 2 & 3 \\ 0 & 2 & 1 & 7 & 0 & 3 & 4 \end{pmatrix}$

(d) $\begin{pmatrix} 0 & 1 & 0 & 2 & 0 & 1 & 0 \\ 0 & 3 & 2 & 6 & 0 & 5 & 4 \\ 0 & 1 & 1 & 2 & 0 & 2 & 2 \\ 0 & 2 & 1 & 4 & 0 & 3 & 2 \end{pmatrix}$

(e) $\begin{pmatrix} 0 & 1 & 0 & 2 & 1 & 1 & 2 \\ 0 & 3 & 2 & 6 & 1 & 5 & 1 \\ 0 & 1 & 1 & 2 & 0 & 2 & 1 \\ 0 & 2 & 1 & 4 & 0 & 3 & 1 \end{pmatrix}$

(7) Recall that two matrices are row equivalent if one can be obtained from the other by a sequence of row operations. When this happens for two matrices A, B, we can write $A \sim B$. Show that $A \sim A$. Next show that if $A \sim B$, then $B \sim A$. Finally show that if $A \sim B$ and $B \sim C$, then $A \sim C$. A relation which has these three properties is called an **equivalence relation**. Part of this is in the chapter. Do it on your own and if you become stuck look at the proof in the chapter. Then finish showing this.

(8) Determine whether the two matrices

$$\begin{pmatrix} 1 & 3 & 5 & 3 \\ 1 & 4 & 6 & 4 \\ 1 & 1 & 3 & 1 \end{pmatrix}, \begin{pmatrix} 1 & 4 & 5 & 2 \\ 1 & 5 & 6 & 3 \\ 1 & 2 & 3 & 0 \end{pmatrix}$$

are row equivalent. **Hint:** Recall the theorem about uniqueness of the row reduced echelon form.

(9) Determine whether the two matrices

$$\begin{pmatrix} 1 & 3 & 5 & 3 \\ 1 & 4 & 6 & 4 \\ 1 & 1 & 3 & 1 \end{pmatrix}, \begin{pmatrix} 1 & 3 & 5 & 3 \\ 1 & 4 & 6 & 4 \\ 2 & 5 & 9 & 5 \end{pmatrix}$$

are row equivalent. **Hint:** Recall the theorem about uniqueness of the row reduced echelon form.

(10) Suppose $AB = AC$ and A is an invertible $n \times n$ matrix. Does it follow that $B = C$? Explain why or why not. What if A were a non invertible $n \times n$ matrix?

(11) Find your own examples:

(a) 2×2 matrices A and B such that $A \neq 0, B \neq 0$ with $AB \neq BA$.
(b) 2×2 matrices A and B such that $A \neq 0, B \neq 0$, but $AB = 0$.
(c) 2×2 matrices A, D, and C such that $A \neq 0, C \neq D$, but $AC = AD$.

(12) Give an example of a matrix A such that $A^2 = I$ and yet $A \neq I$ and $A \neq -I$.

(13) Let

$$A = \begin{pmatrix} 2 & 1 \\ -1 & 3 \end{pmatrix}.$$

Find A^{-1} if possible. If A^{-1} does not exist, determine why.

(14) Let

$$A = \begin{pmatrix} 0 & 1 \\ 5 & 3 \end{pmatrix}.$$

Find A^{-1} if possible. If A^{-1} does not exist, determine why.

(15) Let

$$A = \begin{pmatrix} 2 & 1 \\ 3 & 0 \end{pmatrix}.$$

Find A^{-1} if possible. If A^{-1} does not exist, determine why.

(16) Let

$$A = \begin{pmatrix} 2 & 1 \\ 4 & 2 \end{pmatrix}.$$

Find A^{-1} if possible. If A^{-1} does not exist, determine why.

(17) Let A be a 2×2 matrix which has an inverse. Say $A = \begin{pmatrix} a & b \\ c & d \end{pmatrix}$. Find a formula for A^{-1} in terms of a, b, c, d. Also give a relationship involving a, b, c, d which will ensure that A^{-1} exists.

(18) Let

$$A = \begin{pmatrix} 1 & 2 & 3 \\ 2 & 1 & 4 \\ 1 & 0 & 2 \end{pmatrix}.$$

Find A^{-1} if possible. If A^{-1} does not exist, determine why.

(19) Let

$$A = \begin{pmatrix} 1 & 0 & 3 \\ 2 & 3 & 4 \\ 1 & 0 & 2 \end{pmatrix}.$$

Find A^{-1} if possible. If A^{-1} does not exist, determine why.

(20) Let

$$A = \begin{pmatrix} 1 & 2 & 3 \\ 2 & 1 & 4 \\ 4 & 5 & 10 \end{pmatrix}.$$

Find A^{-1} if possible. If A^{-1} does not exist, determine why.

(21) Let

$$A = \begin{pmatrix} 1 & 2 & 0 & 2 \\ 1 & 1 & 2 & 0 \\ 2 & 1 & -3 & 2 \\ 1 & 2 & 1 & 2 \end{pmatrix}.$$

Find A^{-1} if possible. If A^{-1} does not exist, determine why.

(22) The procedures presented for finding the inverse should work for any field. Let

$$A = \begin{pmatrix} \bar{1} & \bar{2} \\ \bar{2} & \bar{1} \end{pmatrix}$$

Find A^{-1} if possible. Here the entries of this matrix are in F_3 the field of residue classes described earlier.

(23) ↑Let

$$A = \begin{pmatrix} \bar{1} & \bar{2} \\ \bar{1} & \bar{1} \end{pmatrix}$$

Find A^{-1} if possible. Here the entries of this matrix are in F_3 the field of residue classes described earlier.

(24) Using the inverse of the matrix, find the solution to the systems

$$\begin{pmatrix} 1 & 2 & -2 \\ -1 & -1 & 1 \\ 1 & 3 & -2 \end{pmatrix} \begin{pmatrix} x \\ y \\ z \end{pmatrix} = \begin{pmatrix} 1 \\ 2 \\ 3 \end{pmatrix}, \begin{pmatrix} 1 & 2 & -2 \\ -1 & -1 & 1 \\ 1 & 3 & -2 \end{pmatrix} \begin{pmatrix} x \\ y \\ z \end{pmatrix} = \begin{pmatrix} 2 \\ 1 \\ 0 \end{pmatrix},$$

$$\begin{pmatrix} 1 & 2 & -2 \\ -1 & -1 & 1 \\ 1 & 3 & -2 \end{pmatrix} \begin{pmatrix} x \\ y \\ z \end{pmatrix} = \begin{pmatrix} 1 \\ 0 \\ 1 \end{pmatrix}, \begin{pmatrix} 1 & 2 & -2 \\ -1 & -1 & 1 \\ 1 & 3 & -2 \end{pmatrix} \begin{pmatrix} x \\ y \\ z \end{pmatrix} = \begin{pmatrix} 3 \\ -1 \\ -2 \end{pmatrix}.$$

Now give the solution in terms of a, b, and c to

$$\begin{pmatrix} 1 & 0 & 3 \\ 2 & 3 & 4 \\ 1 & 0 & 2 \end{pmatrix} \begin{pmatrix} x \\ y \\ z \end{pmatrix} = \begin{pmatrix} a \\ b \\ c \end{pmatrix}.$$

(25) Using the inverse of the matrix, find the solution to the systems

$$\begin{pmatrix} 1 & 2 & -2 \\ 1 & 3 & -3 \\ 1 & 3 & -2 \end{pmatrix}\begin{pmatrix} x \\ y \\ z \end{pmatrix} = \begin{pmatrix} 1 \\ 2 \\ 3 \end{pmatrix}, \begin{pmatrix} 1 & 2 & -2 \\ 1 & 3 & -3 \\ 1 & 3 & -2 \end{pmatrix}\begin{pmatrix} x \\ y \\ z \end{pmatrix} = \begin{pmatrix} 2 \\ 1 \\ 0 \end{pmatrix},$$

$$\begin{pmatrix} 1 & 2 & -2 \\ 1 & 3 & -3 \\ 1 & 3 & -2 \end{pmatrix}\begin{pmatrix} x \\ y \\ z \end{pmatrix} = \begin{pmatrix} 1 \\ 0 \\ 1 \end{pmatrix}, \begin{pmatrix} 1 & 2 & -2 \\ 1 & 3 & -3 \\ 1 & 3 & -2 \end{pmatrix}\begin{pmatrix} x \\ y \\ z \end{pmatrix} = \begin{pmatrix} 3 \\ -1 \\ -2 \end{pmatrix}.$$

Now give the solution in terms of a, b, and c to

$$\begin{pmatrix} 1 & 0 & 3 \\ 2 & 3 & 4 \\ 1 & 0 & 2 \end{pmatrix}\begin{pmatrix} x \\ y \\ z \end{pmatrix} = \begin{pmatrix} a \\ b \\ c \end{pmatrix}.$$

(26) Using the inverse of the matrix, find the solution to the system

$$\begin{pmatrix} -1 & \frac{1}{2} & \frac{1}{2} & \frac{1}{2} \\ 3 & \frac{1}{2} & -\frac{1}{2} & -\frac{5}{2} \\ -1 & 0 & 0 & 1 \\ -2 & -\frac{3}{4} & \frac{1}{4} & \frac{9}{4} \end{pmatrix}\begin{pmatrix} x \\ y \\ z \\ w \end{pmatrix} = \begin{pmatrix} a \\ b \\ c \\ d \end{pmatrix}.$$

(27) Using the inverse of the matrix, find the solution to the system

$$\begin{pmatrix} 1 & 2 & 2 & -1 \\ -2 & -5 & -3 & 0 \\ -2 & -2 & -5 & 8 \\ 4 & 11 & 6 & 5 \end{pmatrix}\begin{pmatrix} x \\ y \\ z \\ w \end{pmatrix} = \begin{pmatrix} a \\ b \\ c \\ d \end{pmatrix}.$$

(28) Show that if A is an $n \times n$ invertible matrix, and $\mathbf{x}$ is a $n \times 1$ matrix such that $A\mathbf{x} = \mathbf{b}$ for $\mathbf{b}$ an $n \times 1$ matrix, then $\mathbf{x} = A^{-1}\mathbf{b}$.

(29) Prove that if A^{-1} exists, and $A\mathbf{x} = \mathbf{0}$ then $\mathbf{x} = \mathbf{0}$.

(30) Show that if A^{-1} exists for an $n \times n$ matrix, then A^{-1} is unique. That is, if $BA = I$ and $AB = I$, then $B = A^{-1}$. If you are stuck, see the proof of this in the chapter.

(31) Show that if A is an invertible $n \times n$ matrix, then so is A^t and $\left(A^t\right)^{-1} = \left(A^{-1}\right)^t$.

(32) Show that $(AB)^{-1} = B^{-1}A^{-1}$ by verifying that

$$AB\left(B^{-1}A^{-1}\right) = I$$

and $B^{-1}A^{-1}(AB) = I$. **Hint:** Use Problem 30.

(33) Show that $(ABC)^{-1} = C^{-1}B^{-1}A^{-1}$ by verifying that

$$(ABC)\left(C^{-1}B^{-1}A^{-1}\right) = I$$

and

$$\left(C^{-1}B^{-1}A^{-1}\right)(ABC) = I.$$

Generalize to give a theorem which involves an arbitrary finite product of invertible matrices. **Hint:** Use Problem 30.

(34) Using Problem 33, show that an $n \times n$ matrix A is invertible if and only if A is the product of elementary matrices. **Hint:** Recall that from Corollary 3.3, a matrix is invertible if and only if its row reduced echelon form is I. Then recall that every row operation is accomplished by multiplying on the left by an elementary matrix and that the inverse of an elementary matrix is an elementary matrix.

(35) What is the geometric significance of Problem 34 which is implied by Problem 3?

(36) If A is invertible, show that $\left(A^2\right)^{-1} = \left(A^{-1}\right)^2$. **Hint:** Use Problem 30.

(37) If A is invertible, show that $\left(A^{-1}\right)^{-1} = A$. **Hint:** Use Problem 30.

(38) Let the field of scalars be F_3. Do the following operations.

(a) $$\begin{pmatrix} \overline{1} & \overline{2} & \overline{1} \\ \overline{2} & \overline{1} & \overline{2} \\ \overline{0} & \overline{1} & \overline{1} \end{pmatrix}^t + 2 \begin{pmatrix} \overline{1} & \overline{2} & \overline{0} \\ \overline{0} & \overline{2} & \overline{1} \\ \overline{2} & \overline{1} & \overline{0} \end{pmatrix}$$

(b) $$\begin{pmatrix} \overline{1} & \overline{2} & \overline{0} \\ \overline{0} & \overline{2} & \overline{1} \\ \overline{2} & \overline{1} & \overline{0} \end{pmatrix} \begin{pmatrix} \overline{1} & \overline{1} & \overline{2} & \overline{1} \\ \overline{0} & \overline{1} & \overline{2} & \overline{2} \\ \overline{2} & \overline{0} & \overline{1} & \overline{2} \end{pmatrix}$$

(39) Let the field of scalars be F_3 and consider the matrix

$$\begin{pmatrix} \overline{1} & \overline{2} & \overline{1} \\ \overline{2} & \overline{1} & \overline{2} \\ \overline{0} & \overline{1} & \overline{1} \end{pmatrix}$$

Find the row reduced echelon form of this matrix. What is the rank of this matrix? Remember the rank is the number of pivot columns. Deleting the bar on the top of the numbers and letting the field of scalars be $\mathbb{R}$, what is the row reduced echelon form and rank? What if F_3 is changed to F_5?

(40) Let the field of scalars be F_3, and consider the matrix

$$\begin{pmatrix} \overline{1} & \overline{2} & \overline{0} \\ \overline{0} & \overline{2} & \overline{1} \\ \overline{2} & \overline{1} & \overline{0} \end{pmatrix}.$$

Find the row reduced echelon form and rank. Omitting the bar on the numbers and letting the field of scalars be $\mathbb{R}$, find the row reduced echelon form and rank. Remember the rank is the number of pivot columns. Do the same problem, changing F_3 to F_5.

(41) Let the field of scalars be F_3. Find the inverse of

$$\begin{pmatrix} \overline{1} & \overline{1} & \overline{2} \\ \overline{0} & \overline{1} & \overline{2} \\ \overline{2} & \overline{0} & \overline{1} \end{pmatrix}$$

if this inverse exists.

(42) Let the field of scalars be F_3. Find the inverse of

$$\begin{pmatrix} \bar{0} & \bar{2} & \bar{2} \\ \bar{1} & \bar{2} & \bar{0} \\ \bar{2} & \bar{2} & \bar{0} \end{pmatrix}$$

if this inverse exists.

(43) Let the field of scalars be F_5. Find the inverse of

$$\begin{pmatrix} \bar{0} & \bar{4} & \bar{0} \\ \bar{2} & \bar{1} & \bar{2} \\ \bar{1} & \bar{1} & \bar{1} \end{pmatrix}$$

if this inverse exists.

(44) Is there only one LU factorization for a given matrix? **Hint:** Consider the equation

$$\begin{pmatrix} 0 & 1 \\ 0 & 1 \end{pmatrix} = \begin{pmatrix} 1 & 0 \\ 1 & 1 \end{pmatrix} \begin{pmatrix} 0 & 1 \\ 0 & 0 \end{pmatrix}.$$

(45) Find an LU factorization for the matrix $\begin{pmatrix} 1 & 2 & 3 & 4 \\ 2 & 0 & 3 & 2 \\ 1 & 4 & 5 & 3 \end{pmatrix}$.

(46) Find a PLU factorization of $\begin{pmatrix} 1 & 2 & 1 \\ 1 & 2 & 2 \\ 2 & 1 & 1 \end{pmatrix}$.

(47) Find a PLU factorization of $\begin{pmatrix} 1 & 2 & 1 & 2 & 1 \\ 2 & 4 & 2 & 4 & 1 \\ 1 & 2 & 1 & 3 & 2 \end{pmatrix}$.

Chapter 4

Vector spaces

Chapter summary

This chapter is on the general concept of a vector space. This is defined in terms of axioms. We begin by listing these axioms and certain implications. After this we give some simple examples and then discuss linear span and the meaning of independence. Next we discuss the concept of a basis and why any two have the same number of vectors, which shows that the dimension is well defined. This general theory is used to discuss some applications to difference equations.

4.1 The vector space axioms

We now write down the definition of an algebraic structure called a **vector space**. It includes $\mathbb{R}^n$ and $\mathbb{C}^n$ as particular cases. There is a little initial effort in working with a general vector space, but the payoff, due to the wide variety of applications, is incalculably great.

Let F be a given field.

Definition 4.1. A **vector space** over F is a set V containing at least one element which we write as 0. Moreover, there are rules for forming the sum $v+w$, whenever v, w are in V, and the product cv $(c \in F, v \in V)$. We have, for u, v, w in V and a, b in F,

1. $v + w \in V$ and $av \in V$ (**closure**),
2. $v + w = w + v$,
3. $u + (v + w) = (u + v) + w$,
4. $v + 0 = v$,
5. For each v in V, there is a member $-v$ of V such that

$$v + (-v) = 0,$$

6. $1v = v$,
7. $(ab)v = a(bv)$,
8. $a(u + v) = au + av$,

9. $(a+b)u = au + bu$.

This set of axioms has been found to be enough to do calculations in V that resemble those in $\mathbb{R}^n$ or $\mathbb{C}^n$. We refer to the members of V as **vectors** and the members of F as **scalars**.

It is worth noting right away that 0 and $-v$ are unique.

Lemma 4.1. *There is only one* 0 *in* V *and for each* v, *there is only one* $-v$.

Proof. Suppose $0'$ also works. Then since both 0 and $0'$ act as additive identities,

$$0' = 0' + 0 = 0$$

If $w + v = 0$, so that w acts like the additive inverse for v, then adding $-v$ to both sides,

$$w = w + (v + (-v)) = (w + v) + -v = -v.$$

This proves the lemma. □

It is easy to verify that $\mathbb{R}^n$ is a vector space over $\mathbb{R}$. For instance,

$$\begin{aligned} a(\mathbf{u}+\mathbf{v}) &= (au_1 + av_1, \ldots, au_n + av_n) \\ &= a\mathbf{u} + a\mathbf{v}. \end{aligned}$$

The arguments give a more general example.

Definition 4.2. Let F be a field. Then F^n is the set of ordered $n-$tuples $(x_1, \ldots, x_n)$ with each $x_i \in F$, with the following definitions:

$$\begin{aligned} (x_1, \ldots, x_n) + (y_1, \ldots, y_n) &= (x_1 + y_1, \cdots, x_n + y_n), \\ c(x_1, \ldots, x_n) &= (cx_1, \ldots, cx_n) \qquad (c \in F). \end{aligned}$$

It is easy to check that F^n is a vector space. Note that if F is finite, then F^n is a finite vector space.

The following result shows how we use the axioms to reach some conclusions that are used constantly in this book.

Lemma 4.2. *Let* V *be a vector space over* F. *Let* v, w *be in* V *and* $a \in F$.

(i) If $v + w = v$, *then* $w = 0$.
(ii) $a0 = 0$.
(iii) $0v = 0$.
(iv) $(-a)v = -(av)$.
(v) If $v \neq 0$ *and* $av = 0$, *then* $a = 0$.

Be careful with the interpretation of the zeros here. For example, in (iii) the 0 on the left-hand side is in F and the 0 on the right-hand side is in V.

Proof. (We will not cite the uses of the vector space axioms **1**–**9** explicitly.)

(i) Add $-v$ to both sides of

$$v + w = v.$$

Then

$$-v + (v + w) = -v + v = 0.$$

Regroup the left-hand side:

$$(-v + v) + w = 0$$

which reads

$$0 + w = 0.$$

Since $0 + w = w$, this completes the proof.

(ii) Since $a0 = a(0 + 0) = a0 + a0$, we have $a0 = 0$, by (i).

(iii) Since $0v = (0 + 0)v = 0v + 0v$, we have $0v = 0$.

(iv) As $0 = (a + (-a))v$ by (iii), we get

$$0 = av + (-a)v.$$

So $(-a)v$ is the 'additive inverse' of av.

(v) If $av = 0$ and $a \neq 0$, we use (ii) to get $0 = a^{-1}0 = a^{-1}(av) = (a^{-1}a)v = 1v = v$. □

4.2 Some fundamental examples

Our examples in this section are function spaces and sequence spaces. We also define and illustrate the key notation of a **subspace** of a vector space.

Let X, Y be non-empty sets. We recall that a function from X into Y is a rule that assigns a value $f(x) \in Y$ whenever $x \in X$ is given. We write

$$f : X \to Y.$$

We call X the **domain** of f and Y the **codomain** of f. The element $f(x)$ of Y is called the **image** of x under f.

As an example, the function g on $\mathbb{R}^3$ defined by

$$g(\mathbf{x}) = |\mathbf{x}|$$

has domain $\mathbb{R}^3$ and codomain $\mathbb{R}$. Functions are also called **mappings**.

Example 4.1. Let X be a nonempty set and F a field. The set $V(X, F)$ of all functions

$$f : X \to F$$

is a vector space over F.

We have to specify the rules of addition and scalar multiplication. These rules are exactly what you would expect: by definition,

$$(f+g)(x) = f(x) + g(x), (cf)(x) = cf(x)$$

for f, g in $V(X, F)$ and $c \in F$.

It is straightforward to check all laws (1-9) from §3.6. The zero vector is the constant function $f = 0$. The additive inverse of f is $-f$, that is, $(-1)f$.

Definition 4.3. Let V be a vector space over F and let $W \subset V$. We say that W is a **subspace** of V if W is a vector space (using the addition and scalar multiplication defined in V).

Example 4.2. The plane W with equation

$$a_1x_1 + a_2x_2 + a_3x_3 = 0$$

is a subspace of $\mathbb{R}^3$.

The 'reason' for this is that if $\mathbf{v}$ and $\mathbf{u}$ are in W, then $\mathbf{u} + \mathbf{v} \in W$, and $c\mathbf{u} \in W$ whenever c is a scalar. (You can easily verify this.) Because of this, the closure law **1** is obeyed in W, The other laws give little trouble (We get **2,3**, etc. 'free of charge'). It is worth summarizing the idea in a lemma.

Lemma 4.3. *Let W be a subset of the vector space V over F, $0 \in W$. Suppose that for any v, w in W and $c \in F$,*

(i) $v + w \in W$,
(ii) $cv \in W$.

Then W is a subspace of V.

Proof. Law **1.** has been assumed. Laws **2,3,4**, etc. hold because the members of W are in V. Also $-w = (-1)w$ is in W whenever $w \in W$. This gives the remaining law **5**. □

We can put Lemma 4.3 to work in constructing function spaces. Let

$$[a, b] = \{x : a \leq x \leq b\}$$

be an interval in $\mathbb{R}$. The following notations will be used.

$$C[a, b] = \{f : [a, b] \to \mathbb{R} : f \text{ is continuous}\}, \tag{4.1}$$

$$S[a, b] = \{f : [a, b] \to \mathbb{R} : f \text{ has a derivative } f^{(n)} \text{ for } n \geq 1\}. \tag{4.2}$$

(The letter S symbolizes smoothness of the functions.) In the notation used above,

$$S[a, b] \subset C[a, b] \subset V([a, b], \mathbb{R}),$$

the vector space of real valued functions which are defined on $[a, b]$. Moreover, $S[a, b]$ and $C[a, b]$ are subspaces of $V([a, b], \mathbb{R})$, as we now explain.

If f, g are continuous on $[a, b]$, so is $f + g$ and so too is cf $(c \in \mathbb{R})$. In view of Lemma 4.3, this implies that $C[a, b]$ is a subspace of $V([a, b], \mathbb{R})$. Very similar reasoning shows that $S[a, b]$ is a subspace of $V([a, b], \mathbb{R})$. Also $S[a, b]$ is a subspace of $C[a, b]$.

For many applications it is convenient to work with complex functions. We define $C_{\mathbb{C}}[a, b]$, $S_{\mathbb{C}}[a, b]$just as in (4.1), (4.2) with $\mathbb{R}$ replaced by $\mathbb{C}$. Just as above, $C_{\mathbb{C}}[a, b]$ and $S_{\mathbb{C}}[a, b]$ are subspaces of $V([a, b], \mathbb{C})$.

Example 4.3. Let $n \in \mathbb{N}$. Then F^n is the set of all ordered lists of elements of F,

$$\mathbf{x} = (x_1, \cdots, x_n) \tag{4.3}$$

with each $x_j \in F$. This is often denoted as (x_k) where one considers k as a generic index taken fom $\{1, \cdots, n\}$.

Writing $x(k)$ for x_k we see that one way to think of these tuples is as functions

$$x : \{1, \cdots, n\} \to F.$$

Thus F^n is just $V(\{1, \cdots, n\}, F)$, and so F^n is a vector space over F as described in Example 4.1.

Example 4.4. Let F^∞ be the set of all infinite sequences.

$$\mathbf{x} = (x_1, x_2, x_3, \ldots) \tag{4.4}$$

with each $x_j \in F$. Such a sequence is often denoted as (x_n) where one considers n as a generic index taken from $\mathbb{N}$.

Writing $x(n)$ for x_n we see that a sequence is a function $x : \mathbb{N} \to F$, where $\mathbb{N} = \{1, 2, \ldots\}$ is the set of natural numbers. Thus F^∞ is just $V(\mathbb{N}, F)$ and so F^∞ is a vector space over F as described in Example 4.1.

You may think of sequences as signals measured at times $t, 2t, 3t, \ldots$. Then $(x_n) + (y_n)$ is the superposition of two signals and $c(x_n)$ is an amplified signal.

Example 4.5. Let $F^\infty(q)$ be the set of $\mathbf{x} \in F^\infty$ with period q, where q is a given natural number. Thus $\mathbf{x}$ in $F^\infty(q)$ has the property

$$x_{n+q} = x_n \text{ for } n = 1, 2, \ldots$$

One of the exercises for this section is to show that $F^\infty(q)$ is a subspace of F^∞.

Notation 4.1. Let X, Y be sets. The expression $X \cap Y$ means the set of x that are members of both X and Y.

Example 4.6. Let V and W be subspaces of a vector space U over F. Then $V \cap W$ is a subspace of U.

To see this, we note first that $0 \in V \cap W$. Now we apply Lemma 4.3. Let v, w be in $V \cap W$. Then $v + w$ and cv are in V (for given $c \in F$), because v, w are in V. Similarly $v + w$ and cw are in W. So $v + w$ and cv are in $V \cap W$. This shows that $V \cap W$ is a subspace of U.

As a concrete example, let V, W be distinct planes through $\mathbf{0}$ in $U = \mathbb{R}^3$. Then $V \cap W$ is a line through $\mathbf{0}$.

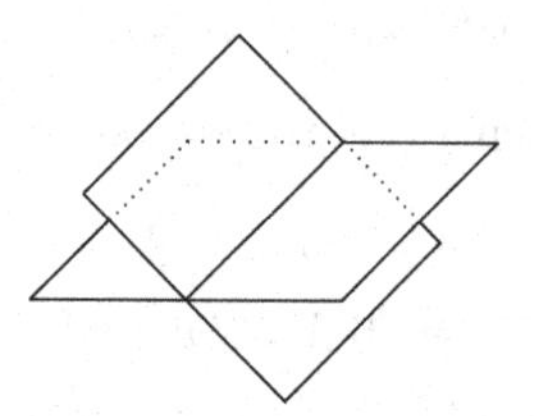

4.3 Exercises

(1) Recall that $F^{\infty}(q)$ consists of periodic sequences in $F^{\infty}, x_{n+q} = x_n$. Show that $F^{\infty}(q)$ is a subspace of F^{∞}.

(2) Let V be a vector space over F a field of scalars. Show that a nonempty subset $W \subseteq V$ is a subspace if and only if, whenever $a, b \in F$ and $u, v \in W$, it follows that $au + bv \in W$.

(3) Let $M = \{\mathbf{u} = (u_1, u_2, u_3, u_4) \in \mathbb{R}^4 : u_3 = u_1 = 0\}$. Is M a subspace? Explain.

(4) Let $M = \{\mathbf{u} = (u_1, u_2, u_3, u_4) \in \mathbb{R}^4 : u_3 \geq u_1\}$. Is M a subspace? Explain.

(5) Let $\mathbf{w} \in \mathbb{R}^4$ and let $M = \{\mathbf{u} = (u_1, u_2, u_3, u_4) \in \mathbb{R}^4 : \langle \mathbf{w}, \mathbf{u} \rangle = 0\}$. Is M a subspace? Explain.

(6) Let $M = \{\mathbf{u} = (u_1, u_2, u_3, u_4) \in \mathbb{R}^4 : u_i \geq 0 \text{ for each } i = 1, 2, 3, 4\}$. Is M a subspace? Explain.

(7) Let $\mathbf{w}, \mathbf{w}_1$ be given vectors in $\mathbb{R}^4$ and define

$$M = \{\mathbf{u} = (u_1, u_2, u_3, u_4) \in \mathbb{R}^4 : \langle \mathbf{w}, \mathbf{u} \rangle = 0 \text{ and } \langle \mathbf{w}_1, \mathbf{u} \rangle = 0\}.$$

Is M a subspace? Explain.

(8) Let $M = \{\mathbf{u} = (u_1, u_2, u_3, u_4) \in \mathbb{R}^4 : |u_1| \leq 4\}$. Is M a subspace? Explain.

(9) Let $M = \{\mathbf{u} = (u_1, u_2, u_3, u_4) \in \mathbb{R}^4 : \sin(u_1) = 1\}$. Is M a subspace? Explain.

(10) Let $M = \{(x, y) : xy = 0\}$. Is M a subspace of $\mathbb{R}^2$?

(11) Let $\mathbf{w} \in \mathbb{R}^3$ be a given vector. Let $M = \{\mathbf{u} \in \mathbb{R}^3 : \mathbf{w} \times \mathbf{u} = \mathbf{0}\}$. Is M a subspace of $\mathbb{R}^3$?

(12) Let $\mathbf{u}, \mathbf{v}$ be given vectors in $\mathbb{R}^3$ and let $M = \{\mathbf{x} \in \mathbb{R}^3 : \langle \mathbf{u} \times \mathbf{v}, \mathbf{x} \rangle = 0\}$. Show that M is a subspace of $\mathbb{R}^3$ and describe M geometrically.

(13) Let M denote all vectors of the form $(x + y, x - y)$ where $x, y \in \mathbb{R}$. Show that M is a subspace of $\mathbb{R}^2$. Next show that M actually equals $\mathbb{R}^2$.

(14) Let $M \subseteq \mathbb{R}^3$ consist of all vectors of the form

$$(x + y - z, x + y + 2z, 2x + 2y + z)$$

where $x, y, z \in \mathbb{R}$. Show that M is a subspace of $\mathbb{R}^3$. Is M equal to $\mathbb{R}^3$? **Hint:** Is $(0,0,1)$ in M?

(15) Suppose $\{x_1, \cdots, x_k\}$ is a set of vectors from a vector space V with field of scalars F. The **span** of these vectors, denoted as span $\{x_1, \cdots, x_k\}$, consists of all expressions of the form $c_1 x_1 + \cdots + c_n x_n$ where the $c_k \in F$. Show that span $\{x_1, \cdots, x_k\}$ is a subspace of V.

(16) Let V be a vector space over the field of scalars F and suppose W, U are two subspaces. Define

$$W + U = \{w + u : u \in U, w \in W\}.$$

Is $W + U$ a subspace?

(17) Show that $\mathbb{R}$ is a vector space over $\mathbb{Q}$, where $\mathbb{Q}$ is the field of scalars consisting of the rational numbers.

(18) Show that the set of numbers of the form $a + b\sqrt{2}$, where $a, b \in \mathbb{Q}$ is a vector space over $\mathbb{Q}$. Show that these numbers form a subspace of $\mathbb{R}$ where $\mathbb{R}$ is a vector space over the field of scalars $\mathbb{Q}$. Show that the set of numbers of this form is also a field.

(19) Consider the set of all polynomials of degree no larger than 4 having real coefficients. With the usual additions and scalar multiplications, show that this set is a vector space over $\mathbb{R}$.

(20) Let F be a field and consider $F[x]$, the set of all polynomials having coefficients in F. Thus a typical thing in $F[x]$ would be an expression of the form

$$a_n x^n + a_{n-1} x^{n-1} + \cdots + a_1 x + a_0$$

where each $a_k \in F$. Define addition and multiplication by scalars in the usual way and show that $F[x]$ is a vector space.

(21) Let V, W be vector spaces over the field F and let $T : V \to W$ be a function which satisfies

$$T(au + bv) = aTu + bTv$$

whenever $a, b \in F$ and $u, v \in V$. Such a function is called a **linear transformation** or **linear mapping**. Let

$$U = \{u \in V : Tu = 0\}.$$

Show that U is a subspace of V. This subspace is called $\ker(T)$. It is referred to as the kernel of T and often as the null space of T. Also define

$$TV = \{Tv : v \in V\}.$$

Show that TV is a subspace of W. This subspace is called the **image** of T and will be denoted as $\operatorname{Im}(T)$.

(22) ↑In the context of the above problem, show that a linear mapping is one-to-one if and only if whenever $T(v) = 0$ it follows that $v = 0$.

(23) Let D denote the derivative. Thus this is defined on $S[a, b]$ by $Df = \frac{df}{dt}$. Show that this is a linear transformation and find $\ker(D)$.

(24) Let V, W be two vector spaces having a field of scalars F. Denote by $\mathcal{L}(V, W)$ the set of linear transformations which map V to W. For S, T two of these, define $T + S$ as follows.

$$(T + S)(u) = T(u) + S(u)$$

and for $c \in F$, $T \in \mathcal{L}(V, W)$ define cT by

$$(cT)(u) = c(Tu).$$

The idea is that if you tell what a function does, you have told what the function is. With this definition, show that $\mathcal{L}(V, W)$ is a vector space over the field of scalars F.

4.4 Linear span and independence

Throughout this chapter, if we mention V without qualification, then V is a vector space over a given field F.

A key idea in describing the 'structure' of V is the *linear span* of a set of vectors. Recall that a **linear combination** of vectors, $v_1, \ldots, v_m$ in V is an expression

$$a_1 v_1 + \cdots + a_m v_m \text{ (each } a_j \in F).$$

Definition 4.4. Let $v_1, \ldots, v_m$ be in V. The set

$$\text{span}\{v_1, \ldots, v_m\} = \{a_1 v_1 + \cdots + a_m v_m : \text{ each } a_j \in F\}$$

is called the **linear span** of $v_1, \ldots, v_m$.

Notation 4.2. Recall from Chapter 1 that we write $\{x_1, \ldots, x_k\}$ for the set whose members are $x_1 \ldots, x_k$.

The definition is most easily digested via simple examples.

Example 4.7. $\text{span}\{0\} = \{0\}$.

For the span consists of all $a0$ $(a \in F)$. This reduces to just one vector, namely 0.

Example 4.8. Let $\mathbf{v} \in \mathbb{R}^3, \mathbf{v} \neq \mathbf{0}$. Then $\text{span}\{\mathbf{v}\}$ is the straight line (infinite in both directions) through $\mathbf{0}$ and $\mathbf{v}$.

Here we are repeating an observation from Chapter 1 since

$$\text{span}\{\mathbf{v}\} = \{t\mathbf{v} : t \in \mathbb{R}\}.$$

Example 4.9. Let $\mathbf{v} \in \mathbb{R}^3, \mathbf{w} \in \mathbb{R}^3, \mathbf{v} \neq \mathbf{0},\ \mathbf{w} \notin \text{span}\{\mathbf{v}\}$. Let P be the plane through the points $\mathbf{0}, \mathbf{v}, \mathbf{w}$. Then $\text{span}\{\mathbf{v}, \mathbf{w}\} = P$.

Observation 4.1. It is obvious that $\text{span}\{\mathbf{v},\mathbf{w}\} \subset P$. To get the 'reverse inclusion' $P \subset \text{span}\{\mathbf{v},\mathbf{w}\}$, we take any $\mathbf{p}$ in P and 'draw' the line $\{\mathbf{p}+t\mathbf{v} : t \in \mathbb{R}\}$. At some point, say $u\mathbf{w}$, this line crosses $\text{span}\{\mathbf{w}\}$. Now $\mathbf{p} = -t\mathbf{v} + u\mathbf{w}$.

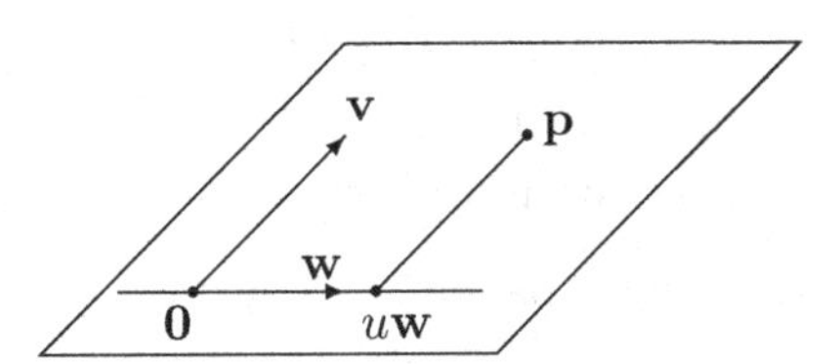

Example 4.10. Let $\mathbf{v} = (1,1,2)$ and $\mathbf{w} = (0,2,1)$ in F_5^3. Both $\mathbf{v}$ and $\mathbf{w}$ lie in the 'plane'

$$P = \{\mathbf{v} \in F_5^3 : x_1 + 2x_2 + x_3 = 0\},$$

which has 25 elements. There is no repetition among the list $a_1\mathbf{v} + a_2\mathbf{w}, a_1, a_2$ in F_5 (otherwise $\mathbf{w}$ would be a scalar multiple of $\mathbf{v}$). So $\text{span}\{\mathbf{v},\mathbf{w}\} \subset P$, and both sets in this 'inclusion' have 25 elements. We get

$$\text{span}\{\mathbf{v},\mathbf{w}\} = P.$$

Example 4.11. The differential equation

$$f'' + f = 0 \tag{4.5}$$

in $S[0, 2\pi]$ has two obvious solutions,

$$f_1(x) = \sin x, \qquad f_2 = \cos x.$$

Clearly $f_2 \notin \text{span}\{f_1\}$ (consider the graphs at 0). As we shall see in Chapter 5, every solution of (4.5) is of the form

$$f(x) = a_1 \sin x + a_2 \cos x.$$

So the set of solutions of (4.5) in $S[0, 2\pi]$ is $\text{span}\{f_1, f_2\}$.

The following simple lemma is often useful.

Lemma 4.4. *Let $v_1, \ldots, v_m$ be vectors in V. Then*

$$W = \text{span}\{v_1, \ldots, v_m\}$$

is a subspace of V.

Proof. We recall that it is only necessary to show that $0 \in W$ (which is obvious) and

(i) $v + w \in W$
(ii) $cv \in W$

for any v, w in W and $c \in F$. Let $v = a_1v_1 + \cdots + a_mv_m$ and $w = b_1w_1 + \cdots + b_mv_m$ with a_j, b_j in F. Then

$$v + w = (a_1 + b_1)v_1 + \cdots + (a_m + b_m)v_m \in \text{span}\{v_1, \ldots, v_m\}.$$
$$cv = (ca_1)v_1 + \cdots + (ca_m)v_m \in \text{span}\{v_1, \ldots, v_m\}.$$

By definition of W, both (i) and (ii) hold. □

When we have the above situation, namely,

$$W = \text{span}\{v_1, \cdots, v_m\},$$

we call $\{v_1, \cdots, v_m\}$ a **spanning set** for W. **Warning:** Some of the vectors in the spanning set may be 'redundant" as we shall see below. It may be that $W = \text{span}\{v_1, \cdots, v_{m-1}\}$, for instance. We also say that the vectors $v_1, \cdots, v_m$ **span** W.

We refer to an equation

$$x_1v_1 + \cdots + x_mv_m = b$$

(where $v_1, \ldots, v_m, b$ are given in V and $x_1, \ldots, x_m$ are 'unknown' scalars) as a **vector equation**. Obviously if $V = F^n$, the vector equation is equivalent to a system of linear equations. For instance, the vector equation

$$x_1 \begin{pmatrix} 1 \\ 2 \end{pmatrix} + x_2 \begin{pmatrix} 3 \\ 4 \end{pmatrix} = \begin{pmatrix} 5 \\ 6 \end{pmatrix}$$

is equivalent to the linear system

$$x_1 + 3x_2 = 5$$
$$2x_1 + 4x_2 = 6.$$

We will describe in Chapter 5how to solve any linear system.

Definition 4.5. Let $v_1, \ldots, v_m$ be in V. We say that $v_1, \ldots, v_m$ is a **linearly independent set** if the only solution of the equation

$$x_1v_1 + \cdots + x_mv_m = 0, \tag{4.6}$$

with $x_1, \ldots, x_m$ in F, is $x_1 = 0, \ldots, x_m = 0$.

We sometimes express this as '(4.6) has only the **trivial solution**'. A set of vectors that is not linearly independent is said to be **linearly dependent**.

The following lemma gives a useful equivalent statement.

Lemma 4.5. *The set of vectors $\{v_1, \ldots, v_m\}$ is linearly independent if and only if $v_1 \neq 0$ and for $j \geq 2$, no v_j is a linear combination of $v_1, \cdots, v_{j-1}$.*

Proof. Suppose first that $\{v_1, \ldots, v_m\}$ is linearly independent. Then $v_1 \neq 0$ since otherwise

$$1v_1 + 0v_2 + \cdots + 0v_m = 0.$$

This violates the linear independence of $\{v_1, \ldots, v_m\}$. If some v_j is a linear combination of $v_1, \cdots, v_{j-1}$, say $v_j = c_1 v_1 + \cdots + c_{j-1} v_{j-1}$, then

$$0 = c_1 v_1 + \cdots + c_{j-1} v_{j-1} + (-1) v_j$$

violating linear independence.

Conversely, suppose $v_1 \neq 0$ and no v_j is a linear combination of $v_1, \cdots, v_{j-1}$. Suppose

$$0 = c_1 v_1 + \cdots + c_m v_m.$$

Then if not all the scalars c_i are zero, let c_k be the last in the list $c_1, c_2, \cdots, c_m$ which is nonzero. Since $v_1 \neq 0$, it follows that $k \geq 2$. Then

$$v_k = -c_k^{-1} (c_1 v_1 + \cdots + c_{k-1} v_{k-1})$$

which violates the second condition that no v_j is a linear combination of the preceding vectors. Hence $c_1 = c_2 = \cdots = c_m = 0$. This proves the lemma. □

Example 4.12. Let $v_1, \ldots, v_k$ be in V. The set

$$0, v_1, \ldots, v_k$$

is linearly dependent.

We need only note that the nontrivial solution

$$1 \cdot 0 + 0 \cdot v_1 + \cdots + 0 \cdot v_k = 0$$

implies the set is dependent.

Note that the order in which a sequence of vectors is listed does not affect the linear independence or linear dependence.

Example 4.13. Let $v \in V, v \neq 0$. The set with one member, v, is a linearly independent set.

For the only solution of the equation

$$x_1 v = 0$$

is $x_1 = 0$.

Example 4.14. Let $v \in V, v \neq 0$ and let $w \in V, w \notin \text{span}\{v\}$. Then $\{v, w\}$ is a linearly independent set.

Consider the equation

$$x_1 v + x_2 w = 0. \tag{4.7}$$

If $x_2 \neq 0$, then $w = (-x_1 x_2^{-1}) v$. This is impossible, since $w \notin \text{span}\{v\}$. If $x_2 = 0$, then $x_1 v = 0$ and $x_1 = 0$. So the only solution of (4.7) is the trivial solution.

To visualize this example, the set $\{\mathbf{v}, \mathbf{w}\}$ in $\mathbb{R}^3$ is linearly independent when $\mathbf{v} \neq 0$ and $\mathbf{w}$ is outside the line $\text{span}\{\mathbf{v}\}$.

Example 4.15. Let $\mathbf{v}_1, \mathbf{v}_2, \mathbf{v}_3$ be vectors in $\mathbb{R}^3, \mathbf{v}_1 \neq \mathbf{0}, \mathbf{v}_2$ is not a multiple of $\mathbf{v}_1$, and $\mathbf{v}_3$ is not in the plane $\text{span}\{\mathbf{v}_1, \mathbf{v}_2\}$. Then $\mathbf{v}_1, \mathbf{v}_2, \mathbf{v}_3$ is a linearly independent set, since the second condition for linear independence of Lemma 4.5 is satisfied.

How do we test the linear independence of a set in F^n, e.g. 4 given vectors in F^5? There is an algorithm for this, **row reduction**, which we have already practiced in Chapter 3. It can be done by hand in simple cases, or using Maple. Here is an example.

Example 4.16. Consider the vectors

$$\begin{pmatrix} 1 & 2 & 0 & 5 \end{pmatrix}^t, \begin{pmatrix} 0 & 2 & 3 & 5 \end{pmatrix}^t, \begin{pmatrix} 1 & 1 & 2 & 6 \end{pmatrix}^t$$

Determine whether these vectors are independent. If they are not independent, find a linear relation which exhibits one of them as a linear combination of the others.

This is easy using Lemma 3.5. Make these three vectors the columns of a 4×3 matrix and row reduce to row reduced echelon form. You can do this by hand or you can use Maple or some other computer algebra system. These computer algebra systems do routine row reductions very well. Lemma 3.5 says that all linear relations are preserved in the row reduced echelon form and this form is such that all linear relations are easily seen. After row operations, the row reduced echelon form is

$$\begin{pmatrix} 1 & 0 & 0 \\ 0 & 1 & 0 \\ 0 & 0 & 1 \\ 0 & 0 & 0 \end{pmatrix},$$

and so the vectors given at the outset are linearly independent, because this is the case with the columns of the row reduced echelon form.

Example 4.17. Here are some vectors.

$$\begin{pmatrix} 2 & 2 & 1 & 1 \end{pmatrix}^t, \begin{pmatrix} 1 & 3 & 1 & 1 \end{pmatrix}^t, \begin{pmatrix} -12 & -8 & -5 & -5 \end{pmatrix}^t, \begin{pmatrix} 9 & 19 & 7 & 7 \end{pmatrix}^t$$

Determine whether they are independent and if they are not, find a linear relation which gives one of the vectors as a linear combination of the others.

Make the vectors the columns of a matrix and row reduce to row reduced echelon form. Then use Lemma 3.5. Using Maple, the row reduced echelon form of this matrix is

$$\begin{pmatrix} 1 & 0 & -7 & 2 \\ 0 & 1 & 2 & 5 \\ 0 & 0 & 0 & 0 \\ 0 & 0 & 0 & 0 \end{pmatrix},$$

and so the third column is -7 times the first plus 2 times the second. This is apparent from the above row reduced echelon form. Thus by Lemma 3.5, the same is true of the columns of the original matrix.

$$-7 \begin{pmatrix} 2 \\ 2 \\ 1 \\ 1 \end{pmatrix} + 2 \begin{pmatrix} 1 \\ 3 \\ 1 \\ 1 \end{pmatrix} = \begin{pmatrix} -12 \\ -8 \\ -5 \\ -5 \end{pmatrix}$$

You should note that there is also a linear relation for the fourth column as a linear combination of the first two.

Example 4.18. Maple gives the row reduced form of

$$\begin{pmatrix} 1 & 2 & 3 & 2 \\ 3 & -1 & 1 & 1 \\ 5 & 2 & 6 & 2 \\ 1 & 0 & 4 & 2 \\ 2 & 1 & 4 & 0 \end{pmatrix}$$

as

$$\begin{pmatrix} 1 & 0 & 0 & 0 \\ 0 & 1 & 0 & 0 \\ 0 & 0 & 1 & 0 \\ 0 & 0 & 0 & 1 \\ 0 & 0 & 0 & 0 \end{pmatrix}.$$

So $(1,3,5,1,2), (2,-1,2,0,1), (3,1,6,4,4), (2,1,2,2,0)$ is a linearly independent set.

Proposition 4.1. A set of n vectors in F^m, with $n > m$, is linearly dependent.

Proof. Let the vectors be columns of an $m \times n$ matrix. Then as pointed out earlier, the number of nonzero rows in its row reduced echelon form equals the number of pivot columns. However, there are at most m nonzero rows, and so, not all these columns can be pivot columns. In reading from left to right, pick the first one which is not a pivot column. By the description of the row reduced echelon form, this column is a linear combination of the preceding columns. By Lemma 4.5, the set of columns is not linearly independent. This proves the proposition. □

Example 4.19. Consider

$$\begin{pmatrix} 1 & 1 & 1 & 8 \\ 1 & 2 & 1 & 15 \\ 1 & 1 & 2 & 6 \end{pmatrix}.$$

By Proposition 4.1, the columns cannot be linearly independent. By Lemma 4.5, some column is a linear combination of the preceding columns. Find such a column and exhibit it as a linear combination of the preceding columns.

The row reduced echelon form is

$$\begin{pmatrix} 1 & 0 & 0 & 3 \\ 0 & 1 & 0 & 7 \\ 0 & 0 & 1 & -2 \end{pmatrix}.$$

Therefore,

$$-3\,(\text{first column}) + (-7)\,(\text{second column})$$

$$+2\,(\text{ third column}) + (\text{fourth column}) = \mathbf{0}.$$

This is obvious in the row reduced echelon form. By Lemma 3.5 this is also true of the columns in the original matrix. Thus

$$-3\begin{pmatrix}1\\1\\1\end{pmatrix}+(-7)\begin{pmatrix}1\\2\\1\end{pmatrix}+2\begin{pmatrix}1\\1\\2\end{pmatrix}+\begin{pmatrix}8\\15\\6\end{pmatrix}=\begin{pmatrix}0\\0\\0\end{pmatrix}.$$

You might not have seen this non trivial linear relationship right away, but the row reduced echelon form makes it immediately clear.

4.5 Exercises

(1) Here are three vectors. Determine whether they are linearly independent or linearly dependent.

$$\begin{pmatrix}1\\2\\0\end{pmatrix},\begin{pmatrix}2\\0\\1\end{pmatrix},\begin{pmatrix}3\\0\\0\end{pmatrix}$$

(2) Here are three vectors. Determine whether they are linearly independent or linearly dependent.

$$\begin{pmatrix}4\\2\\0\end{pmatrix},\begin{pmatrix}2\\2\\1\end{pmatrix},\begin{pmatrix}3\\0\\1\end{pmatrix}$$

(3) Here are three vectors. Determine whether they are linearly independent or linearly dependent.

$$\begin{pmatrix}1\\2\\3\end{pmatrix},\begin{pmatrix}4\\5\\1\end{pmatrix},\begin{pmatrix}3\\1\\0\end{pmatrix}$$

(4) Here are four vectors. Determine whether they span $\mathbb{R}^3$. Are these vectors linearly independent?

$$\begin{pmatrix}1\\2\\3\end{pmatrix},\begin{pmatrix}4\\3\\3\end{pmatrix},\begin{pmatrix}3\\1\\0\end{pmatrix},\begin{pmatrix}2\\4\\6\end{pmatrix}$$

(5) Here are four vectors. Determine whether they **span** $\mathbb{R}^3$. (That is, find whether the span of the vectors is $\mathbb{R}^3$.) Are these vectors linearly independent?

$$\begin{pmatrix}1\\2\\3\end{pmatrix},\begin{pmatrix}4\\3\\3\end{pmatrix},\begin{pmatrix}3\\2\\0\end{pmatrix},\begin{pmatrix}2\\4\\6\end{pmatrix}$$

(6) Here are three vectors in $\mathbb{R}^3$ and a fourth vector.

$$\begin{pmatrix}1\\1\\2\end{pmatrix}\begin{pmatrix}0\\1\\1\end{pmatrix}\begin{pmatrix}7\\5\\12\end{pmatrix},\begin{pmatrix}2\\1\\0\end{pmatrix}$$

Determine whether the fourth vector is in the linear span of the first three.

(7) Here are three vectors in $\mathbb{R}^3$ and a fourth vector.

$$\begin{pmatrix} 1 \\ 1 \\ 2 \end{pmatrix} \begin{pmatrix} 0 \\ 1 \\ 1 \end{pmatrix} \begin{pmatrix} 7 \\ 5 \\ 11 \end{pmatrix}, \begin{pmatrix} 2 \\ 1 \\ 2 \end{pmatrix}$$

Determine whether the fourth vector is in the linear span of the first three.

(8) Here are three vectors in $\mathbb{R}^3$ and a fourth vector.

$$\begin{pmatrix} 1 \\ 0 \\ 2 \end{pmatrix} \begin{pmatrix} 1 \\ 1 \\ 1 \end{pmatrix} \begin{pmatrix} 1 \\ -1 \\ 3 \end{pmatrix}, \begin{pmatrix} 5 \\ 6 \\ 4 \end{pmatrix}$$

Determine whether the fourth vector is in the span of the first three.

(9) The following are 3×4 matrices and so you know by Proposition 4.1 that the columns are not independent. Find a subset of these columns which is independent, which is as large as possible, and write one of the columns as a linear combination of the others.

(a)

$$\begin{pmatrix} 3 & 2 & -4 & 4 \\ 1 & 1 & -1 & 1 \\ 1 & 1 & -1 & 2 \end{pmatrix}$$

(b)

$$\begin{pmatrix} 3 & 6 & 2 & 4 \\ 2 & 4 & 2 & 3 \\ 1 & 2 & 1 & 2 \end{pmatrix}$$

(c)

$$\begin{pmatrix} 3 & 6 & 2 & 11 \\ 2 & 4 & 2 & 8 \\ 1 & 2 & 1 & 4 \end{pmatrix}$$

(10) The following are 3×5 matrices and so you know by Proposition 4.1 that the columns are not independent. Find a subset of these columns which is independent, which is also as large as possible. Write one of the columns as a linear combination of the others.

(a)

$$\begin{pmatrix} 3 & 2 & -4 & 3 & -4 \\ 2 & -3 & -7 & -11 & -7 \\ 1 & 1 & -1 & 2 & -1 \end{pmatrix}$$

(b)

$$\begin{pmatrix} 3 & 2 & -4 & 3 & 4 \\ 2 & -3 & -7 & -11 & 3 \\ 1 & 1 & -1 & 2 & 2 \end{pmatrix}$$

(c)

$$\begin{pmatrix} 3 & 2 & -4 & 4 & -7 \\ 2 & -3 & -7 & 3 & -13 \\ 1 & 1 & -1 & 2 & -1 \end{pmatrix}$$

(11) Let B be an $m \times n$ matrix. Show that

$$\ker(B) = \{\mathbf{x} \in \mathbb{R}^n : B\mathbf{x} = \mathbf{0}\}$$

is a subspace of $\mathbb{R}^n$. Now suppose that A is an $m \times m$ invertible matrix. Show that $\ker(AB) = \ker(B)$.

(12) Consider the functions $\{1, x, x^2, x^3, x^4\}$ in the vector space of functions defined on $(0, 1)$. Show that these vectors are linearly independent. **Hint:** You need to consider $a + bx + cx^2 + dx^3 + ex^4 = 0$ and show that all the constants equal 0. A fun way to do this is to take derivatives of both sides several times.

(13) Consider the functions$\{1 + x^2, 1 + x + x^2, 1 + 2x^2\}$ in the vector space of functions defined on $(0, 1)$. Show that these vectors are linearly independent.

(14) Suppose $\{\mathbf{x}_1, \cdots, \mathbf{x}_m\}$ are some vectors in $\mathbb{R}^n$ such that

$$\langle \mathbf{x}_k, \mathbf{x}_j \rangle = \delta_{kj}$$

where is the Kronecker delta

$$\delta_{kj} = \begin{cases} 1 \text{ if } k = j \\ 0 \text{ if } k \neq j \end{cases}$$

Show that this must be an independent set of vectors. **Hint:** Suppose

$$c_1\mathbf{x}_1 + \cdots + c_m\mathbf{x}_m = \mathbf{0}$$

and take the inner product of both sides with $\mathbf{x}_k$. Explain why the properties of the inner product imply $c_k = 0$. Then note that k was arbitrary.

(15) Let the vector space $\mathbb{Q}(\sqrt{3})$ consist of all numbers of the form $a + b\sqrt{3}$ where a, b are rational numbers, and let the field of scalars be $\mathbb{Q}$ with the usual rules of addition and multiplication. Show that this is a vector space, that $\text{span}\{1, \sqrt{3}\} = \mathbb{Q}(\sqrt{3})$, and that the vectors $1, \sqrt{3}$ are linearly independent.

(16) Let $a \neq b$ with neither a nor b equal to 0. Consider the two vectors in $\mathbb{R}^2$,

$$\begin{pmatrix} a \\ a^2 \end{pmatrix}, \begin{pmatrix} b \\ b^2 \end{pmatrix}.$$

Show that these vectors are linearly independent. Generalize this to the case of three distinct points, each non-zero.

(17) Suppose that

$$\begin{pmatrix} x \\ y \end{pmatrix}, \begin{pmatrix} a \\ b \end{pmatrix}$$

are two vectors in $\mathbb{R}^2$. Show that this set of vectors is linearly independent if and only if $xb - ay \neq 0$.

(18) For a vector in F^{∞}

$$\mathbf{x} = (x_1, x_2, \cdots),$$

denote by $P_n \mathbf{x}$ the vector in F^n given by

$$P_n \mathbf{x} = (x_1, \cdots, x_n).$$

Suppose that $\mathbf{x}_1, \cdots, \mathbf{x}_n$ are each vectors in F^{∞} and $\{P_n \mathbf{x}_1, \cdots, P_n \mathbf{x}_n\}$ is an independent set in F^n. Show that then $\mathbf{x}_1, \cdots, \mathbf{x}_n$ are linearly independent in F^{∞}.

(19) Let V consist of all continuous functions which are periodic of period 2π. Show that this is a vector space. Next consider the functions

$$\{\sin x, \sin 2x, \sin 3x, \cdots\}.$$

Show that this is a linearly independent set of vectors in V in the sense that any finite subset of these is linearly independent. **Hint:** You might want to make use of the identity

$$\int_{-\pi}^{\pi} \sin(mx) \sin(nx)\, dx = 0$$

if n, m are distinct positive integers.

(20) Show that a set of n vectors with $n \geq 2$ is linearly independent if and only if none of the vectors in the set is a linear combination of the others.

(21) Suppose$f(x)$ and $g(x)$ are two differentiable functions defined on an interval (a, b). Thus they are vectors in the vector space of functions defined on (a, b). Show that they are linearly independent if $(f/g)'(x) \neq 0$ for some $x \in (a, b)$.

(22) Here are three vectors in F_3^3

$$\begin{pmatrix} \bar{1} \\ \bar{2} \\ \bar{0} \end{pmatrix}, \begin{pmatrix} \bar{2} \\ \bar{1} \\ \bar{0} \end{pmatrix}, \begin{pmatrix} \bar{1} \\ \bar{0} \\ \bar{1} \end{pmatrix}$$

Are these three vectors linearly independent? Recall that F_3 refers to the field of residue classes. What if you considered the three vectors in $\mathbb{R}^3$,

$$\begin{pmatrix} 1 \\ 2 \\ 0 \end{pmatrix}, \begin{pmatrix} 2 \\ 1 \\ 0 \end{pmatrix}, \begin{pmatrix} 1 \\ 0 \\ 1 \end{pmatrix}$$

Are these vectors linearly independent?

(23) Here is a matrix whose entries are in F_3. Find its row reduced echelon form.

$$\begin{pmatrix} \bar{1} & \bar{0} & \bar{1} \\ \bar{0} & \bar{1} & \bar{1} \\ \bar{1} & \bar{1} & \bar{0} \end{pmatrix}$$

(24) In Example 4.10 explain why the 'plane' P has exactly 25 elements.

4.6 Basis and dimension

A non zero vector space V over F is said to be **finite-dimensional** if

$$V = \operatorname{span}\{v_1, \ldots, v_n\} \tag{4.8}$$

for some vectors $v_1, \ldots, v_n$ in V. This set of vectors is called a **spanning set.** The **dimension**[1] of V is the smallest possible number n such that there exists a spanning set having n vectors. With such a choice, $v_1, \ldots, v_n$ *is a linearly independent set.* Clearly $v_1 \neq 0$ since otherwise

$$V = \operatorname{span}\{v_2, \ldots, v_n\}.$$

Moreover, $v_j \notin \operatorname{span}\{v_1, \ldots, v_{j-1}\}$. Otherwise, in an expression

$$v = a_1 v_1 + \cdots + a_j v_j + \cdots + a_n v_n,$$

we could replace v_j with $b_1 v_1 + \cdots + b_{j-1} v_{j-1}$, leading to, say,

$$v = c_1 v_1 + \cdots + c_{j-1} v_{j-1} + a_{j+1} v_{j+1} + \cdots + a_n v_n$$

and to $V = \operatorname{span}\{v_1, \ldots, v_{j-1}, v_{j+1}, \ldots, v_n\}$.

Definition 4.6. Suppose that

$$V = \operatorname{span}\{v_1, \ldots, v_n\}$$

and that $v_1, \ldots, v_n$ is a linearly independent set. We say that $v_1, \ldots, v_n$ is a **basis** of V.

We shall see shortly that any two bases[2] of V have the same number of elements, so that dimension, which is written as $\dim V$, is just the number of elements in any basis of V.

Example 4.20. Clearly in $\mathbb{R}^3$ a line $L = \operatorname{span}\{\mathbf{v}_1\}, \mathbf{v}_1 \neq 0$ has dimension 1 ($\mathbf{v}_1$ is a basis of L). A plane $P = \operatorname{span}\{\mathbf{v}_1, \mathbf{v}_2\}$ has dimension 2 (here $\mathbf{v}_1 \neq 0, \mathbf{v}_2 \notin \operatorname{span} \mathbf{v}_1$). It is obvious that no basis would have 1 element, and therefore $\mathbf{v}_1, \mathbf{v}_2$ is a basis of P.

Example 4.21. A basis for F^n is the set

$$\mathbf{e}_j = (1, 0, \ldots, 0), \ldots, \mathbf{e}_n = (0, \ldots, 0, 1).$$

Certainly

$$F^n = \operatorname{span}\{\mathbf{e}_1, \ldots, \mathbf{e}_n\}$$

[1] If V consists of only the zero vector, it isn't very interesting, but it is convenient in this case to say that the dimension is 0.

[2] Bases is the plural of basis. It is much easier to say than basiss which involves a lot of hissing. There are other words which end in an s or an s sound for which the plural is not obtained by simply adding another s. Hippopotamus is one which comes to mind. So is mouse.

since any $(x_1, \ldots, x_n)$ in F^n can be written

$$\mathbf{x} = x_1\mathbf{e}_1 + \cdots + x_n\mathbf{e}_n.$$

The linear independence of $\mathbf{e}_1, \ldots, \mathbf{e}_n$ is obvious. We call this particular basis the **standard basis** of F^n.

If vectors $v_1, \ldots, v_m$ are given and $V = \text{span}\{v_1, \ldots, v_m\}$, the **discard process** means that we discard v_1 if $v_1 = 0$ and for $j > 1$, we discard v_j if $v_j \in \text{span}\{v_1, \ldots, v_{j-1}\}$. Beginning with a spanning set of vectors, this process can be applied till a basis is obtained. This is because every time a vector is discarded, those vectors which remain still span V.

Lemma 4.6. *If $v_1, \ldots, v_n$ is a basis of V and $w_1, \ldots, w_m$ is a linearly independent set in V, then $m \leq n$.*

Proof. Since $w_m \in \text{span}\{v_1, \ldots, v_n\}$, the set $w_m, v_1, \ldots, v_n$ is linearly dependent, and its linear span is V. Applying the discard process to the above list results in $w_m, v_{i_1}, \ldots, v_{i_k}$, with $k \leq n-1$ a basis of V. Repeat the procedure with $w_{m-1}, w_m, v_{i_1}, \ldots, v_{i_k}$. The discard process gives a basis $w_{m-1}, w_m, v_{j_1}, \ldots, v_{j_s}$ of $V, s \leq n-2$. After further steps we reach a basis $w_2, \ldots, w_{m-1}, w_m, v_a, v_b, \ldots$ Since w_1 is not a combination of $w_2, \ldots, w_{m-1}$, at least one v_j is still left. We have discarded at least $m-1$ of the v_j. So $m-1 \leq n-1$.

This is such an important result that we give another proof of it. Suppose to the contrary that $m > n$. Then by assumption, there exist scalars c_{ji} such that $w_i = \sum_{j=1}^{n} c_{ji}v_j$, $i = 1, \cdots, m$. Thus the matrix whose ji^{th} entry is c_{ji} has more columns than rows. It follows from Corollary 3.2 that there exists a non-zero vector $\mathbf{x} \in F^m$ such that $\sum_{i=1}^{m} c_{ji}x_i = 0$ for each j. Hence

$$\sum_{i=1}^{m} x_i w_i = \sum_{i=1}^{m} x_i \sum_{j=1}^{n} c_{ji}v_j = \sum_{j=1}^{n} \left(\sum_{i=1}^{m} c_{ji}x_i \right) v_j = \sum_{j=1}^{n} 0v_j = 0$$

which contradicts the linear independence of the vectors $w_1, \ldots, w_m$. Therefore, $m \leq n$ as claimed. □

Note that it was only necessary to assume that $v_1, \cdots, v_n$ is a spanning set. Thus in words, the above says roughly that spanning sets have at least as many vectors as linearly independent sets.

Corollary 4.1. *Any two bases of a finite-dimensional vector space V contain the same number of elements.*

Proof. Let $v_1, \ldots, v_n$ and $w_1, \ldots, w_m$ be two bases of V. By Lemma 4.6, $m \leq n$. Reversing roles of the v_j and the w_k, we obtain $n \leq m$. Hence $n = m$. □

An informal description of the above corollary is as follows:

$$\overset{\text{spanning set}}{\{v_1, \ldots, v_n\}}, \ \overset{\text{independent}}{\{w_1, \ldots, w_m\}} \Longrightarrow m \leq n$$

$$\overset{\text{spanning set}}{\{w_1, \ldots, w_m\}}, \ \overset{\text{independent}}{\{v_1, \ldots, v_n\}} \Longrightarrow n \leq m$$

Definition 4.7. The **dimension** of a finite-dimensional vector space V, written $\dim V$, is the number of vectors in any basis of V. By convention, the space $\{0\}$ has dimension 0.

Of course $\dim F^n = n$ (we already know a basis with n elements). See Example 4.21. The following is a useful lemma when dealing with bases.

Lemma 4.7. *If $\{v_1, \cdots, v_m\}$ is a linearly independent set of vectors and if w is not in* span $\{v_1, \cdots, v_m\}$, *then $\{v_1, \cdots, v_m, w\}$ is also a linearly independent set of vectors.*

Proof. Suppose

$$c_1 v_1 + \cdots + c_m v_m + dw = 0.$$

If $d = 0$, then each $c_j = 0$ because of linear independence of $v_1, \cdots, v_m$. But if $d \neq 0$, a contradiction is obtained immediately because then it is clear that $w \in \text{span}\{v_1, \cdots, v_m\}$. Hence $c_1 = \cdots = c_m = d = 0$. □

Corollary 4.2. *Let W be a nonzero subspace of an n dimensional vector space V. Then W has a basis. Furthermore, every linearly independent set of vectors in W can be extended to a basis of W.*

Proof. Let W be a nonzero subspace of an n dimensional vector space V. Pick $w_1 \in W$ such that $w_1 \neq 0$. Then $\{w_1\}$ is linearly independent. Suppose you have obtained $\{w_1, \cdots, w_k\}$ where $k \geq 1$, and the set of vectors is independent. Then if

$$\text{span}\{w_1, \cdots, w_k\} = W,$$

stop. This is the basis. If $\text{span}\{w_1, \cdots, w_k\} \neq W$, there exists w_{k+1} in W, $w_{k+1} \notin \text{span}\{w_1, \cdots, w_k\}$. By Lemma 4.7 $\{w_1, \cdots, w_k, w_{k+1}\}$ is linearly independent. Continue this way till for some $m \geq 1$, $\text{span}\{w_1, \cdots, w_m\} = W$. The process must terminate in finitely many steps because, if it did not do so, you could obtain an independent set of vectors having more than n vectors, which cannot occur by Theorem 4.1. The last assertion of the corollary follows from the above procedure. Simply start with the given linearly independent set of vectors and repeat the argument. This proves the corollary. □

We often wish to find a basis for a subspace. A typical case is $\ker(A)$ where A is an $m \times n$ matrix. See Problem 4 on Page 120.

Definition 4.8. For A an $m \times n$ matrix, $\ker(A)$ consists of all the vectors $\mathbf{x}$ such that $A\mathbf{x} = \mathbf{0}$. To find $\ker(A)$, solve the resulting system of equations.

Example 4.22. Here is a matrix.

$$A = \begin{pmatrix} 2 & 2 & 16 & 6 & 14 \\ 3 & 3 & 24 & 9 & 21 \\ 1 & 1 & 8 & 3 & 7 \\ 2 & 4 & 26 & 8 & 18 \end{pmatrix}$$

Find a basis for $\ker(A)$.

You want to find all the vectors $\mathbf{x}$ such that $A\mathbf{x} = \mathbf{0}$. The augmented matrix for the system of equations to be solved is then

$$\begin{pmatrix} 2 & 2 & 16 & 6 & 14 & 0 \\ 3 & 3 & 24 & 9 & 21 & 0 \\ 1 & 1 & 8 & 3 & 7 & 0 \\ 2 & 4 & 26 & 8 & 18 & 0 \end{pmatrix}.$$

The row reduced echelon form is

$$\begin{pmatrix} 1 & 0 & 3 & 2 & 5 & 0 \\ 0 & 1 & 5 & 1 & 2 & 0 \\ 0 & 0 & 0 & 0 & 0 & 0 \\ 0 & 0 & 0 & 0 & 0 & 0 \end{pmatrix}$$

and so you see there are three free variables. The solution to the equation $A\mathbf{x} = \mathbf{0}$ is then

$$x_1 = -3x_3 - 2x_4 - 5x_5,\ x_2 = -5x_3 - x_4 - 2x_5$$

You can write the above equations as

$$\begin{pmatrix} x_1 \\ x_2 \\ x_3 \\ x_4 \\ x_5 \end{pmatrix} = \begin{pmatrix} -3x_3 - 2x_4 - 5x_5 \\ -5x_3 - x_4 - 2x_5 \\ x_3 \\ x_4 \\ x_5 \end{pmatrix}.$$

Written as a span, the right side is of the form

$$x_3 \begin{pmatrix} -3 \\ -5 \\ 1 \\ 0 \\ 0 \end{pmatrix} + x_4 \begin{pmatrix} -2 \\ -1 \\ 0 \\ 1 \\ 0 \end{pmatrix} + x_5 \begin{pmatrix} -5 \\ -2 \\ 0 \\ 0 \\ 1 \end{pmatrix}, x_3, x_4, x_5 \in \mathbb{R}$$

$$= \text{span} \left\{ \begin{pmatrix} -3 \\ -5 \\ 1 \\ 0 \\ 0 \end{pmatrix}, \begin{pmatrix} -2 \\ -1 \\ 0 \\ 1 \\ 0 \end{pmatrix}, \begin{pmatrix} -5 \\ -2 \\ 0 \\ 0 \\ 1 \end{pmatrix} \right\}.$$

Note that the above three vectors are linearly independent and so these vectors are a basis for $\ker(A)$.

Corollary 4.3. *Let* $\dim V = n$. *Any linearly independent set* $v_1, \ldots, v_n$ *in* V *is a basis of* V.

Proof. Suppose not. Then there exists $v \in V, v \notin \text{span}\{v_1, \ldots, v_n\}$. Now by Lemma 4.7 we have a linearly independent set $v_1, \ldots, v_n, v$ with $n+1$ elements, which contradicts Lemma 4.6. $\square$

The following is a useful algorithm for extending a linearly independent set of vectors in $\mathbb{R}^n$ to form a basis for $\mathbb{R}^n$.

Procedure 4.1. An algorithm for finding a basis starting with a linearly independent set in the case that $V = \mathbb{R}^n$ is to form the matrix

$$\begin{pmatrix} \mathbf{u}_1 & \cdots & \mathbf{u}_m & \mathbf{e}_1 & \cdots & \mathbf{e}_n \end{pmatrix}$$

which has the indicated vectors as columns. Then $\mathbb{R}^n$ equals the span of these columns. Obtain the row reduced echelon form of this matrix. The first m columns become $\mathbf{e}_1, \cdots, \mathbf{e}_m$ respectively, thanks to Lemma 3.5, because they are linearly independent. Thus none of $\mathbf{u}_1, \cdots, \mathbf{u}_m$ can be a linear combination of the others, and so they will all be pivot columns. Then the pivot columns of the above matrix, which include the $\mathbf{u}_k$, yield a basis for $\mathbb{R}^n$.

Example 4.23. Here is a linearly independent set of vectors in $\mathbb{R}^4$.

$$\begin{pmatrix} 1 \\ 2 \\ 1 \\ 0 \end{pmatrix}, \begin{pmatrix} 2 \\ 2 \\ 1 \\ 0 \end{pmatrix}.$$

Extend this set to a basis for $\mathbb{R}^n$.

As indicated above, you can do this by finding the row reduced echelon form for

$$\begin{pmatrix} 1 & 2 & 1 & 0 & 0 & 0 \\ 2 & 2 & 0 & 1 & 0 & 0 \\ 1 & 1 & 0 & 0 & 1 & 0 \\ 0 & 0 & 0 & 0 & 0 & 1 \end{pmatrix}.$$

Doing the necessary row operations or letting a computer algebra system do it for us, this row reduced echelon form is

$$\begin{pmatrix} 1 & 0 & -1 & 0 & 2 & 0 \\ 0 & 1 & 1 & 0 & -1 & 0 \\ 0 & 0 & 0 & 1 & -2 & 0 \\ 0 & 0 & 0 & 0 & 0 & 1 \end{pmatrix}$$

which is sufficient to identify the pivot columns. Therefore, a basis for $\mathbb{R}^4$ is

$$\begin{pmatrix} 1 \\ 2 \\ 1 \\ 0 \end{pmatrix}, \begin{pmatrix} 2 \\ 2 \\ 1 \\ 0 \end{pmatrix}, \begin{pmatrix} 0 \\ 1 \\ 0 \\ 0 \end{pmatrix}, \begin{pmatrix} 0 \\ 0 \\ 0 \\ 1 \end{pmatrix}.$$

A basis of V can be used as a **frame of reference**, in the following sense. Let $\beta = \{v_1, \ldots, v_n\}$ be a basis of V. Then every $v \in V$ can be written in *one and only one way* as

$$v = a_1 v_1 + \cdots + a_n v_n \qquad (a_j \in F). \tag{4.9}$$

We already know that such $a_1, \ldots, a_n$ exist in F. If

$$v = b_1 v_1 + \cdots + b_n v_n,$$

then

$$(a_1 - b_1)v_1 + \cdots + (a_n - b_n)v_n = 0.$$

By linear independence, $a_1 = b_1, \ldots, a_n = b_n$. There is no 'alternative' to the field elements a_j used in (4.9). So we may write

$$[v]_\beta = \begin{pmatrix} a_1 \\ \vdots \\ a_n \end{pmatrix}$$

and call $[v]_\beta$ the **coordinate vector of v in the basis β.**

Example 4.24. Let P be a plane in $\mathbb{R}^3$, $P = \text{span}\{\mathbf{u}, \mathbf{w}\}$. If $\mathbf{v} \in P$, then $\mathbf{v} = x_1\mathbf{u} + x_2\mathbf{w}$, say, and v has coordinate vector

$$[\mathbf{v}]_\beta = \begin{pmatrix} x_1 \\ x_2 \end{pmatrix}$$

in the basis $\beta = \{\mathbf{u}, \mathbf{w}\}$.

We shall see later that for a particular application of linear algebra there is a basis that best fits the application.

The earliest example is perhaps Euler's choice of a frame of reference in a rotating body to be a set of axes *fixed in the body*; see Johns (2005).

Definition 4.9. A vector space that is not finite-dimensional is said to be **infinite-dimensional**. In view of Lemma 4.6, V is infinite-dimensional if we can construct a linearly independent set $w_1, \ldots, w_m$ in V with arbitrarily large m.

As an example of an infinite-dimensional vector space, consider the following.

Example 4.25. $S[a, b]$ is infinite-dimensional whenever $a < b$.

For the vectors $f_0(x) = 1, f_1(x) = x, \ldots, f_m(x) = x^m$ form a linearly independent set. Whenever

$$a_0 f_0 + a_1 f_1 + \cdots + a_m f_m = 0,$$

the polynomial $a_0 + a_1 x + \cdots + a_m x^m$ is 0 for every x in (a, b). This contradicts the fact that there are at most m roots to this equation if any of the coefficients $a_0, a_1, \ldots, a_m$ is nonzero.

4.7 Linear difference equations

In this section we consider a subspace of the sequence space F^∞. Let $a_1, \ldots, a_m$ be given elements of $F, a_m \neq 0$. Let V be the set of all sequences (x_n) which satisfy the **linear difference equation** (for every $k \geq 1$)

$$x_{m+k} + a_1 x_{m-1+k} + \cdots + a_{m-1} x_{k+1} + a_m x_k = 0. \tag{4.10}$$

Example 4.26. $x_1 = x_2 = 1$ and for $m \geq 1$,

$$x_{m+2} = x_{m+1} + x_m.$$

A sequence of this kind with any choice of x_1 and x_2 is called a *Fibonacci sequence.*

Lemma 4.8. *V is an m-dimensional subspace of F^∞.*

Proof. It is clear that if (x_n) and (y_n) are in V, then so too are $(x_n + y_n)$ and (cx_n), whenever $c \in F$.

We now exhibit a basis. Let

$$\mathbf{v}_j = (0, \ldots, 0, 1, 0, \ldots, 0, v_{j,m+1}, v_{j,m+2}, \ldots)$$

where the 1 is in j^{th} place and the entries from $v_{j,m+1}$ onwards are constructed successively according to the rule (4.10). Clearly $\mathbf{v}_1, \ldots, \mathbf{v}_m$ is a linearly independent set in V. See Problem 18 on Page 109. Now let

$$\mathbf{z} = (z_1, z_2, \ldots)$$

be any element of V, and

$$\mathbf{z}' = \mathbf{z} - z_1 \mathbf{v}_1 - \cdots - z_m \mathbf{v}_m \in V.$$

Since $z_n' = 0$ for $n = 1, \ldots, m$, we find from (4.10) that $z_{m+1}' = 0$, then $z_{m+2}' = 0$, and so on. So $\mathbf{z}' = \mathbf{0}$ and $\mathbf{z} = z_1 \mathbf{v}_1 + \cdots + z_m \mathbf{v}_m$. This proves that $\mathbf{v}_1, \ldots, \mathbf{v}_m$ is a basis of V. □

A basis that lets us write the members of V in a neater way can be given at once if the polynomial

$$P(z) = z^m + a_1 z^{m-1} + \cdots + a_2 z^2 + \cdots + a_{m-1} z + a_m$$

has distinct zeros $z_1, \ldots, z_m$ in F. In this case, we can show that the vectors

$$\mathbf{w}_j = (1, z_j, z_j^2, z_j^3, \ldots) \qquad (j = 1, \ldots, m) \tag{4.11}$$

form a basis of V. The vector w_j is in V because z_j satisfies $P(z) = 0$. Thus

$$z_j^m + a_1 z_j^{m-1} + \cdots + a_2 z_j^2 + \cdots + a_{m-1} z_j + a_m = 0,$$

so z_j^m is the correct term to place in the mth slot. Also

$$\begin{aligned} & z_j^{m+1} + a_1 z_j^m + \cdots + a_2 z_j^3 + \cdots + a_{m-1} z_j^2 + a_m z_j \\ = \; & z_j \left(z_j^m + a_1 z_j^{m-1} + \cdots + a_2 z_j^2 + \cdots + a_{m-1} z_j + a_m \right) = 0, \end{aligned}$$

so z_j^{m+1} is the correct entry for the $m+1$ position. Continuing this way, we see that $\mathbf{w}_j \in V$ as claimed.

In order to show that $\mathbf{w}_1, \ldots, \mathbf{w}_m$ are a basis, it suffices to verify that they are linearly independent and then apply Corollary 4.3. The columns of the $m \times m$ matrix

$$\begin{pmatrix} 1 & \cdots & 1 \\ z_1 & \cdots & z_m \\ z_1^2 & \cdots & z_m^2 \\ \vdots & & \vdots \\ z_1^{m-1} & \cdots & z_m^{m-1} \end{pmatrix}$$

form a linearly independent set. (We shall deduce this from the theory of determinants in Chapter 7. See also Problem 26 below on Page 124.) For $m = 2$ or 3 it is also easy to verify directly by finding the row reduced echelon form of the matrix. Clearly then, $\mathbf{w}_1, \ldots, \mathbf{w}_m$ is a linearly independent set in V. (Problem 18 on Page 109).

It follows that if $\mathbf{u} \in V$, then $\mathbf{u}$ must be a linear combination of the $\mathbf{w}_j$. Thus

$$u_k = C_1 z_1^{k-1} + C_2 z_2^{k-1} + \cdots + C_m z_m^{k-1} \tag{4.12}$$

Example 4.27. Let $F = \mathbb{R}$,

$$x_1 = 1, x_2 = 1, x_{k+2} = x_{k+1} + x_k \ (k \geq 1).$$

Then for $k \geq 1$,

$$x_k = \frac{1}{\sqrt{5}}\left(\frac{1+\sqrt{5}}{2}\right)^k - \frac{1}{\sqrt{5}}\left(\frac{1-\sqrt{5}}{2}\right)^k. \tag{4.13}$$

The formula for x_k does not obviously generate integers!

Here is why the above formula results. In the above notation, the polynomial $P(z)$ is $z^2 - z - 1$ with zeros

$$z_1 = \frac{1+\sqrt{5}}{2}, z_2 = \frac{1-\sqrt{5}}{2}.$$

Thus, with $\mathbf{w}_1, \mathbf{w}_2$ as in (4.11),

$$\mathbf{x} = c_1\mathbf{w}_1 + c_2\mathbf{w}_2$$

for certain reals c_1, c_2. The given values $x_1 = 1, x_2 = 1$ impose on c_1, c_2 the requirement to be solutions of

$$1 = c_1 + c_2,$$

$$1 = c_1\left(\frac{1+\sqrt{5}}{2}\right) + c_2\left(\frac{1-\sqrt{5}}{2}\right).$$

It follows from solving this system of equations that

$$\begin{aligned} c_1 &= \frac{1}{2} + \frac{1}{10}\sqrt{5} = \frac{1}{\sqrt{5}}\left(\frac{1+\sqrt{5}}{2}\right), \\ c_2 &= \frac{1}{2} - \frac{1}{10}\sqrt{5} = -\frac{1}{\sqrt{5}}\left(\frac{1-\sqrt{5}}{2}\right). \end{aligned}$$

It follows that the solution is

$$\begin{aligned} x_k &= \frac{1}{\sqrt{5}}\left(\frac{1+\sqrt{5}}{2}\right)\left(\frac{1+\sqrt{5}}{2}\right)^{k-1} - \frac{1}{\sqrt{5}}\left(\frac{1-\sqrt{5}}{2}\right)\left(\frac{1-\sqrt{5}}{2}\right)^{k-1} \\ &= \frac{1}{\sqrt{5}}\left(\frac{1+\sqrt{5}}{2}\right)^{k} - \frac{1}{\sqrt{5}}\left(\frac{1-\sqrt{5}}{2}\right)^{k}. \end{aligned}$$

We began with a linear difference equation

$$x_{m+k} + a_1 x_{m-1+k} + \cdots + a_{m-1}x_{k+1} + a_m x_k = 0$$

and if the polynomial

$$P(x) = x^m + a_1 x^{m-1} + \cdots + a_{m-1}x + a_m \tag{4.14}$$

corresponding to this has m distinct roots, we were able to obtain solutions to the difference equation as described above.

It suffices to consider, instead of (4.12), the following expression as a formula for u_k provided none of the $z_i = 0$.

$$u_k = C_1 z_1^k + C_2 z_2^k + \cdots + C_m z_m^k$$

This is because

$$\begin{aligned} u_k &= C_1 z_1^{k-1} + C_2 z_2^{k-1} + \cdots + C_m z_m^{k-1} = \frac{C_1}{z_1} z_1^k + \cdots + \frac{C_m}{z_m} z_m^k \\ &= C_1' z_1^k + \cdots + C_m' z_m^k \end{aligned}$$

and finding the C_i is equivalent to finding the C_i'. If you apply this to the above example, you may find the desired formula appears more easily and with less algebraic manipulations.

This illustrates the following procedure for solving linear difference equations in the case where the roots to $P(x)$ are distinct.

Procedure 4.2. Suppose $\mathbf{x} \in F^\infty$ satisfies the linear difference equation

$$x_{m+k} + a_1 x_{m-1+k} + \cdots + a_{m-1}x_{k+1} + a_m x_k = 0,$$

and also suppose that there exist m distinct solutions $z_1, \cdots, z_m$ to the polynomial equation

$$x^m + a_1 x^{m-1} + \cdots + a_{m-1}x + a_m = 0.$$

Then there exist constants $c_1, \cdots, c_m$ such that

$$x_k = c_1 z_1^{k-1} + \cdots + c_m z_m^{k-1}$$

If $a_m \neq 0$, then there exist constants $c_1, \cdots, c_m$ such that

$$x_k = c_1 z_1^k + \cdots + c_m z_m^k$$

These constants c_i only need to be chosen in such a way that the given values for x_k for $k = 1, 2, \cdots, m$ are achieved. This may be accomplished by solving a system of equations.

4.8 Exercises

(1) Consider the following sets of vectors in $\mathbb{R}^3$. Explain why none of them is linearly independent before doing any computations. Next exhibit some vector as a linear combination of the others and give a basis for the span of these vectors.

(a)

$$\begin{pmatrix} 1 \\ -1 \\ 2 \end{pmatrix}, \begin{pmatrix} -1 \\ 1 \\ -2 \end{pmatrix}, \begin{pmatrix} 1 \\ 0 \\ 1 \end{pmatrix}, \begin{pmatrix} 5 \\ -3 \\ 8 \end{pmatrix}, \begin{pmatrix} 3 \\ -1 \\ 4 \end{pmatrix}$$

(b)

$$\begin{pmatrix} 1 \\ -1 \\ 2 \end{pmatrix}, \begin{pmatrix} 1 \\ 0 \\ 1 \end{pmatrix}, \begin{pmatrix} 2 \\ -1 \\ 3 \end{pmatrix}, \begin{pmatrix} 5 \\ -3 \\ 8 \end{pmatrix}, \begin{pmatrix} 5 \\ -3 \\ 6 \end{pmatrix}$$

(c)

$$\begin{pmatrix} 1 \\ 1 \\ 2 \end{pmatrix}, \begin{pmatrix} 1 \\ 1 \\ 1 \end{pmatrix}, \begin{pmatrix} 2 \\ 2 \\ 7 \end{pmatrix}, \begin{pmatrix} 5 \\ 3 \\ 6 \end{pmatrix}, \begin{pmatrix} 8 \\ 8 \\ 19 \end{pmatrix}$$

(2) Consider the following sets of vectors in $\mathbb{R}^4$. Explain why none of them is linearly independent before doing any computations. Next exhibit some vector as a linear combination of the others and give a basis for the span of these vectors.

(a)

$$\begin{pmatrix} 1 \\ 1 \\ 2 \\ 0 \end{pmatrix}, \begin{pmatrix} 1 \\ 0 \\ 2 \\ 1 \end{pmatrix}, \begin{pmatrix} 2 \\ 3 \\ 4 \\ -1 \end{pmatrix}, \begin{pmatrix} 1 \\ 3 \\ 2 \\ -2 \end{pmatrix}, \begin{pmatrix} 1 \\ 1 \\ 2 \\ 1 \end{pmatrix}$$

(b)

$$\begin{pmatrix} 1 \\ 0 \\ 1 \\ 1 \end{pmatrix}, \begin{pmatrix} 3 \\ 1 \\ 2 \\ 0 \end{pmatrix}, \begin{pmatrix} 14 \\ 4 \\ 10 \\ 2 \end{pmatrix}, \begin{pmatrix} -12 \\ -5 \\ -7 \\ 3 \end{pmatrix}, \begin{pmatrix} 1 \\ 0 \\ 2 \\ 1 \end{pmatrix}$$

(c)

$$\begin{pmatrix} 1 \\ 1 \\ 1 \\ 1 \end{pmatrix}, \begin{pmatrix} 3 \\ 3 \\ 2 \\ 0 \end{pmatrix}, \begin{pmatrix} 1 \\ 1 \\ 3 \\ 7 \end{pmatrix}, \begin{pmatrix} 1 \\ 2 \\ 2 \\ 1 \end{pmatrix}, \begin{pmatrix} 19 \\ 20 \\ 13 \\ -2 \end{pmatrix}$$

(d)

$$\begin{pmatrix} 1 \\ 2 \\ 1 \\ 1 \end{pmatrix}, \begin{pmatrix} 3 \\ 6 \\ 2 \\ 0 \end{pmatrix}, \begin{pmatrix} 1 \\ 2 \\ 3 \\ 7 \end{pmatrix}, \begin{pmatrix} 6 \\ 12 \\ 5 \\ 3 \end{pmatrix}, \begin{pmatrix} 21 \\ 42 \\ 15 \\ 3 \end{pmatrix}$$

(3) The following are independent sets of $r < k$ vectors in $\mathbb{R}^k$ for some k. Extend each collection of vectors to obtain a basis for $\mathbb{R}^k$.

(a)

$$\begin{pmatrix} 1 \\ 2 \\ 1 \\ 0 \end{pmatrix}, \begin{pmatrix} 0 \\ 1 \\ 0 \\ 1 \end{pmatrix}$$

(b)

$$\begin{pmatrix} 1 \\ 2 \\ 1 \end{pmatrix} \begin{pmatrix} 0 \\ 1 \\ 1 \end{pmatrix}$$

(c)

$$\begin{pmatrix} 0 \\ 0 \\ 1 \\ 2 \end{pmatrix}, \begin{pmatrix} 1 \\ 0 \\ 1 \\ 0 \end{pmatrix}, \begin{pmatrix} 0 \\ 1 \\ 0 \\ 1 \end{pmatrix}$$

(4) For A an $m \times n$ matrix, recall that

$$\ker(A) = \{\mathbf{x} \in \mathbb{R}^n : A\mathbf{x} = \mathbf{0}\}.$$

Show that $\ker(A)$ is always a subspace. Now for each of the following matrices, find a basis for $\ker(A)$ and give a description of $\ker(A)$. Then find a basis for the image of A, defined as $\{A\mathbf{x} : \mathbf{x} \in \mathbb{R}^n\}$, which is the same as the span of the columns. (Why?)

(a)

$$\begin{pmatrix} 1 & 2 & 1 & 0 & 0 \\ 2 & 3 & 1 & 1 & 3 \\ 2 & 1 & -3 & 1 & 2 \\ 1 & 1 & -2 & 0 & 1 \end{pmatrix}$$

(b)

$$\begin{pmatrix} 2 & 2 & 2 & 4 & 22 \\ 3 & 3 & 3 & 3 & 27 \\ 1 & 1 & 1 & 3 & 13 \\ 2 & 2 & 4 & 4 & 26 \end{pmatrix}$$

(c)

$$\begin{pmatrix} 2 & 2 & 2 & 16 & 14 \\ 3 & 3 & 3 & 24 & 21 \\ 1 & 1 & 1 & 8 & 7 \\ 2 & 2 & 4 & 20 & 18 \end{pmatrix}$$

(5) Let V denote all polynomials having real coefficients which have degree no more than 2. Show that this is a vector space over $\mathbb{R}$. Next show that $\{1, x, x^2\}$ is a basis for V. Now determine whether $\{1, x + x^2, 1 + x^2\}$ is a basis for this vector space.

(6) Let V be all polynomials having real coefficients, which have degree no more than 3. Show that $\{1, x, x^2, x^3\}$ is a basis. Now determine whether $\{1 + x^2 + 4x^3, 1 + x + x^2 + 6x^3, x + x^2 + 3x^3, 1 + 4x^3\}$ is a basis. **Hint:** Take a linear combination, set equal to 0, differentiate, and then let $x = 0$.

(7) Let V consist of all polynomials of degree no larger than n which have coefficients in $\mathbb{R}$. Show that this is a vector space over $\mathbb{R}$ and find its dimension by giving a basis. Tell why what you have identified as a basis is a basis.

(8) In the situation of Problem 7 where V is the set of polynomials of degree no larger than n, suppose there are n distinct points in $\mathbb{R}$ $\{a_1, a_2, \cdots, a_n\}$ and for $p, q \in V$ define

$$\langle p, q \rangle \equiv p(a_1) q(a_1) + \cdots + p(a_n) q(a_n).$$

Show that this is an inner product on V in the sense that the above product satisfies the axioms (1.1) - (1.4) on Page 4.

(9) Let U, W be subspaces of a vector space V. Then $U + W$ is defined to be all sums of the form $u + w$ where $u \in U$ and $w \in W$. Show that $U + W$ is a subspace and also that the dimension of this subspace is no larger than the sum of the dimensions of U and W. We recall that the dimension of a subspace U is denoted as $\dim(U)$. Thus you wish to show that

$$\dim(U + W) \leq \dim(U) + \dim(W).$$

Hint: At one point you might want to use the theorem that every linearly independent set can be extended to a basis.

(10) We recall that whenever U, W are subspaces, then so is $U \cap W$. In the context of Problem 9, show that

$$\dim(U + W) = \dim(U) + \dim(W)$$

if and only if $U \cap W = \{0\}$.

(11) Consider the set of $n \times n$ matrices having complex entries. Show that this is a vector space over $\mathbb{C}$. Find its dimension by describing a basis.

(12) Let A be any $n \times n$ matrix with entries in any field F. Show that there exists a polynomial

$$p(\lambda) = \lambda^{n^2} + a\lambda^{n^2-1} + \cdots + c\lambda + d$$

such that

$$p(A) = A^{n^2} + aA^{n^2-1} + \cdots + cA + dI = 0$$

where the 0 is the zero matrix. (In fact, there exists a polynomial of degree n such that this happens but this is another topic which will be presented later.) **Hint:** Have a look at Problem 11.

(13) Given two vectors in $\mathbb{R}^2$,

$$\begin{pmatrix} a \\ b \end{pmatrix}, \begin{pmatrix} x \\ y \end{pmatrix}$$

show that they are a basis for $\mathbb{R}^2$ if and only if one is not a multiple of the other. Is this still true in F^2 where F is an arbitrary field?

(14) Show that in any vector space, a set of two vectors is independent if and only if neither vector is a multiple of the other.

(15) Give an example of three vectors in $\mathbb{R}^3$ which is dependent, but none of the vectors is a multiple of any of the others.

(16) Given two vectors in $\mathbb{R}^2$,

$$\begin{pmatrix} a \\ b \end{pmatrix}, \begin{pmatrix} x \\ y \end{pmatrix},$$

show that they are a basis for $\mathbb{R}^2$ if and only if $ay - bx \neq 0$. Is this still true in F^2 where F is an arbitrary field?

(17) The set of complex numbers $\mathbb{C}$ is a vector space over $\mathbb{R}$, the real numbers. What is the dimension of this vector space?

(18) Consider the set of all complex numbers of the form $a + ib$ where $a, b \in \mathbb{Q}$, the field of rational numbers. Show that this set of complex numbers is a vector space over the rational numbers and find a basis for this vector space.

(19) Let F be a field and let $p(\lambda)$ be a monic irreducible polynomial. This means that $p(\lambda)$ is of the form $\lambda^n + a_{n-1}\lambda^{n-1} + \cdots + a_1\lambda + a_0$, and the only polynomials having coefficients in F which divide $p(\lambda)$ are constants and constant multiples of $p(\lambda)$. For example you could have $x^7 - 11$ with field of scalars equal to the rational numbers. Suppose now that $a \in G$, a possibly larger field and that $p(a) = 0$. For example you could have, in the given example, $a = \sqrt[7]{11}$. Show that the numbers $F_1 \equiv \{g(a) : g \text{ is a polynomial having coefficients in } F\}$ is a vector space over F of dimension n. Next show that F_1 is also a field. You can consider the next two problems for the existence of the larger field G.

(20) Suppose $p(\lambda)$ and $q(\lambda)$ are two polynomials having coefficients in a field F. The greatest common divisor of $p(\lambda), q(\lambda)$ sometimes denoted as $(p(\lambda), q(\lambda))$ is defined to be the monic polynomial $g(\lambda)$ such that $g(\lambda)$ divides both $p(\lambda)$ and $q(\lambda)$ and if $l(\lambda)$ is any other polynomial, which divides both $p(\lambda)$ and $q(\lambda)$, then $l(\lambda)$ divides $g(\lambda)$. Show that there exist polynomials $m(\lambda), n(\lambda)$ such that $(p, q)(\lambda) = n(\lambda)p(\lambda) + m(\lambda)q(\lambda)$.

(21) Suppose $p(\lambda)$ is a monic polynomial of degree n for $n > 1$ having coefficients in F a field of scalars. Suppose also that $p(\lambda)$ is irreducible. Consider the question whether there exists a larger field G such that a root of $p(\lambda)$ is found in G. Show that such a field must exist and can be identified with the polynomials $q(\lambda)$ of degree less than n having coefficients in F. Next show that there exists a field G, possibly larger than F, such that $p(\lambda)$ can be written in the form $\prod_{i=1}^{n}(\lambda - a_i)$, the $a_i \in G$, possibly not all distinct. Next show that if $q(\lambda)$ is any polynomial, having coefficients in G, which divides $p(\lambda)$, then $q(\lambda)$ must be the product of linear factors as well. **Hint:** Define multiplication as follows.

$$q_1(\lambda)\, q_2(\lambda) = r(\lambda)$$

where $q_1(\lambda)\, q_2(\lambda) = k(\lambda)\, p(\lambda) + r(\lambda)$ with the degree of $r(\lambda)$ less than the degree of $p(\lambda)$ or else $r(\lambda) = 0$. This larger field is called a field extension.

(22) A geometric sequence satisfies $a_{n+1} = ra_n$ where r is the common ratio. Find a formula for a_n given that $a_1 = 1$.

(23) Consider the following diagram which illustrates payments equal to P made at equally spaced intervals of time.

0 P P P P P P P P P P

The 0 at the far left indicates this is at time equal to 0. The payments are made at the ends of equally spaced intervals of time as suggested by the picture. This situation is known as an ordinary annuity. Suppose the bank, into which these payments are made, pays an interest rate of r per payment period. Thus if there is an amount A_n at the end of the nth payment period, then at the end of the next payment period there will be

$$A_{n+1} = (1+r)\, A_n + P$$

while at the end of the first payment period, there is $A_1 = P$. Find a formula for A_n which gives the amount in the bank at the end of the nth payment period. **Hint:** This is a nonhomogeneous linear difference equation. To solve this difference equation, write

$$A_{n+2} = (1+r)\, A_{n+1} + P$$

and subtract to obtain the sort of problem discussed above,

$$A_{n+2} - (2+r)\, A_{n+1} + (1+r)\, A_n = 0.$$

along with the initial conditions $A_1 = P$, $A_2 = P(2+r)$.

(24) Let A be an $m \times n$ matrix and let $\mathbf{b}$ be a $m \times 1$ vector. Explain why there exists a solution $\mathbf{x}$ to the system

$$A\mathbf{x} = \mathbf{b}$$

if and only if the rank of A equals the rank of $(A|\mathbf{b})$. Recall that the rank of a matrix is the number of pivot columns.

(25) Let A be an $m \times n$ matrix. Col(A) will denote the column space of A which is defined as the span of the columns. Row(A) will denote the row space of A and is defined as the span of the rows. Show that a basis for Row(A) consists of the nonzero rows in the row reduced echelon form of A while a basis of Col(A) consists of the pivot columns of A. Explain why the dimension of Row(A) equals the dimension of Col(A). Also give an example to show that the pivot columns of the row reduced echelon form of A cannot be used as a basis for Col(A).

(26) Suppose you are given ordered pairs $(x_1, y_1), \cdots, (x_n, y_n)$ where the x_i are distinct and the y_i are arbitrary. The Lagrange interpolating polynomial is a polynomial of degree $n-1$ defined as

$$p(x) = \sum_{j=1}^{n} y_j \frac{\prod_{i,i\neq j}(x - x_i)}{\prod_{i,i\neq j}(x_j - x_i)}.$$

Here $\prod_{i,i\neq j}(x - x_i)$ means to take the product of all $(x - x_i)$ for all $i \neq j$. Thus j is fixed and you are considering the product of all these terms for all $i \neq j$. The symbol $\prod_{i,i\neq j}(x_j - x_i)$ means to take the product of all $(x_j - x_i)$ for all $i \neq j$. For example, if $n = 3$, this polynomial is

$$y_1 \frac{(x - x_2)(x - x_3)}{(x_1 - x_2)(x_1 - x_3)} + y_2 \frac{(x - x_1)(x - x_3)}{(x_2 - x_1)(x_2 - x_3)} + y_3 \frac{(x - x_2)(x - x_1)}{(x_3 - x_2)(x_3 - x_1)}.$$

Show that the polynomial passes through the given points. Now observe that the existence of a degree $n-1$ polynomial

$$y = a_{n-1}x^{n-1} + \cdots + a_1 x + a_0$$

passing through the given points is equivalent to the existence of a solution to the system of equations

$$\begin{pmatrix} x_1^{n-1} & x_1^{n-2} & \cdots & 1 \\ x_2^{n-1} & x_2^{n-2} & \cdots & 1 \\ \vdots & \vdots & & \vdots \\ x_n^{n-1} & x_n^{n-2} & \cdots & 1 \end{pmatrix} \begin{pmatrix} a_{n-1} \\ a_{n-1} \\ \vdots \\ a_0 \end{pmatrix} = \begin{pmatrix} y_1 \\ y_2 \\ \vdots \\ y_n \end{pmatrix}$$

Explain why this requires that the rank of the $n \times n$ matrix above must be n. Hence it has an inverse. Consequently, its transpose also is invertible. (Recall $(A^t)^{-1} = (A^{-1})^t$ from the definition of the inverse.)

Chapter 5

Linear mappings

Chapter summary

We define linear mappings and give examples. Then we show how every linear mapping on F^n can be considered in terms of a matrix. After this, rank and nullity are described in terms of vector space ideas and a fundamental lemma due to Sylvester is proved about the rank and nullity of a product of linear transformations. This is used to give a discussion of linear differential equations.

5.1 Definition and examples

Most applications of linear algebra ultimately depend on the idea of a linear mapping, which we now define.

Definition 5.1. Let V and W be vector spaces over the field F. Let $T : V \to W$ be a mapping having the property that

$$T(a_1v_1 + a_2v_2) = a_1T(v_1) + a_2T(v_2) \tag{5.1}$$

whenever v_1, v_2 are in V and a_1, a_2 are in F. We say that T is a **linear mapping**.

The property (5.1) is called **linearity** of T. In this book we write Tv as an abbreviation for $T(v)$ (unless v is a compound expression, such as the case $v = a_1v_1 + a_2v_2$). Note that $T0 = 0$ from (5.1).

Example 5.1. Let $T : \mathbb{R}^2 \to \mathbb{R}^2$ be a **rotation** through angle θ. That is, $T\mathbf{x}$ is obtained by rotating $\mathbf{v}$ by θ anticlockwise. (The word 'anticlockwise' is left implicit from now on.)

Rotations are important in computer graphics. Here is a picture to illustrate the result of rotating the vectors from the origin to points of a picture anticlockwise through an angle of 45 degrees.

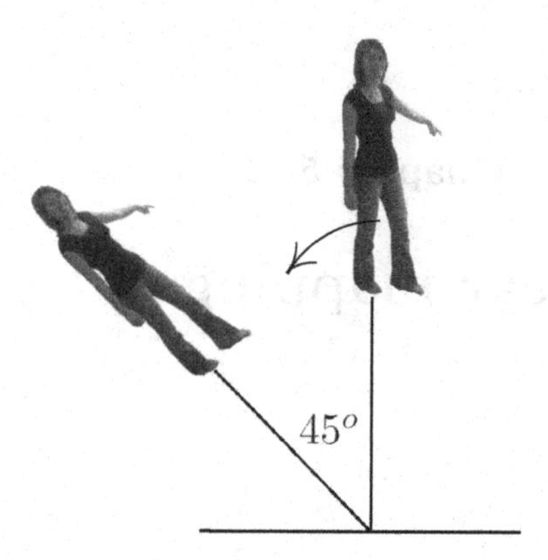

In fact, rotations are linear as shown in the following picture which shows the effect of rotation on a sum of vectors.

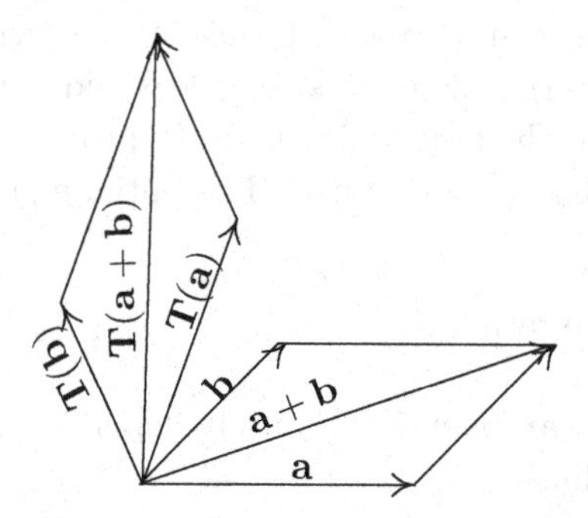

Linearity can always be seen as two properties:

$$T(\mathbf{v}_1 + \mathbf{v}_2) = T\mathbf{v}_1 + T\mathbf{v}_2,\ T(a\mathbf{v}) = aT\mathbf{v}.$$

In the case of a rotation, these equations say that a parallelogram remains a parallelogram under rotation, and that scaling a vector before or after rotation has the same outcome.

Example 5.2. Let $n > 1$. Let $P : F^n \to F^n$ be the **projection**

$$P(x_1, \ldots, x_n) = (x_1, \ldots, x_{n-1}, 0).$$

(You should be able to visualize P easily for $F = \mathbb{R},\ n = 2, 3$.) There is no difficulty in verifying (5.1).

Example 5.3. Let $V = S[a, b]$ or $V = S_{\mathbb{C}}[a, b]$. The **differentiation mapping** $D : V \to V$,

$$Df = f'$$

is linear. The verification uses the simple properties

$$(f + g)' = f' + g',\ (kf)' = kf'$$

of differentiation, where k is a constant.

Example 5.4. Let V be as in Example 5.3. For a fixed c in $[a,b]$, let

$$(Uf)(x) = \int_c^x f(t)dt.$$

It is not difficult to verify that Uf is in V and that $U : V \to V$ is linear.

Another important observation is that the composition of linear transformations is a linear transformation.

Definition 5.2. Let V, W, Z be vector spaces and let $T : V \to W$ and $S : W \to Z$ be linear mappings. Then ST denotes the mapping $ST : V \to Z$ defined by

$$STv = S(Tv)$$

Proposition 5.1. Let T, S be linear mappings as defined above, then ST is also a linear mapping.

Proof. This follows from

$$\begin{aligned} ST(a_1v_1 + a_2v_2) &= S(T(a_1v_1 + a_2v_2)) = S(a_1Tv_1 + a_2Tv_2) \\ &= a_1S(Tv_1) + a_2S(Tv_2) = a_1ST(v_1) + ST(v_2). \ \square \end{aligned}$$

5.2 The matrix of a linear mapping from F^n to F^m

In a finite-dimensional space, it is convenient to show the action of linear mappings using matrices. Consider a linear mapping $T : F^n \to F^m$ and let

$$A = \begin{pmatrix} T\mathbf{e}_1 & \cdots & T\mathbf{e}_n \end{pmatrix}. \tag{5.2}$$

This $m \times n$ matrix is called the **matrix of** T.

Example 5.5. In Example 5.1, a simple sketch shows that

$$T\mathbf{e}_1 = \begin{pmatrix} \cos\theta \\ \sin\theta \end{pmatrix}, T\mathbf{e}_2 = \begin{pmatrix} -\sin\theta \\ \cos\theta \end{pmatrix}.$$

So the matrix of T is

$$A(\theta) = \begin{pmatrix} \cos\theta & -\sin\theta \\ \sin\theta & \cos\theta \end{pmatrix}. \tag{5.3}$$

Here is a discussion of the details.

Let $\mathbf{e}_1 \equiv \begin{pmatrix} 1 \\ 0 \end{pmatrix}$ and $\mathbf{e}_2 \equiv \begin{pmatrix} 0 \\ 1 \end{pmatrix}$. These identify the geometric vectors which point along the positive x axis and positive y axis as shown.

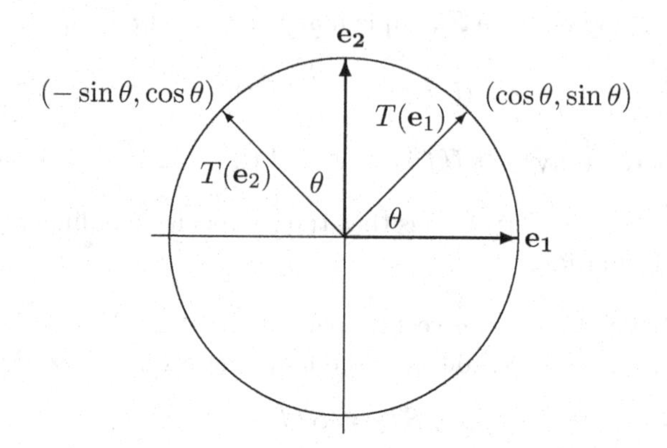

From the above, you only need to find $T\mathbf{e}_1$ and $T\mathbf{e}_2$, the first being the first column of the desired matrix A and the second being the second column. From the definition of the cos, sin the coordinates of $T(\mathbf{e}_1)$ are as shown in the picture. The coordinates of $T(\mathbf{e}_2)$ also follow from simple trigonometry. Thus

$$T\mathbf{e}_1 = \begin{pmatrix} \cos\theta \\ \sin\theta \end{pmatrix}, T\mathbf{e}_2 = \begin{pmatrix} -\sin\theta \\ \cos\theta \end{pmatrix}.$$

Therefore,

$$A = \begin{pmatrix} \cos\theta & -\sin\theta \\ \sin\theta & \cos\theta \end{pmatrix}.$$

For those who prefer a more algebraic approach, the definition of

$$(\cos\theta, \sin\theta)$$

is as the x and y coordinates of a point on the unit circle resulting from rotating the point at $(1,0)$ through an angle of θ. Now the point of the vector from $(0,0)$ to $(0,1)$ $\mathbf{e}_2$ is exactly $\pi/2$ further along the unit circle from $\mathbf{e}_1$. Therefore, when $\mathbf{e}_2$ is rotated through an angle of θ, the x and y coordinates are given by

$$(x,y) = (\cos(\theta+\pi/2), \sin(\theta+\pi/2)) = (-\sin\theta, \cos\theta).$$

Example 5.6. The matrix of the projection in Example 5.2 is

$$A = \begin{pmatrix} 1 & \cdots & 0 & 0 \\ 0 & \ddots & & 0 \\ \vdots & & 1 & \vdots \\ 0 & \cdots & 0 & 0 \end{pmatrix}.$$

Let A be an $m \times n$ matrix over F, $A = (a_{ij})$. Let

$$\mathbf{x} = \begin{pmatrix} x_1 \\ \vdots \\ x_n \end{pmatrix} \in F^n.$$

We recall that

$$Ax = \begin{pmatrix} a_{11}x_1 + & \cdots & +a_{1n}x_n \\ \vdots & & \vdots \\ a_{m1}x_1 + & \cdots & +a_{mn}x_n \end{pmatrix}.$$

or equivalently

$$Ax = x_1\mathbf{a}_1 + \cdots + x_n\mathbf{a}_n. \tag{5.4}$$

Lemma 5.1. *Let $T : F^n \to F^m$ and suppose the matrix of T is A. Then $T\mathbf{x}$ is (the column vector) $A\mathbf{x}$.*

Proof.

$$T\mathbf{x} = T\left(\sum_{i=1}^{n} x_i\mathbf{e}_i\right)$$

$$= \sum_{i=1}^{n} x_i T(\mathbf{e}_i) \text{ by linearity} = \begin{pmatrix} T\mathbf{e}_1 & \cdots & T\mathbf{e}_n \end{pmatrix} \begin{pmatrix} x_1 \\ \vdots \\ x_n \end{pmatrix} = A\mathbf{x}$$

the third step coming from (5.4). □

Thus to find $T\mathbf{x}$, it suffices to multiply $\mathbf{x}$ on the left by A.

Note that we can start with a given $m \times n$ matrix A and *define* $T : F^n \to F^m$ by $T\mathbf{x} = A\mathbf{x}$. The matrix of T is A, since $T\mathbf{e}_j = A\mathbf{e}_j =$ column j of A.

Throughout the book, in any expression $A\mathbf{x}$ (or $B\mathbf{y}$, for that matter) the capital letter denotes an $m \times n$ matrix; and the bold lower case letter, a column vector of length n.

Example 5.7. Let T be the rotation in Example 5.1. Then

$$T\mathbf{x} = \begin{pmatrix} \cos\theta & -\sin\theta \\ \sin\theta & \cos\theta \end{pmatrix} \begin{pmatrix} x_1 \\ x_2 \end{pmatrix} = \begin{pmatrix} (\cos\theta)\,x_1 - (\sin\theta)\,x_2 \\ (\sin\theta)\,x_1 + (\cos\theta)\,x_2 \end{pmatrix}.$$

Example 5.8. Let P be the projection in Example 5.2. Then

$$P\mathbf{x} = \begin{pmatrix} 1 & \cdots & 0 & 0 \\ 0 & \ddots & & 0 \\ \vdots & & 1 & \vdots \\ 0 & \cdots & 0 & 0 \end{pmatrix} \begin{pmatrix} x_1 \\ \vdots \\ x_{n-1} \\ x_n \end{pmatrix}.$$

(as we know already).

We observe then that $T : F^n \to F^m$ is a linear transformation if and only if there exists an $m \times n$ matrix A such that for all $\mathbf{x} \in F^n, T\mathbf{x} = A\mathbf{x}$.

In other words, linear transformations (mappings) from F^n to F^m may always be obtained by multiplying on the left by an appropriate $m \times n$ matrix.

Now here is an example which involves the composition of two linear mappings.

Example 5.9. A linear transformation is defined by first rotating all vectors in $\mathbb{R}^2$ through an angle of $\pi/4$ and then reflecting across the line $y = x$. Find the result of doing these two transformations to the vector $(1, 2)^t$.

The effect of rotating through an angle of $\pi/4$ is accomplished by multiplying by

$$\begin{pmatrix} \cos(\pi/4) & -\sin(\pi/4) \\ \sin(\pi/4) & \cos(\pi/4) \end{pmatrix}$$

After this, we must multiply by the matrix which accomplishes the reflection across the line $y = x$. Drawing a picture you see that if S is this linear transformation,

$$S\mathbf{e}_1 = \mathbf{e}_2, \ S\mathbf{e}_2 = \mathbf{e}_1$$

Therefore, the matrix of this transformation is

$$\begin{pmatrix} 0 & 1 \\ 1 & 0 \end{pmatrix}$$

It follows that the desired vector is

$$\begin{pmatrix} 0 & 1 \\ 1 & 0 \end{pmatrix} \begin{pmatrix} \cos(\pi/4) & -\sin(\pi/4) \\ \sin(\pi/4) & \cos(\pi/4) \end{pmatrix} \begin{pmatrix} 1 \\ 2 \end{pmatrix}$$

$$= \begin{pmatrix} \frac{1}{2}\sqrt{2} & \frac{1}{2}\sqrt{2} \\ \frac{1}{2}\sqrt{2} & -\frac{1}{2}\sqrt{2} \end{pmatrix} \begin{pmatrix} 1 \\ 2 \end{pmatrix} = \begin{pmatrix} \frac{3}{2}\sqrt{2} \\ -\frac{1}{2}\sqrt{2} \end{pmatrix}.$$

The matrix of the linear transformation which is the composition of the two described is

$$\begin{pmatrix} \frac{1}{2}\sqrt{2} & \frac{1}{2}\sqrt{2} \\ \frac{1}{2}\sqrt{2} & -\frac{1}{2}\sqrt{2} \end{pmatrix}.$$

Here is another very interesting example.

Example 5.10. A linear transformation first rotates all vectors through an angle of θ and then rotates all vectors through an angle of ϕ. Find the matrix of the linear transformation.

First of all, we know that this matrix is of the form

$$\begin{pmatrix} \cos(\theta+\phi) & -\sin(\theta+\phi) \\ \sin(\theta+\phi) & \cos(\theta+\phi) \end{pmatrix}.$$

On the other hand, this linear transformation is also the composition of the two which come from rotation through the two given angles. Thus it is also equal to

$$\begin{pmatrix} \cos(\phi) & -\sin(\phi) \\ \sin(\phi) & \cos(\phi) \end{pmatrix} \begin{pmatrix} \cos(\theta) & -\sin(\theta) \\ \sin(\theta) & \cos(\theta) \end{pmatrix}$$

$$= \begin{pmatrix} \cos\theta\cos\phi - \sin\theta\sin\phi & -\cos\theta\sin\phi - \cos\phi\sin\theta \\ \cos\theta\sin\phi + \cos\phi\sin\theta & \cos\theta\cos\phi - \sin\theta\sin\phi \end{pmatrix}.$$

It follows that

$$\begin{aligned}
&\begin{pmatrix} \cos(\theta+\phi) & -\sin(\theta+\phi) \\ \sin(\theta+\phi) & \cos(\theta+\phi) \end{pmatrix} \\
&= \begin{pmatrix} \cos\theta\cos\phi - \sin\theta\sin\phi & -\cos\theta\sin\phi - \cos\phi\sin\theta \\ \cos\theta\sin\phi + \cos\phi\sin\theta & \cos\theta\cos\phi - \sin\theta\sin\phi \end{pmatrix}.
\end{aligned}$$

This implies that

$$\begin{aligned}
\cos(\theta+\phi) &= \cos\theta\cos\phi - \sin\theta\sin\phi \\
\sin(\theta+\phi) &= \cos\theta\sin\phi + \cos\phi\sin\theta
\end{aligned}$$

which are the familiar identities for the trigonometric functions of the sum of two angles. You may recall that these were fairly difficult to obtain in the context of trigonometry, but the above computation shows that linear algebra delivers these identities without any trouble.

5.3 Exercises

(1) Find the matrix for the linear transformation which rotates every vector in $\mathbb{R}^2$ through an angle of $\pi/3$.

(2) Find the matrix for the linear transformation which rotates every vector in $\mathbb{R}^2$ through an angle of $\pi/4$.

(3) Find the matrix for the linear transformation which rotates every vector in $\mathbb{R}^2$ through an angle of $-\pi/3$.

(4) Find the matrix for the linear transformation which rotates every vector in $\mathbb{R}^2$ through an angle of $2\pi/3$.

(5) Find the matrix for the linear transformation which rotates every vector in $\mathbb{R}^2$ through an angle of $\pi/12$. **Hint:** Note that $\pi/12 = \pi/3 - \pi/4$.

(6) Find the matrix for the linear transformation which rotates every vector in $\mathbb{R}^2$ through an angle of $2\pi/3$ and then reflects across the x axis.

(7) Find the matrix for the linear transformation which rotates every vector in $\mathbb{R}^2$ through an angle of $\pi/3$ and then reflects across the x axis.

(8) Find the matrix for the linear transformation which rotates every vector in $\mathbb{R}^2$ through an angle of $\pi/4$ and then reflects across the x axis.

(9) Find the matrix for the linear transformation which rotates every vector in $\mathbb{R}^2$ through an angle of $\pi/6$ and then reflects across the x axis followed by a reflection across the y axis.

(10) Find the matrix for the linear transformation which reflects every vector in $\mathbb{R}^2$ across the x axis and then rotates every vector through an angle of $\pi/4$.

(11) Find the matrix for the linear transformation which reflects every vector in $\mathbb{R}^2$ across the y axis and then rotates every vector through an angle of $\pi/4$.

(12) Find the matrix for the linear transformation which reflects every vector in $\mathbb{R}^2$ across the x axis and then rotates every vector through an angle of $\pi/6$.

(13) Find the matrix for the linear transformation which reflects every vector in $\mathbb{R}^2$ across the line $y = x$ and then rotates every vector through an angle of $\pi/6$.

(14) Find the matrix for the linear transformation which rotates every vector through an angle of $\pi/6$ and then reflects every vector in $\mathbb{R}^2$ across the line $y = x$.

(15) Find the matrix for the linear transformation which reflects every vector in $\mathbb{R}^2$ across the y axis and then rotates every vector through an angle of $\pi/6$.

(16) Find the matrix for the linear transformation which rotates every vector in $\mathbb{R}^2$ through an angle of $5\pi/12$. **Hint:** Note that $5\pi/12 = 2\pi/3 - \pi/4$.

(17) Find the matrix of the linear transformation which rotates every vector in $\mathbb{R}^3$ anticlockwise about the z axis when viewed from the positive z axis through an angle of 30° and then reflects through the xy plane.

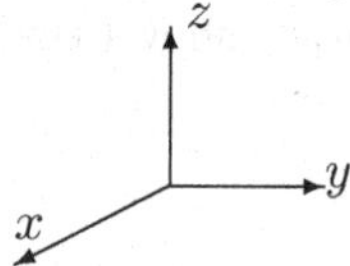

(18) For $\mathbf{u}, \mathbf{v} \in \mathbb{R}^n, \mathbf{u} \neq \mathbf{0}$, let

$$\operatorname*{proj}_{\mathbf{u}}(\mathbf{v}) = \left\langle \mathbf{v}, \frac{\mathbf{u}}{|\mathbf{u}|} \right\rangle \frac{\mathbf{u}}{|\mathbf{u}|}.$$

Show that the mapping $\mathbf{v} \to \operatorname{proj}_{\mathbf{u}}(\mathbf{v})$ is a linear transformation. Find the matrices $\operatorname{proj}_{\mathbf{u}}$ for the following.

(a) $\mathbf{u} = (1, -2, 3)^t$.
(b) $\mathbf{u} = (1, 5, 3)^t$.
(c) $\mathbf{u} = (1, 0, 3)^t$.

(19) Show that the function $T_{\mathbf{u}}$ defined by $T_{\mathbf{u}}(\mathbf{v}) \equiv \mathbf{v} - \operatorname{proj}_{\mathbf{u}}(\mathbf{v})$ is also a linear transformation.

(20) Show that $\langle \mathbf{v} - \operatorname{proj}_{\mathbf{u}}(\mathbf{v}), \mathbf{u} \rangle = 0$. For a given nonzero vector $\mathbf{u}$ in $\mathbb{R}^n$, conclude that every vector in $\mathbb{R}^n$ can be written as the sum of two vectors, one which is perpendicular to $\mathbf{u}$ and one which is parallel to $\mathbf{u}$.

(21) Show that if T is a linear transformation mapping a vector space V to a vector space W that T is one-to-one if and only if, whenever $Tv = 0$, it follows that $v = 0$.

(22) Let T be a linear mapping from V to W where V and W are vector spaces over a field F. Then

$$\ker(T) \equiv \{v \in V : Tv = 0\}.$$

Show directly from the definition, that $\ker(T)$ is a subspace of V. Also let

$$\begin{aligned} \operatorname{Im}(T) &\equiv \{w \in W : w = Tv \text{ for some } v \in V\} \\ &= \{Tv : v \in V\}. \end{aligned}$$

Show directly from the definition that $\operatorname{Im}(T)$ is a subspace of W.

(23) ↑Show that in the situation of Problem 22, where V and W are finite dimensional vector spaces and $T \neq 0$, that there exists a basis for $\operatorname{Im}(T)$

$$\{Tv_1, \cdots, Tv_m\}$$

and that in this situation,

$$\{v_1, \cdots, v_m\}$$

is linearly independent.

(24) ↑In the situation of Problem 23 in which $\ker(T) \neq \{0\}$, show that there exists $\{z_1, \cdots, z_n\}$ a basis for $\ker(T)$. Now for an arbitrary $Tv \in \operatorname{Im}(T)$, explain why

$$Tv = a_1 Tv_1 + \cdots + a_m Tv_m$$

and why this implies that

$$v - (a_1 v_1 + \cdots + a_m v_m) \in \ker(T).$$

Then explain why $V = \operatorname{span}\{v_1, \cdots, v_m, z_1, \cdots, z_n\}$.

(25) ↑In the situation of the above problem, show that

$$\{v_1, \cdots, v_m, z_1, \cdots, z_n\}$$

is a basis for V and therefore, $\dim(V) = \dim(\ker(T)) + \dim(\operatorname{Im}(T))$. In case that T is one-to-one or the zero transformation, give the correct definition of $\dim(\{0\})$ so that this will still be so.

(26) ↑Let A be a linear transformation from V to W and let B be a linear transformation from W to U, where V, W, U are all finite dimensional vector spaces over a field F. Using the definition of ker in Problem 22, Explain why

$$A(\ker(BA)) \subseteq \ker(B), \ \ker(A) \subseteq \ker(BA).$$

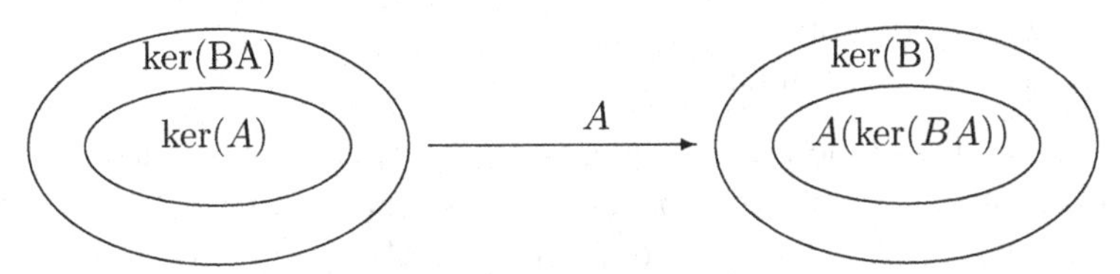

(27) ↑Let $\{x_1, \cdots, x_n\}$ be a basis of $\ker(A)$ and let $\{Ay_1, \cdots, Ay_m\}$ be a basis of $A(\ker(BA))$. Let $z \in \ker(BA)$. Show the following:

(a) $Az \in \operatorname{span}\{Ay_1, \cdots, Ay_m\}$

(b) There exist scalars a_i such that

$$A(z - (a_1 y_1 + \cdots + a_m y_m)) = 0.$$

(c) Next explain why it follows that

$$z - (a_1 y_1 + \cdots + a_m y_m) \in \operatorname{span}\{x_1, \cdots, x_n\}.$$

(d) Now explain why
$$\ker(BA) \subseteq \operatorname{span}\{x_1, \cdots, x_n, y_1, \cdots, y_m\}.$$
(e) Explain why
$$\dim(\ker(BA)) \leq \dim(\ker(B)) + \dim(\ker(A)). \tag{5.5}$$
This important inequality is due to Sylvester.

(f) Show that strict inequality must hold if $A(\ker(BA))$ is a proper subset of $\ker(B)$. **Hint:** On this last part, there must exist a vector w in $\ker(B)$ which is not in the span of $\{Ay_1, \cdots, Ay_m\}$. What does this mean about independence of the set
$$\{Ay_1, \cdots, Ay_m, w\}?$$

(28) ↑In the above problem, an inequality of the form
$$\dim(\ker(BA)) \leq \dim(\ker(B)) + \dim(\ker(A))$$
was established. Strict inequality holds if $A(\ker(BA))$ is a proper subset of $\ker(B)$. Show that if
$$A(\ker(BA)) = \ker(B),$$
then equality holds in Sylvester's inequality. Thus with the previous problem, (5.5) holds and equality occurs if and only if $A(\ker(BA)) = \ker(B)$. **Hint:** As above, let $\{x_1, \cdots, x_n\}$ be a basis for $\ker(A)$. Thus $\dim(\ker(A)) = n$. Next let $\{Ay_1, \cdots, Ay_m\}$ be a basis for $A(\ker(BA)) = \ker(B)$ so $\dim(\ker(B)) = m$. Now repeat the argument of the above problem, letting $z \in \ker(BA)$.

(29) Generalize the result of the previous problem to any finite product of linear mappings.

(30) Here are some descriptions of functions mapping $\mathbb{R}^n$ to $\mathbb{R}^n$.

(a) T multiplies the jth component of $\mathbf{x}$ by a nonzero number b.

(b) T replaces the ith component of $\mathbf{x}$ with b times the jth component added to the ith component.

(c) T switches two components.

Show that these functions are linear and describe their matrices.

(31) In Problem 30, sketch the effects of the linear transformations on the unit square $[0,1] \times [0,1]$ in $\mathbb{R}^2$. From Problem 34 on Page 91 give a geometric description of the action of an arbitrary invertible matrix in terms of products of these special matrices in Problem 30.

(32) Let $\mathbf{u} = (a, b)$ be a unit vector in $\mathbb{R}^2$. Find the matrix which reflects all vectors across this vector.

u

Hint: You might want to notice that $(a, b) = (\cos\theta, \sin\theta)$ for some θ. First rotate through $-\theta$. Next reflect through the x axis which is easy. Finally rotate through θ.

(33) Any matrix Q which satisfies $Q^tQ = QQ^t = I$ is called an **orthogonal** matrix. Show that the linear transformation determined by an orthogonal $n \times n$ matrix always preserves the length of a vector in $\mathbb{R}^n$. **Hint:** First either recall, depending on whether you have done Problem 47 on Page 56, or show that for any matrix A,
$$\langle A\mathbf{x}, \mathbf{y} \rangle = \langle \mathbf{x}, A^t\mathbf{y} \rangle$$
Next $||\mathbf{x}||^2 = \langle Q^tQ\mathbf{x}, \mathbf{x} \rangle \cdots$.

(34) ↑Let $\mathbf{u}$ be a unit vector. Show that the linear transformation of the matrix $I - 2\mathbf{u}\mathbf{u}^t$ preserves all distances and satisfies
$$\left(I - 2\mathbf{u}\mathbf{u}^t\right)^t \left(I - 2\mathbf{u}\mathbf{u}^t\right) = I.$$
This matrix is called a Householder reflection and it is an important example of an orthogonal matrix.

(35) ↑Suppose $|\mathbf{x}| = |\mathbf{y}|$ for $\mathbf{x}, \mathbf{y} \in \mathbb{R}^n$. The problem is to find an orthogonal transformation Q, (see Problem 33) which has the property that $Q\mathbf{x} = \mathbf{y}$ and $Q\mathbf{y} = \mathbf{x}$. Show that
$$Q \equiv I - 2\frac{\mathbf{x} - \mathbf{y}}{|\mathbf{x} - \mathbf{y}|^2} (\mathbf{x} - \mathbf{y})^t$$
does what is desired.

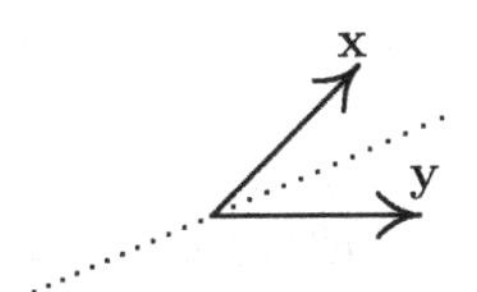

(36) Let $\mathbf{a}$ be a fixed nonzero vector. The function $T_{\mathbf{a}}$ defined by $T_{\mathbf{a}}\mathbf{v} = \mathbf{a} + \mathbf{v}$ has the effect of translating all vectors by adding $\mathbf{a}$. Show that this is **not** a linear transformation. Explain why it is not possible to realize $T_{\mathbf{a}}$ in $\mathbb{R}^3$ by multiplying by a 3×3 matrix.

(37) ↑In spite of Problem 36 we can represent both translations and rotations by matrix multiplication at the expense of using higher dimensions. This is done by the homogeneous coordinates. We will illustrate in $\mathbb{R}^3$ where most interest in this is found. For each vector $\mathbf{v} = (v_1, v_2, v_3)^t$, consider the vector in $\mathbb{R}^4$ $(v_1, v_2, v_3, 1)^t$. What happens when you do
$$\begin{pmatrix} 1 & 0 & 0 & a_1 \\ 0 & 1 & 0 & a_2 \\ 0 & 0 & 1 & a_3 \\ 0 & 0 & 0 & 1 \end{pmatrix} \begin{pmatrix} v_1 \\ v_2 \\ v_3 \\ 1 \end{pmatrix}?$$
Describe how to consider both rotations and translations all at once by forming appropriate 4×4 matrices.

(38) Find the matrix which rotates all vectors about the positive z axis through an angle of θ where the rotation satisfies a right hand rule with respect to the given vector. That is, if the thumb of the right hand is pointing in the direction of $\mathbf{k} = \mathbf{e}_3$, then the fingers of the right hand move in the direction of rotation as the hand is closed.

(39) ↑Let $\mathbf{u} = (a, b, c)$ in $\mathbf{R}^3$, $(a, b) \neq (0, 0)$ where $\mathbf{u}$ is a given unit vector. Find the matrix which rotates all vectors through an angle of θ where the rotation satisfies a right hand rule with respect to the given vector. That is, if the thumb of the right hand is pointing in the direction of $\mathbf{u}$, then the fingers of the right hand move in the direction of rotation as the hand is closed. (This is a hard problem. It will be presented from a different point of view later. To see it worked, see the supplementary exercises on the web page.) **Hint:** First find a rotation which will move $\mathbf{u}$ to $\mathbf{e}_3$. Then rotate counter clockwise about $\mathbf{e}_3$ as in the above problem. Finally multiply by the inverse of the transformation which achieved the first rotation. You might want to make use of the cross product and its properties. You should get a 3×3 matrix whose columns are

$$\begin{pmatrix} \cos\theta - a^2\cos\theta + a^2 \\ -ba\cos\theta + ba + c\sin\theta \\ -(\sin\theta)\, b - (\cos\theta)\, ca + ca \end{pmatrix}, \begin{pmatrix} -ba\cos\theta + ba - c\sin\theta \\ -b^2\cos\theta + b^2 + \cos\theta \\ (\sin\theta)\, a - (\cos\theta)\, cb + cb \end{pmatrix},$$

$$\begin{pmatrix} (\sin\theta)\, b - (\cos\theta)\, ca + ca \\ -(\sin\theta)\, a - (\cos\theta)\, cb + cb \\ (1 - c^2)\cos\theta + c^2 \end{pmatrix}.$$

5.4 Rank and nullity

We noted above that, if $T : F^n \to F^m$, then

$$\operatorname{span}\{T\mathbf{e}_1, \ldots, T\mathbf{e}_n\} = \operatorname{Im} T.$$

Recall why this is. If $T\mathbf{x}$ is something in $\operatorname{Im} T$, then for $\mathbf{x} = x_1\mathbf{e}_1 + \cdots + x_n\mathbf{e}_n$,

$$T\mathbf{x} = T(x_1\mathbf{e}_1 + \cdots + x_n\mathbf{e}_n) = x_1 T\mathbf{e}_1 + \cdots + x_n T\mathbf{e}_n$$

which is in $\operatorname{span}\{T\mathbf{e}_1, \ldots, T\mathbf{e}_n\}$.

So if T has matrix $A = (\mathbf{a}_1 \ldots \mathbf{a}_n) = (T\mathbf{e}_1 \ldots T\mathbf{e}_n)$,

$$\operatorname{span}\{\mathbf{a}_1, \ldots, \mathbf{a}_n\} = \operatorname{span}\{T\mathbf{e}_1, \ldots, T\mathbf{e}_n\} = \operatorname{Im} T.$$

It is customary to refer to $\operatorname{span}\{\mathbf{a}_1, \ldots, \mathbf{a}_n\}$ as the **column space** of A, written $\operatorname{Col}(A)$, so that

$$\operatorname{Col}(A) = \operatorname{Im} T.$$

Definition 5.3. The **rank** $R(T)$of a linear mapping $T : F^n \to F^m$ is the dimension of $\operatorname{Im} T$. The **rank** $R(A)$of a matrix A is the dimension of $\operatorname{Col}(A)$.

Of course this implies that T and the matrix of T have the same rank.

Proposition 5.2. Let A be an $n \times n$ matrix. Then A is invertible if and only if $R(A) = n$. Also A is invertible if and only if the row reduced echelon form of A is I. In any case, the pivot columns are a basis for $\operatorname{Im} A = \operatorname{Col}(A)$.

Proof. By Corollary 3.3 A is invertible if and only if every column of A is a pivot column; which happens if and only if the row reduced echelon form of A is I. This in turn happens if and only if the columns of A are linearly independent, which occurs if and only if the columns of A are a basis for F^n. Finally note that the pivot columns are a basis for $\text{Col}(A)$ by Proposition 3.2. This proves the proposition. $\square$

Note that this implies a singular $n \times n$ matrix has rank $< n$. The proposition also shows that the rank defined here coincides with the rank defined in Definition 3.11, which was given as the number of pivot columns. Thus there is a routine way to find the rank and a basis for $\text{Col}(A)$. You just get the row reduced echelon form and count the number of pivot columns (nonzero rows in the row reduced echelon form). This gives the rank. A basis for $\text{Col}(A)$ is this set of pivot columns. Of course, in finding the row reduced echelon form, it is a good idea to let a computer algebra system do the busy work.

Example 5.11. The following is a matrix followed by its row reduced echelon form which was computed by Maple.

$$A = \begin{pmatrix} 1 & 4 & 5 & -3 & 1 \\ 3 & 5 & 8 & -2 & 11 \\ 6 & 7 & 13 & -1 & 9 \\ 9 & 1 & 10 & -8 & 7 \\ 2 & 6 & 8 & -4 & 0 \end{pmatrix} \sim \begin{pmatrix} 1 & 0 & 1 & 0 & 0 \\ 0 & 1 & 1 & 0 & 0 \\ 0 & 0 & 0 & 1 & 0 \\ 0 & 0 & 0 & 0 & 1 \\ 0 & 0 & 0 & 0 & 0 \end{pmatrix}.$$

Hence A has rank 4 and a basis of $\text{Col}(A)$ is $\mathbf{a}_1, \mathbf{a}_2, \mathbf{a}_4, \mathbf{a}_5$, where

$$A = \begin{pmatrix} \mathbf{a}_1 & \mathbf{a}_2 & \mathbf{a}_3 & \mathbf{a}_4 & \mathbf{a}_5 \end{pmatrix}.$$

Definition 5.4. If we write a matrix A in terms of row vectors,

$$A = \begin{pmatrix} \mathbf{r}_1 \\ \vdots \\ \mathbf{r}_m \end{pmatrix},$$

then $\text{span}\{\mathbf{r}_1, \ldots, \mathbf{r}_m\}$ is called the **row space** of A and is written $\text{Row}(A)$.

Lemma 5.2. *The row space of A has dimension $R(A)$. Also, a basis for the row space is the set of nonzero rows in the row reduced echelon form.*

Proof. This is obvious if A is row reduced, since $R(A)$ = number of pivot columns = number of nonzero rows, and it is clear from the description of the row reduced echelon form that the first entry of a nonzero row is to the right of the first entry of any row above it. If for some $i_1 < i_2 < \cdots < i_r$,

$$c_1 \mathbf{r}_{i_1} + c_2 \mathbf{r}_{i_2} + \cdots + c_r \mathbf{r}_{i_r} = \mathbf{0}$$

and not all the c_j equal 0, you could pick the first nonzero c_j and solve for that row as a linear combination of other rows below it. As just noted, this is not possible.

Hence we need only show that row operations do not change the row space. This is almost obvious: sets such as

$$\mathbf{r}_1 + k\mathbf{r}_2, \mathbf{r}_2, \ldots, \mathbf{r}_m$$

or

$$c\mathbf{r}_1, \mathbf{r}_2, \ldots, \mathbf{r}_m \text{ (where } c \neq 0)$$

have the same linear span as $\mathbf{r}_1, \ldots, \mathbf{r}_m$. □

Example 5.12. Let $A = \begin{pmatrix} 2 & 1 \\ 2 & 1 \end{pmatrix}$. Row reduction gives

$$A \sim \begin{pmatrix} 1 & \frac{1}{2} \\ 0 & 0 \end{pmatrix} = B.$$

A and B both have the same row space, the line in $\mathbb{R}^2$ with equation

$$x_1 - 2x_2 = 0.$$

However $\mathrm{Col}\,(A)$ is the line with equation $x_1 = x_2$ and $\mathrm{Col}\,(B)$ is the line with equation $x_2 = 0$. They are distinct subspaces with the same dimension.

Remember that in finding $\mathrm{Col}\,(A)$ you must use the columns in the original matrix and in finding the row space, it is sufficient to consider the rows in the row reduced echelon form.

An important subspace for any linear mapping $T : V \to W$ is the **kernel** of T,

$$\ker T = \{v \in V : Tv = 0\}.$$

(It is easily checked that $\ker T$ satisfies the closure requirement of Lemma 4.3 so that $\ker T$ is a subspace of V.)

Definition 5.5. We write $N(T)$ for the dimension of $\ker T$ (if $\ker T$ is a finite-dimensional space) and refer to $N(T)$ as the **nullity** of T. If $T : F^n \to F^m$ has matrix A, then $\ker\,(A)$ is defined to be $\ker\,(T)$. The nullity of A, is defined to be $N(T)$, the nullity of T. (Note that $\ker\,(A)$ is sometimes called the **null space** of A.)

Example 5.13. Let $D : S[a, b] \to S[a, b]$ be the differentiation mapping. Then $\ker D = \mathrm{span}\{1\}$, since the functions with derivative 0 are the constants.

Example 5.14. The row reduced echelon form for the matrix A shown below is the matrix B

$$A = \begin{pmatrix} 1 & 4 & 5 \\ 2 & 5 & 7 \\ 3 & 6 & 9 \end{pmatrix} \sim \begin{pmatrix} 1 & 0 & 1 \\ 0 & 1 & 1 \\ 0 & 0 & 0 \end{pmatrix} = B.$$

We noted earlier that the solutions of $A\mathbf{x} = 0$ are the same as the solutions of $B\mathbf{x} = 0$. The equation $B\mathbf{x} = \mathbf{0}$ is equivalent to $x_1 + x_3 = 0, x_2 + x_3 = 0$ or $\mathbf{x} = (-x_3, -x_3, x_3)$; that is $\ker A = \mathrm{span}\{(-1, -1, 1)\}$ and $N(A) = 1$.

There is a simple relationship between the rank and nullity of a matrix. This is a consequence of

Theorem 5.1. *Let* $T : V \to W$ *be a linear mapping and suppose* V *is finite-dimensional. Then*

$$N(T) + R(T) = \dim V.$$

Proof. We can dismiss the easy case $T = 0$, $N(T) = \dim V$, $R(T) = 0$. Suppose $T \neq 0$. Let $v_1, \ldots, v_k$ be a basis of $\ker T$. We make the convention that $k = 0$ if $\ker T = 0$. By Corollary 4.2, there exist vectors $v_{k+1}, \ldots, v_m$, so that $v_1, \ldots, v_m$ is a basis of V. We claim that

$$Tv_{k+1}, \ldots, Tv_m$$

is a basis of $\operatorname{Im} T$. In fact, we recall that

$$\operatorname{Im} T = \operatorname{span}\{Tv_1, \ldots, Tv_m\} = \operatorname{span}\{Tv_{k+1}, \ldots, Tv_m\}$$

on discarding zero vectors. It remains to show that $Tv_{k+1}, \ldots, Tv_m$ is a linearly independent set. Suppose that

$$x_{k+1}Tv_{k+1} + \cdots + x_m Tv_m = 0.$$

Then

$$T\left(x_{k+1}v_{k+1} + \cdots + x_m v_m\right) = 0,$$

and so $x_{k+1}v_{k+1} + \cdots + x_m v_m \in \ker T$. This implies that there exist scalars $y_1, \cdots, y_m$ such that

$$x_{k+1}v_{k+1} + \cdots + x_m v_m = y_1 v_1 + \cdots + y_k v_k.$$

Since $\{v_1, \ldots, v_m\}$ is a basis, all the scalars in the above equation are 0. Now the theorem is true since $k + (m - k) = m$. □

It follows at once that for an $m \times n$ matrix A,

$$N(A) + R(A) = n.$$

Also, it is clear at this point that for A an $m \times n$ matrix, $N(A)$ equals the number of non pivot columns. This is because it has already been observed that $R(A)$ is the number of pivot columns. Thus it is very easy to compute the rank and nullity of a matrix.

Example 5.15. Here is a matrix.

$$\begin{pmatrix} 1 & 2 & 0 & 4 & 1 \\ 1 & 2 & 2 & 10 & 1 \\ 1 & 2 & 2 & 10 & 2 \\ 1 & 2 & 1 & 7 & 1 \end{pmatrix}$$

Find its rank and nullity.

Using Maple or simply doing it by hand, the row reduced echelon form is

$$\begin{pmatrix} 1 & 2 & 0 & 4 & 0 \\ 0 & 0 & 1 & 3 & 0 \\ 0 & 0 & 0 & 0 & 1 \\ 0 & 0 & 0 & 0 & 0 \end{pmatrix}.$$

Therefore, there are three pivot columns and two nonpivot columns, so the rank is 3 and the nullity is 2.

5.5 Rank and nullity of a product

We note some useful results of Sylvester.

Lemma 5.3. *Let $T : V \to W$ and $U : W \to Z$ be linear mappings, with finite dimensional spaces V, W.*

(i) $R(UT) \leq \min\{R(U), R(T)\}$.
(ii) If T is a bijection, then

$$R(UT) = R(U).$$

(iii) If U is a bijection, then

$$R(UT) = R(T).$$

Proof. (i) Clearly

$$\operatorname{Im} UT \subseteq \operatorname{Im} U.$$

So $\operatorname{Im} UT$ has dimension no more than the dimension of $\operatorname{Im} U$ and

$$R(UT) \leq R(U).$$

Let $v_1, \ldots, v_m$ be the vectors constructed in the proof of Theorem 5.1. Clearly

$$\operatorname{Im} UT = \{UTv : v \in V\} \tag{5.6}$$

$$= \operatorname{span}\{UTv_{k+1,\ldots,}UTv_m\},$$

$$R(UT) \leq m - k = R(T).$$

(ii) If T is a bijection then

$$\operatorname{Im} UT = \{UTv : v \in V\} = \{Uw : w \in W\}$$
$$= \operatorname{Im} U,$$

so of course $R(UT) = R(U)$.

(iii) Let $v_1, \ldots, v_m$ be as above. If U is a bijection, then

$$UTv_{k+1}, \cdots, UTv_m$$

is a linearly independent set. For an equation

$$x_{k+1}UTv_{k+1} + \cdots + x_m UTv_m = 0$$

leads to an equation

$$x_{k+1}Tv_{k+1} + \cdots + x_m Tv_m = 0$$

and thence to $x_{k+1} = \cdots = x_m = 0$. Now (5.6) gives

$$R(UT) = m - k = R(T). \ \Box$$

Lemma 5.4. *Let $T : V \to W$ where $\ker T$ is finite-dimensional, and let Z be a finite-dimensional subspace of W. Then*

$$V_Z = \{v \in V : Tv \in Z\}$$

is a subspace of V having dimension $\leq \dim Z + N(T)$.

Proof. It is easy to check that V_Z is a subspace of V using the closure property of Lemma 4.3. We '**restrict**' T to V_Z, that is consider the mapping $T' : V_Z \to Z$ defined by $T'v = Tv$ $(v \in V_Z)$. Clearly

$$\operatorname{Im} T' \subset Z,\ R(T') \leq \dim Z,\ \ker T' \subset \ker T,\ N(T') \leq N(T).$$

So

$$\dim V_Z = R(T') + N(T') \leq \dim Z + N(T). \ \square$$

Lemma 5.5. *Let $T : V \to W$, $U : W \to Z$ be linear mappings, where* $\ker T$ *and* $\ker U$ *are finite-dimensional. Then*

$$N(UT) \leq N(U) + N(T).$$

Proof.

$$\begin{aligned}\ker UT &= \{v \in V : UTv = 0\} \\ &= \{v \in V : Tv \in \ker U\}.\end{aligned}$$

By Lemma 5.4, the dimension of $\ker UT$ is at most

$$\dim(\ker U) + N(T) = N(U) + N(T). \ \square$$

Of course this result holds for any finite product of linear transformations by induction. One way this is quite useful is in the case of a finite product of linear transformations $\prod_{i=1}^{l} L_i$. Then

$$\dim\left(\ker \prod_{i=1}^{l} L_i\right) \leq \sum_{i=1}^{l} \dim\left(\ker L_i\right).$$

So, if you can find a linearly independent set of vectors in $\ker\left(\prod_{i=1}^{l} L_i\right)$ of size

$$p = \sum_{i=1}^{l} \dim\left(\ker L_i\right),$$

then this linearly independent set must be a basis for $\ker\left(\prod_{i=1}^{l} L_i\right)$. To see this, suppose $\{v_1, \cdots, v_p\}$ is linearly independent in $\dim\left(\ker \prod_{i=1}^{l} L_i\right)$ where

$$p = \sum_{i=1}^{l} \dim\left(\ker L_i\right) \geq \dim\left(\ker \prod_{i=1}^{l} L_i\right)$$

Then, since the vectors are linearly independent, $\dim\left(\ker \prod_{i=1}^{l} L_i\right) \geq p$ and so the inequality in the above is an equality. Therefore, the given set of vectors must span $\ker \prod_{i=1}^{l} L_i$ because if there exists $w \notin \operatorname{span}\{v_1, \cdots, v_p\}$, then $\{v_1, \cdots, v_p, w\}$ is independent and the above inequality would be violated.

Example 5.16. Let A, B be $n \times n$ matrices over F, with A invertible. Then AB is invertible if, and only if, B is invertible.

To see this, use Lemma 5.3 (ii):

$$AB \text{ invertible } \Leftrightarrow R(AB) = n \Leftrightarrow R(B) = n \Leftrightarrow B \text{ invertible}.$$

($\Leftrightarrow$ is an abbreviation for 'if and only if'.)

We end the section with a few remarks on powers of linear mappings and the transpose of a matrix. For any linear mapping $T : V \to V$ we can form powers

$$T^2 = TT,\ T^3 = TTT$$

and so on. Thus $T^h = T \cdots T$ with h factors. Of course we can define A^h for an $n \times n$ matrix in just the same way. It is simple to see that

$$T^{h+k} = T^h T^k, \quad A^{h+k} = A^h A^k.$$

Example 5.17. Let $A = \begin{pmatrix} 0 & 1 \\ 0 & 0 \end{pmatrix}$. Then $A^2 = 0$. Hence $A^h = 0$ for $h = 3, 4, \ldots$.

For linear mapping $T : V \to V$ we can form polynomials in T. If $P(z) = a_0 z^n + \cdots + a_n$, then $P(T)$ is defined to be $a_0 T^n + \cdots + a_{n-1} T + a_n I$. Evidently $P(T)Q(T) = Q(T)P(T)$ for any polynomials P and Q.

5.6 Linear differential equations with constant coefficients

Given $g_1(x), \ldots, g_n(x)$ in $S_{\mathbb{C}}[a, b]$, the linear mapping $T : S_{\mathbb{C}}[a, b] \to S_{\mathbb{C}}[a, b]$ given by

$$Tf = f^{(n)} + g_1 f^{(n-1)} + \cdots + g_n f \tag{5.7}$$

is a **linear differential operator**. In this section we only treat the case that all the g_k are constant. See Collins (2006) for the very important questions raised by the general case, and also some introductory material on *PDEs*. Suppose now that

$$Tf = f^{(n)} + a_1 f^{(n-1)} + \cdots a_n f \tag{5.8}$$

where $a_1, \ldots, a_n$ are in $\mathbb{C}$. We factorize the polynomial $P(z) = z^n + a_1 z^{n-1} + \cdots + a_n$ as

$$P(z) = (z - c_1)^{h(1)} \cdots (z - c_k)^{h(k)} \tag{5.9}$$

where $c_1, \ldots, c_k$ are distinct complex numbers. The positive integer $h(j)$ is the **multiplicity** of c_j as a zero of P.

To begin with, we consider a very easy special case of the above.

Lemma 5.6. *Let a be a real number. Then the solution to the initial value problem*

$$y'(t) = ay(t),\ y(0) = 1$$

is $y(t) = e^{at}$.

Proof. First you can verify that this function works using elementary calculus. Why is it the only one which does so? Suppose $y_1(t)$ is a solution. Then $y(t) - y_1(t) = z(t)$ solves the initial value problem

$$z'(t) = az(t),\ z(0) = 0.$$

Thus $z' - az = 0$. Multiply by e^{-at} on both sides. By the chain rule,

$$e^{-at}(z' - az) = \frac{d}{dt}\left(e^{-at}z(t)\right) = 0,$$

and so there is a constant C such that $e^{-at}z(t) = C$. Since $z(0) = 0$, this constant is 0, and so $z(t) = y(t) - y_1(t) = 0$. □

We want to define $e^{(a+ib)t}$ in such a way that

$$\frac{d}{dt}e^{(a+ib)t} = (a + ib)\, e^{(a+ib)t}.$$

Also, to conform to the real case, we require $e^{(a+ib)0} = 1$. Thus it is desired to find a function $y(t)$ which satisfies the following two properties.

$$y(0) = 1, \ y'(t) = (a + ib)\, y(t). \tag{5.10}$$

Proposition 5.3. Let $y(t) = e^{at}(\cos(bt) + i\sin(bt))$. Then $y(t)$ is a solution to (5.10) and furthermore, this is the only function which satisfies the conditions of (5.10).

Proof. It is easy to see that if $y(t)$ is as given above, then $y(t)$ satisfies the desired conditions. First

$$y(0) = e^0(\cos(0) + i\sin(0)) = 1.$$

Next

$$\begin{aligned} y'(t) &= ae^{at}(\cos(bt) + i\sin(bt)) + e^{at}(-b\sin(bt) + ib\cos(bt)) \\ &= ae^{at}\cos bt - e^{at}b\sin bt + i\left(ae^{at}\sin bt + e^{at}b\cos bt\right). \end{aligned}$$

On the other hand,

$$(a + ib)\left(e^{at}(\cos(bt) + i\sin(bt))\right)$$
$$= ae^{at}\cos bt - e^{at}b\sin bt + i\left(ae^{at}\sin bt + e^{at}b\cos bt\right)$$

which is the same thing. Remember $i^2 = -1$.

It remains to verify that this is the only function which satisfies (5.10). Suppose $y_1(t)$ is another function which works. Then, letting $z(t) \equiv y(t) - y_1(t)$, it follows that

$$z'(t) = (a + ib)\, z(t), \ z(0) = 0.$$

Now $z(t)$ has a real part and an imaginary part, $z(t) = u(t) + iv(t)$. Then $\overline{z}(t) \equiv u(t) - iv(t)$ and

$$\overline{z}'(t) = (a - ib)\,\overline{z}(t), \ \overline{z}(0) = 0.$$

Then $|z(t)|^2 = z(t)\,\overline{z}(t)$ and by the product rule,

$$\begin{aligned} \frac{d}{dt}|z(t)|^2 &= z'(t)\,\overline{z}(t) + z(t)\,\overline{z}'(t) \\ &= (a + ib)\, z(t)\,\overline{z}(t) + (a - ib)\, z(t)\,\overline{z}(t) \\ &= (a + ib)\,|z(t)|^2 + (a - ib)\,|z(t)|^2 \\ &= 2a\,|z(t)|^2, \ |z(0)|^2 = 0. \end{aligned}$$

It follows from Lemma 5.6 that the solution is

$$|z(t)|^2 = 0e^{2at} = 0.$$

Thus $z(t) = 0$, and so $y(t) = y_1(t)$, proving the uniqueness assertion of the proposition. □

Note that the function $e^{(a+ib)t}$ is never equal to 0. This is because its absolute value is e^{at} (Why?).

With this, it is possible to describe the solutions to $Tf = 0$ in terms of linear algebra.

Lemma 5.7. *Let T be as in (5.8). Let $s \in \{0, 1, \ldots, h(j) - 1\}$. The function f in $S_{\mathbb{C}}[a, b]$ defined by*

$$f(x) = x^s e^{c_j x}$$

is a solution of the differential equation

$$Tf = 0.$$

Proof. With D the differentiation mapping and I the identity mapping on $S_{\mathbb{C}}[a, b]$, we have

$$(D - qI)x^l e^{cx} = lx^{l-1}c^{cx} + (c - q)x^l e^{cx} \tag{5.11}$$

for $l \in \{0, 1, 2, \ldots\}$ and $q \in \mathbb{C}$. In particular,

$$(D - cI)x^l e^{cx} = lx^{l-1}e^{cx}.$$

Repeated application of this gives

$$(D - cI)^h x^s e^{cx} = 0 \text{ if } h > s. \tag{5.12}$$

We can factorize T in the following way:

$$T = (D - c_1 I)^{h(1)} \cdots (D - c_k I)^{h(k)}, \tag{5.13}$$

the order of the factors being of no importance. In particular,

$$T = M(D - c_j I)^{h(j)}$$

for a linear mapping $M : S_{\mathbb{C}}[a, b] \to S_{\mathbb{C}}[a, b]$. So $Tf = 0$ follows from (5.12) since $s < h(j)$. □

We summarize the observations (5.11) and (5.12) in the following lemma.

Lemma 5.8. *For each $j \geq 1$*

$$(D - cI)^j x^j e^{cx} = Ke^{cx}$$

for some nonzero K. Also if $n > j$

$$(D - cI)^n x^j e^{cx} = 0$$

and if $q \neq c$,

$$(D - qI)^m x^j e^{cx} = p(x) e^{cx}$$

where $p(x)$ has degree j.

Lemma 5.9. *The n functions $x^s e^{c_j x}$ $(1 \le j \le k,\ 0 \le s < h(j))$ are linearly independent in $S_{\mathbb{C}}[a,b]$.*

Proof. Consider a linear combination of $x^s e^{c_j x}$ for $1 \le j \le k$ and $s < h(j)$. This is of the form

$$\sum_{r=1}^{M} a_r f_r(x) + A_0 e^{c_l x} + \cdots + A_{h(l)-1} x^{h(l)-1} e^{c_l x} = 0 \tag{5.14}$$

where the functions f_r are each of the form $f_r(x) = x^s e^{c_j x}, c_j \ne c_l,\ s < h(j)$ and $h(l)$ is the largest of all the $h(j)$ for $j \le k$. Renumbering if necessary, we can assume $l = k$. Apply the operator

$$(D - c_1 I)^{h(1)} \cdots (D - c_{k-1} I)^{h(k-1)}$$

to both sides of the above equation. By Lemma 5.8, this sends all terms in the sum on the left to 0 and yields an expression of the form

$$A_0 p_0(x) e^{cx} + \cdots + A_{h-1} p_{h-1}(x) e^{cx} = 0,\ h = h(l)$$

for polynomials $p_j(x)$, each having degree j. Explicitly, $p_j(x) = a_j x^j + a_{j-1} x^{j-1} + \cdots, a_j \ne 0$. We want to conclude that all the $A_j = 0$. If not, suppose A_j is the last nonzero one. Then apply $(D - cI)^j$ to both sides, which yields $CA_j e^{cx} = 0$ for some $C \ne 0$. This is a contradiction. It follows that (5.14) is of the form

$$\sum_{r=1}^{M'} a_r f_r(x) + B_0 e^{c_{l-1} x} + \cdots + B_{h(l-1)-1} x^{h(l-1)-1} e^{c_{l-1} x} = 0$$

where $h(l-1)$ is the largest of the remaining $h(j)$. Now a repeat of the above reasoning shows that these $B_j = 0$ also. Continuing this way, it follows the given functions are linearly independent as claimed.This proves the lemma. □

This lemma gives most of the proof of the following theorem.

Theorem 5.2. *Let $T : S_{\mathbb{C}}(a,b) \to S_{\mathbb{C}}(a,b)$ be given in (5.8),*

$$Tf = f^{(n)} + a_1 f^{(n-1)} + \cdots + a_n f.$$

Then $N(\ker(T)) = n$ and $\ker T$ has a basis of functions of the form $x^s e^{c_j x}$, $1 \le s \le h(j) - 1$, where the polynomial

$$z^n + a_1 z^{n-1} + \cdots + a_n$$

factorizes as

$$(z - c_1)^{h(1)} \cdots (z - c_k)^{h(k)}.$$

Proof. We have identified n linearly independent solutions of $Tf = 0$. This gives a basis of $\ker T$. For, writing

$$P(z) = (z - z_1) \cdots (z - z_n), D_j = (D - z_j I),$$

where the z_j may be repeated, we have $T = D_1 \cdots D_n$. If $D_j f = 0$, then

$$(f(x)e^{-c_j x})' = 0;$$

$f(x) = \text{const. } e^{c_j x}$.

The kernel of D_j has dimension 1. By Lemma 5.5, the dimension of $\ker T$ is at most n. Hence we have found a basis of $\ker T$. This proves the theorem. □

Example 5.18. Give the general solution of

$$f^{(4)} - 11f^{(3)} + 44f^{(2)} - 76f' + 48f = 0.$$

Solution (for any interval of $\mathbb{R}$). Using Maple,

$$z^4 - 11z + 44z^2 - 76z + 48 = (z-2)^2(z-3)(z-4).$$

According to the above results, the general solution is

$$a_1 e^{2x} + a_2 x e^{2x} + a_3 e^{3x} + a_4 e^{4x}.$$

Here we are working in $S_{\mathbb{C}}[a,b]$, so $a_j \in \mathbb{C}$.

Example 5.19. Show that the subspace W of $S[a,b]$ consisting of solutions of

$$f'' + f = 0 \tag{5.15}$$

is $\text{span}\{\sin x, \cos x\}$.

As already noted, $\text{span}\{\sin x, \cos x\} \subset W$. To get the reverse inclusion, we first work in $S_{\mathbb{C}}[a,b]$. The polynomial $P(z) = z^2 + 1$ has zeros $-i, i$, and the general solution of (5.15) in $S_{\mathbb{C}}[a,b]$ is

$$\begin{aligned} f(x) &= ae^{ix} + be^{-ix} \\ &= (a_1 + ia_2)(\cos x + i \sin x) + (b_1 + ib_2)(\cos x - \sin x), \end{aligned}$$

and so f is in span $\{\sin x, \cos x\}$. This gives the required result.

5.7 Exercises

(1) For each of the following matrices A, find a basis for $\text{Col}(A)$, the row space of A, $\text{Row}(A)$, and $\ker(A)$, and give the rank and nullity of A.

(a) $\begin{pmatrix} 1 & 2 & 3 & 1 & 1 \\ 1 & 2 & 3 & 2 & 3 \\ 1 & 2 & 3 & 1 & 2 \end{pmatrix}$

(b) $\begin{pmatrix} 1 & 1 & 4 & 3 & 1 \\ 1 & 2 & 5 & 4 & 3 \\ 1 & 1 & 4 & 3 & 2 \end{pmatrix}$

(c) $\begin{pmatrix} 1 & 1 & 4 & 3 & 3 \\ 1 & 2 & 5 & 4 & 5 \\ 1 & 1 & 4 & 3 & 3 \end{pmatrix}$

(d) $\begin{pmatrix} 1 & 1 & 1 & 6 & 2 \\ 1 & 2 & 3 & 13 & 5 \\ 1 & 1 & 2 & 9 & -1 \end{pmatrix}$

(2) For each of the following matrices A, find a basis for $\mathrm{Col}(A)$, the row space of A, and $\ker(A)$, and give the rank and nullity of A.

(a) $\begin{pmatrix} 1 & 1 & 1 & 1 & 3 \\ 1 & 2 & 2 & 2 & 0 \\ 1 & 2 & 3 & 4 & 7 \\ 1 & 1 & 1 & 2 & 1 \end{pmatrix}$

(b) $\begin{pmatrix} 1 & 1 & 1 & 6 & 6 \\ 1 & 2 & 2 & 10 & 9 \\ 1 & 2 & 3 & 11 & 11 \\ 1 & 1 & 1 & 6 & 6 \end{pmatrix}$

(c) $\begin{pmatrix} 1 & 2 & 1 & 1 & 6 & 8 \\ 1 & 2 & 2 & 2 & 9 & 15 \\ 1 & 2 & 2 & 3 & 11 & 20 \\ 1 & 2 & 1 & 1 & 6 & 8 \end{pmatrix}$

(d) $\begin{pmatrix} 1 & 1 & 1 & 6 & 6 \\ 1 & 3 & 2 & 13 & 10 \\ 1 & 2 & 3 & 11 & 11 \\ 1 & 1 & 1 & 6 & 6 \end{pmatrix}$

(e) $\begin{pmatrix} 1 & 1 & 1 \\ 1 & 2 & 2 \\ 1 & 2 & 3 \\ 1 & 1 & 1 \end{pmatrix}$

(f) $\begin{pmatrix} 1 & 1 & 10 \\ 1 & 2 & 13 \\ 1 & 2 & 13 \\ 1 & 1 & 10 \end{pmatrix}$

(3) Explain why if A is an $m \times n$ matrix with $n > m$, then $\ker(A) \neq \{\mathbf{0}\}$.

(4) Explain why $S_{\mathbb{C}}(a, b)$ is a vector space but does not have finite dimension.

(5) Find all solutions to the differential equation $y'' - 3y' + 2y = 0$.

(6) Find all solutions to the differential equation $y''' + 2y'' - 15y' - 36y = 0$. Describe the way in which the general solution is the kernel of a linear transformation.

(7) Find the solution to the initial value problem consisting of the differential equation

$$y''' - 8y'' + 20y' - 16y = 0$$

along with the initial conditions $y(0) = 2, y'(0) = -1, y''(0) = 2$. **Hint:** Find the solutions to the differential equation and then choose the constants correctly to satisfy the initial conditions.

(8) Find the general solution to the differential equation $y^{(4)} - 4y^{(3)} + 6y'' - 4y' + y = 0$.

(9) Find the solution to the initial value problem consisting of the equation $y^{(4)} - 2y'' + y = 0$ and the initial conditions, $y(0) = 0, y'(0) = 1, y''(0) = 0, y'''(0) = 0$.

(10) Let $L : V \to W$ where L is linear and V, W are vector spaces. Suppose you want to find all solutions v to the equation $Lv = w$ for a fixed w. Show that if you have a single solution to this last equation v_p, then all solutions are of the form $v_p + v$ for some $v \in \ker(L)$. Sometimes this is written as $v_p + \ker(L)$. The solution v_p is called a particular solution.

(11) ↑Consider the equation $y''' - 8y'' + 20y' - 16y = t^2 + t$. It happens that $y_p(t) = -\frac{27}{128} - \frac{7}{32}t - \frac{1}{16}t^2$ is a solution to this equation. Find the general solution.

(12) Consider the equation $w' + a(t) w = f(t)$ for $t \in [a, b]$. This is an ordinary differential equation in which there is a non-constant coefficient $a(t)$. Show the solutions to this equation are of the form $w(t) = e^{-A(t)} \int_a^t e^{A(s)} f(s)\, ds + Ce^{-A(t)}$ where $A'(t) = a(t)$. **Hint:** Multiply on both sides by $e^{A(t)}$. Then observe that by the chain rule, the left side is the derivative of the expression $\left(e^{A(t)} w(t)\right)$.

(13) ↑To solve $y'' + ay' + by = f$, show that it suffices to find a single solution y_p and then add to it the general solution of $y'' + ay' + by = 0$, which is easy to find. One way to find the single solution is to take a solution y to the second equation and look for a particular solution to the first equation in the form $y(t) v(t)$. This is called the method of reduction of order. It **always works.** Show that when yv is plugged in to the given equation, you obtain the following equation for v'.

$$v'' + \left(\frac{2y' + ay}{y}\right) v' = \frac{f}{y}$$

Then you solve this using the methods of the above problem.

(14) There is another method for finding a particular solution (see Problem 10) to $Ty = f$ where T is a constant coefficient differential operator and f is of a special form which often occurs in applications. It works whenever f is of the form $p(t) e^{\alpha t}$ where complex values are allowed as coefficients of $p(t)$ and α. As described in the above discussion of differential equations, such functions are solutions to $\hat{T}y = 0$ where $\hat{T}$ is a constant coefficient differential operator of the sort described above. One first finds the general solution to the **homogeneous equation** $Ty = 0$ as described above. Next one looks for a particular solution of the form $q(t) e^{\alpha t}$ where $q(t)$ is the most general polynomial which has the same degree as $p(t)$. This is not unreasonable because T will take functions of this form and deliver other functions of this form. Then observe the terms of this proposed particular solution. If any of them is a solution to the homogeneous equation, multiply the proposed solution by t. Continue with

this process until no term in the proposed solution yields a solution to the homogeneous equation $Ty = 0$. Then stop. There will exist coefficients for $q(t)$ such that $t^m q(t) e^{\alpha t}$ is a particular solution. For example, consider $y'' + 2y' + y = e^{-t}$. The solution to the homogeneous equation is $C_1 e^{-t} + C_2 t e^{-t}$. Now you look for a particular solution y_p in the form

$$y_p(t) = (A + Bt) e^{-t}$$

This is doomed to failure because e^{-t} is a solution to the homogeneous equation. Therefore, multiply by t. The new proposed particular solution is of the form

$$y_p(t) = \left(At + Bt^2\right) e^{-t}$$

One of the terms is a solution to the homogeneous equation and so you must multiply by t again. Then the next proposed particular solution is

$$y_p(t) = \left(At^2 + Bt^3\right) e^{-t}$$

Now no term is a solution to the homogeneous equation so there will be a particular solution of this form. In fact, a particular solution is $y_p(t) = \frac{1}{2} t^2 e^{-t}$. Go through the details and verify that this is the case. It will involve solving some simple linear equations. Then from Problem 10, the general solution is of the form

$$\frac{1}{2} t^2 e^{-t} + C_1 e^{-t} + C_2 t e^{-t}$$

At this point, you can solve for C_i using whatever initial conditions are given in the problem. This is called the method of undetermined coefficients and is a standard way to find particular solutions although it is mathematically less general than some other approaches. All methods up till now will be included later in the book when we consider first order systems.

(15) Use the method of undetermined coefficients of Problem 14 to find the general solution to the following differential equations.

  (a) $y'' - 2y' + y = te^{-t}$
  (b) $y''' - y'' - y' + y = (1 + t) e^t$
  (c) $y'' - y = e^{-t} + e^t$ **Hint:** Divide and conquer. First find y_{p_1} such that $y''_{p_1} - y_{p_1} = e^{-t}$ and then find y_{p_2} such that $y''_{p_2} - y_{p_2} = e^t$. Explain why the sum of these two particular solutions yields a particular solution to the given differential equation. This illustrates the **principle of superposition**.
  (d) $y'' + y = \cos t$ **Hint:** The easy way to do this one is to consider instead $z'' + z = e^{it}$ and then explain why the real part of z is the desired particular solution. Also explain why the imaginary part of z is a particular solution to the equation $y'' + y = \sin t$. This illustrates a general principle.

(16) Find the solution to the initial value problem consisting of the equation $y'' - 5y' + 4y = e^t$ and the initial conditions $y(0) = 0, y'(0) = 1$.

(17) The principle of superposition for a linear differential operator T as discussed above has two forms. First if $Ty_k = 0$ for $k = 1, 2, \cdots, n$ and if c_k are scalars in $\mathbb{C}$, then $T\left(\sum_{k=1}^{n} c_k y_k\right) = 0$ also. That is, linear combinations of solutions to the homogeneous equation are solutions to the homogeneous equations. The second form of this principle is this. If $Ty_{p_k} = f_k$, and if the c_k are scalars in $\mathbb{C}$, then

$$T\left(\sum_{k=1}^{n} c_k y_{p_k}\right) = \sum_{k=1}^{n} c_k f_k.$$

Does this principle have anything to do with the linear map being a differential operator or is it really just an algebraic property of linear maps? If you give the correct discussion of the first part, you will see that this is just a property of algebra. A generalized verion of this principle is used to analyse vibrations in which a periodic forcing function is written as a generalized sum of simpler functions involving cosines and sines.

(18) If W is a subspace of V, a finite dimensional vector space and if $\dim(W) = \dim(V)$, show that $W = V$.

(19) Let V be a vector space over a field F and let $V_1, \cdots, V_m$ be subspaces. We write

$$V = V_1 \oplus \cdots \oplus V_m \tag{5.16}$$

if every $v \in V$ can be written in a unique way in the form

$$v = v_1 + \cdots + v_m$$

where each $v_i \in V_i$. This is called a direct sum. If this uniqueness condition does not hold, we simply write

$$V_1 + \cdots + V_m,$$

and this symbol means all vectors of the form

$$v_1 + \cdots + v_m, \ v_j \in V_j \text{ for each } j.$$

Show that (5.16) is equivalent to saying that if

$$0 = v_1 + \cdots + v_m, \ v_j \in V_j \text{ for each } j,$$

then each $v_j = 0$. Next show that in the situation of (5.16), if $\beta_i = \{u_1^i, \cdots, u_{m_i}^i\}$ is a basis for V_i, then the union of $\beta_1, \cdots, \beta_m$ is a basis for V.

(20) ↑Suppose you have finitely many linear mappings $L_1, L_2, \cdots, L_m$ which map V to V and suppose that they commute. That is, $L_i L_j = L_j L_i$ for all i, j. Also suppose L_k is one-to-one on $\ker(L_j)$ whenever $j \neq k$. Letting P denote the product of these linear transformations, $P = L_1 L_2 \cdots L_m$, first show that

$$\ker(L_1) + \cdots + \ker(L_m) \subseteq \ker(P)$$

Next show that

$$\ker(L_1) + \cdots + \ker(L_m) = \ker(L_1) \oplus \cdots \oplus \ker(L_m).$$

Using Sylvester's theorem, and the result of Problem 18, show that

$$\ker(P) = \ker(L_1) \oplus \cdots \oplus \ker(L_m).$$

Hint: By Sylvester's theorem and the above problem,

$$\begin{aligned}\dim(\ker(P)) &\leq \sum_i \dim(\ker(L_i)) \\ &= \dim(\ker(L_1) \oplus \cdots \oplus \ker(L_m)) \leq \dim(\ker(P)).\end{aligned}$$

Now consider Problem 18.

(21) Let $\mathcal{M}(F^n, F^n)$ denote the set of all $n \times n$ matrices having entries in F. With the usual operations of matrix addition and scalar multiplications, explain why this is a vector space. See Chapter 2if you don't recall why this is. Now show that the dimension of this vector space is n^2. If $A \in \mathcal{M}(F^n, F^n)$, explain why there exists a monic polynomial of smallest possible degree of the form

$$p(\lambda) = \lambda^k + a_{k-1}\lambda^{k-1} + \cdots + a_1\lambda + a_0$$

such that

$$p(A) = A^k + a_{k-1}A^{k-1} + \cdots + a_1 A + a_0 I = 0.$$

This is called the minimial polynomial of A. **Hint:** Consider the matrices $I, A, A^2, \cdots, A^{n^2}$. There are $n^2 + 1$ of these matrices. Can they be linearly independent? Now consider all polynomials satisfied by A and pick one of smallest degree. Then divide by the leading coefficient.

(22) ↑Suppose A is an $n \times n$ matrix. From the preceding problem, suppose the minimal polynomial factors as

$$(\lambda - \lambda_1)^{r_1}(\lambda - \lambda_2)^{r_2} \cdots (\lambda - \lambda_k)^{r_k}$$

where r_j is the algebraic multiplicity of λ_j. Thus

$$(A - \lambda_1 I)^{r_1}(A - \lambda_2 I)^{r_2} \cdots (A - \lambda_k I)^{r_k} = 0,$$

and so, letting

$$P = (A - \lambda_1 I)^{r_1}(A - \lambda_2 I)^{r_2} \cdots (A - \lambda_k I)^{r_k}$$

and $L_j = (A - \lambda_j I)^{r_j}$, apply the result of Problem 20 to verify that

$$\mathbb{C}^n = \ker(L_1) \oplus \cdots \oplus \ker(L_k)$$

and that $A : \ker(L_j) \to \ker(L_j)$. In this context, $\ker(L_j)$ is called the generalized eigenspace for λ_j. You need to verify that the conditions of the result of this problem hold.

(23) In the context of Problem 22, show that there exists a nonzero vector $\mathbf{x}$ such that $(A - \lambda_j I)\,\mathbf{x} = \mathbf{0}$. This $\mathbf{x}$ is called an eigenvector. The λ_j is called an eigenvalue. **Hint:** There must exist a vector $\mathbf{z}$ such that

$$(A - \lambda_1 I)^{r_1} (A - \lambda_2 I)^{r_2} \cdots (A - \lambda_j I)^{r_j - 1} \cdots (A - \lambda_k I)^{r_k} \mathbf{y} = \mathbf{z} \neq \mathbf{0}$$

Why? Now what happens if you apply $(A - \lambda_j I)$ to $\mathbf{z}$?

(24) Let A be an $n \times n$ matrix and let B be invertible. Show that $B^{-1}AB$ has the same rank and nullity as A.

(25) Consider the operators $(D - c_j)^{h(j)} : S_{\mathbb{C}}(a, b) \to S_{\mathbb{C}}(a, b)$ for the c_k distinct complex numbers and it is desired to find $\ker\left((D - c_1)^{h(1)} \cdots (D - c_k)^{h(k)}\right)$. Show the following.

(a) The operators $(D - c_j)^{h(j)}$ commute. That is,

$$(D - c_j)^{h(j)} (D - c_k)^{h(k)} = (D - c_k)^{h(k)} (D - c_j)^{h(j)}$$

This follows from simply examining carefully the meaning of these operators.

(b) Show next that $(D - b)$ is one to one on $\ker(D - a)^m$. To do this, suppose $y \in \ker(D - a)^m$ and $(D - b)\,y = 0$. You want to show that $y = 0$. Let k be the smallest index such that $(D - a)^k y = 0$. Then since $(D - b)\,y = 0$,

$$(D - a)\,y + (a - b)\,y = 0$$

Now multiply on the left by the linear map $(D - a)^{k-1}$. Get a contradiction.

(c) Explain why $(D - c_j)^{h(j)}$ is one to one on $\ker(D - c_k)^{h(k)}$ whenever $k \neq j$. Now apply the result of Problem 20 above to conclude that

$$\begin{aligned} &\ker\left((D - c_1)^{h(1)} \cdots (D - c_k)^{h(k)}\right) \\ = &\ker\left((D - c_1)^{h(1)}\right) \oplus \cdots \oplus \ker\left((D - c_k)^{h(k)}\right) \end{aligned}$$

(d) Next use direct computations to describe $\ker\left((D - c_j)^{h(j)}\right)$ as a linear combination of the functions of the form $x^r e^{c_j x}$ for $r < h(j)$. This gives another proof of the main result on linear ordinary differential equations, Theorem 5.2. To carry out this description, you only need to use the theory of first order differential equations presented earlier. See Problem 12 for example for more than is needed. Another application of Problem 20 will occur in the section on the Jordan Canonical form presented later.

(26) Let $Tf = f^{(n)} + a_1 f^{(n-1)} + \cdots a_n f$ as in Section 5.6. The Cauchy problem is the following.

$$Ty = 0,\ y(a) = y_0,\ y'(a) = y_1,\ \cdots,\ y^{(n-1)}(a) = y_{n-1}.$$

It can be shown that the Cauchy problem **always** has a **unique** solution for any choice of constants $y_0, y_1, \cdots, y_{n-1}$ and for any a. It was shown in Section

5.6 that there exist n functions $\{y_1, \cdots, y_n\}$ which are linearly independent and such that for each $j, Ty_j = 0$ and that every solution y to $Ty = 0$ must be a linear combination of these basis functions. Now consider the matrix

$$\begin{pmatrix} y_1(x) & y_2(x) & \cdots & y_n(x) \\ y_1'(x) & y_2'(x) & \cdots & y_n'(x) \\ \vdots & \vdots & & \vdots \\ y_1^{(n-1)}(x) & y_2^{(n-1)}(x) & \cdots & y_n^{(n-1)}(x) \end{pmatrix}$$

Show that this matrix has rank equal to n for any x.

(27) Let V, W be finite dimensional vector spaces with field of scalars F and let T be a linear transformation which maps V to W. Recall from Problem 22 on Page 99 or Problem 21 on Page 132 that T is one-to-one if and only if $Tv = 0$ implies $v = 0$. Show that in this situation, T maps a basis for V to a basis for $\text{Im}(T)$. Also show that if T is one-to-one and has a right inverse $S : \text{Im}(T) \to V$, then the right inverse is one-to-one and linear.

(28) ↑In the situation of the above problem, suppose T is a one-to-one linear transformation which maps onto W. Now suppose $S : W \to V$ is a right inverse for T,

$$TS = I.$$

Show that $ST = I$ also. Here I is the identity map on either V or W.

(29) Let T be a linear transformation which maps V to W. Recall that

$$\ker(T) = \{v \in V : Tv = 0\}$$

For $v, u \in V$, let v be similar to u, written as $v \sim u$ if $v - u \in \ker(T)$. Show that $\sim$ is an example of an **equivalence relation**. This means that

(a) $u \sim u$.
(b) If $u \sim v$, then $v \sim u$.
(c) If $u \sim v$ and $v \sim y$, then $u \sim y$.

The equivalence class, denoted by $[u]$ consists of all vectors v which are similar to u. Show that for any set which has such an equivalence relation, $[u] = [v]$ or else the two sets $[u]$ and $[v]$ have nothing in common. Thus the set of equivalence classes partitions the set into mutually disjoint nonempty subsets.

(30) ↑In the situation of the above problem, where the equivalence class is given there, and V is a finite dimensional vector space having field of scalars F, define $[u] + [v]$ as $[u + v]$ and for a a scalar, $a[u] = [au]$. Show that these operations are well defined and that with these operations the set of equivalence classes, denoted by $V/\ker(T)$, is a vector space with field of scalars F. Now define

$$\widetilde{T} : V/\ker(T) \to W$$

by $\widetilde{T}([v]) = T(v)$. Verify $\widetilde{T}$ is a well defined linear one-to-one mapping onto $\text{Im}(T)$. Recall that from Problem 22 on Page 99, a linear transformation T is one-to-one if and only if whenever $Tv = 0$, it follows that $v = 0$.

(31) ↑In the above situation, the rank of T equals the dimension of $\text{Im}(T)$. Explain why this must equal the dimension of $V/\ker(T)$. Letting s be the dimension of $\ker(T)$, explain why $\dim(V) - s = \dim(V/\ker(T))$. **Hint:** Letting $\{Tv_1, \cdots, Tv_r\}$ be a basis for $\text{Im}(T)$, explain why $\{[v_1], \cdots, [v_r]\}$ is a basis for $V/\ker(T)$. If $v \in V$, then

$$[v] = \sum_{k=1}^{r} c_k [v_k]$$

for some scalars c_k. Then explain why

$$v - \sum_{k=1}^{r} c_k v_k \in \ker(T)$$

If $\{z_1, \cdots, z_s\}$ is a basis for $\ker(T)$, explain why $\{z_1, \cdots, z_s, v_1, \cdots, v_r\}$ is a basis for V.

Chapter 6

Inner product spaces

Chapter summary

We define what is meant by an inner product space and discuss the concept of orthogonal and orthonormal sets. Next we give a method for constructing these sets from a given basis and use this to define what is meant by an orthogonal projection. This is used to develop the method of least squares. This method is used in various applications such as the method of least squares regression lines and Fourier series.

6.1 Definition and examples

As promised in Chapter 1, we now set up an axiomatic version of the inner product which we originally described in $\mathbb{R}^3$, and then extended to $\mathbb{R}^n$ and $\mathbb{C}^n$.

Definition 6.1. Let V be a vector space over F, where F is either $\mathbb{R}$ or $\mathbb{C}$. Suppose that for any u, v in V there is defined a number $\langle u, v\rangle \in F$, the **inner product** of u and v, with the following properties for u, v, w in V and $c \in F$:

(1) $\langle u+v, w\rangle = \langle u, w\rangle + \langle v, w\rangle$
(2) $\langle cu, v\rangle = c\langle u, v\rangle$
(3) $\langle u, v\rangle = \overline{\langle v, u\rangle}$
(4) $\langle u, u\rangle > 0$ for $u \neq 0$.

Then V is an **inner product space**. We say that V is a real inner product space if $F = \mathbb{R}$, a complex inner product space if $F = \mathbb{C}$.

Note that complex conjugation can be omitted in (3) if V is a real inner product space.

Example 6.1. $V = \mathbb{R}^n$ is a real inner product space with

$$\langle u, v\rangle = u_1 v_1 + \cdots + u_n v_n.$$

Example 6.2. $V = \mathbb{C}^n$ is a complex inner product space with

$$\langle u, v\rangle = u_1 \bar{v}_1 + \cdots + u_n \bar{v}_n.$$

For these basic examples, most of the details for the verification were noted in Chapter 1.

Example 6.3. Let h be a given positive continuous function on $[a, b]$. For f and g in $C_{\mathbb{C}}[a, b]$, let

$$\langle f, g\rangle = \int_a^b f(x)\overline{g(x)}h(x)dx. \tag{6.1}$$

With this definition, $C_{\mathbb{C}}[a, b]$ becomes a complex inner product space. We call h the **weight function** for the space. Verifying (1)–(3) is an easy matter. As for (4), if $f \neq 0$,

$$\langle f, f\rangle = \int_a^b |f(x)|^2 h(x)dx.$$

The integrand is ≥ 0, and is positive on some interval since the functions are continuous and $|f|^2 h \neq 0$. Clearly we have $\langle f, f\rangle > 0$.

We now prove some geometrical facts about V.

Definition 6.2. Let V be an inner product space. Let u, v be in V. The **length** of u is

$$\|u\| = \langle u, u\rangle^{1/2}.$$

The **distance** from u to v is

$$d(u, v) = \|u - v\|.$$

In Example 6.3, the distance from f to g is

$$\sqrt{\int_a^b |f(x) - g(x)|^2 h(x)dx}.$$

This is a 'mean square distance'. By drawing a sketch, you will see that $d(f, g)$ can be arbitrarily small even if $|f(x) - g(x)| > 1$ for some values of x.

Theorem 6.1. *Let V be an inner product space over F. Let u, v, w be in $V, c \in F$. Then*

(i) $\|cv\| = |c|\|v\|$;
(ii) $|\langle u, v\rangle| \leq \|u\|\|v\|$ *(Cauchy-Schwarz inequality);*
(iii) $\|u + v\| \leq \|u\| + \|v\|$ *(triangle inequality);*
(iv) $d(u, w) \leq d(u, v) + d(v, w)$.

Often (iv) is called the triangle inequality.

Proof. We begin by observing that

$$\langle 0, v\rangle = 0. \tag{6.2}$$

To see this, we use

$$\langle 0, v\rangle = \langle 0+0, v\rangle = \langle 0, v\rangle + \langle 0, v\rangle.$$

We also have

$$\langle u, cv\rangle = \bar{c}\langle u, v\rangle. \tag{6.3}$$

To see this,

$$\langle u, cv\rangle = \overline{\langle cv, u\rangle} = \overline{c\langle v, u\rangle} = \bar{c}\overline{\langle v, u\rangle} = \bar{c}\langle u, v\rangle.$$

Now we proceed to prove (i)–(iv).

(i) $\|cv\| = \langle cv, cv\rangle^{1/2} = (c\bar{c}\langle v, v\rangle)^{1/2} = |c|\,\|v\|$.

(ii) This is obvious from (6.2) if $v = 0$. If $v \neq 0$, we use the properties of inner product to get, for any $t \in F$.

$$\begin{aligned} 0 &\leq \langle u - tv, u - tv\rangle \\ &= \langle u, u\rangle - t\langle v, u\rangle - \bar{t}\langle u, v\rangle + t\bar{t}\langle v, v\rangle. \end{aligned} \tag{6.4}$$

The most favorable choice of t is

$$t = \frac{\langle u, v\rangle}{\langle v, v\rangle}.$$

(We will see why later on). With this choice, (6.4) gives

$$\begin{aligned} 0 &\leq \langle u, u\rangle - \frac{\langle u, v\rangle\langle v, u\rangle}{\langle v, v\rangle} - \frac{\overline{\langle u, v\rangle}\langle u, v\rangle}{\langle v, v\rangle} + \frac{\langle u, v\rangle\overline{\langle u, v\rangle}}{\langle v, v\rangle} \\ &= \langle u, u\rangle - \frac{|\langle u, v\rangle|^2}{\langle v, v\rangle}. \end{aligned}$$

Rearranging, we obtain (ii).

(iii) Since $\langle u, v\rangle + \langle v, u\rangle = \langle u, v\rangle + \overline{\langle u, v\rangle} = 2\Re\langle u, v\rangle$, we have

$$\begin{aligned} \|u + v\|^2 &= \langle u, u\rangle + 2\Re\langle u, v\rangle + \langle v, v\rangle \\ &\leq \langle u, u\rangle + 2\,|\langle u, v\rangle| + \langle v, v\rangle \\ &= ||u||^2 + 2\,||u||\,||v|| + ||v||^2 \\ &= (\|u\| + \|v\|)^2 \end{aligned}$$

which gives (iii) at once. Recall that $\Re\,(x + iy) = x$. Thus $\Re z$ denotes the real part of z and so $|z| \geq \Re z$.

(iv) We have

$$\|u - w\| = \|u - v + v - w\| \leq \|u - v\| + \|v - w\|$$

by (iii). □

6.2 Orthogonal and orthonormal sets

Vectors u and v in an inner product space are **orthogonal** if $\langle u, v\rangle = 0$. A set $v_1, \ldots, v_m$ of nonzero vectors in an inner product space V is said to be an **orthogonal set** if

$$\langle v_j, v_k\rangle = 0 \text{ whenever } j \neq k.$$

We think of the vectors as perpendicular to each other. An **orthonormal set** is an orthogonal set of vectors each having length 1.

Orthogonal sets have some very convenient properties.

Lemma 6.1. *An orthogonal set $v_1, \ldots, v_m$ is a linearly independent set.*

Proof. Suppose that

$$x_1 v_1 + \cdots + x_m v_m = 0$$

for scalars $x_1, \ldots, x_m$. Taking the inner product of both sides with v_j,

$$\langle x_1 v_1 + \cdots + x_m v_m, v_j\rangle = 0.$$

However,

$$\begin{aligned}&\langle x_1 v_1 + \cdots + x_m v_m, v_j\rangle \\ &\quad = x_1\langle v_1, v_j\rangle + \cdots + x_m\langle v_m, v_j\rangle \\ &\quad = x_j\langle v_j, v_j\rangle.\end{aligned}$$

The other terms vanish by orthogonality. We conclude that $x_j\langle v_j, v_j\rangle = 0$. Since $v_j \neq 0$, we deduce that $x_j = 0$. Since j is arbitrary, linear independence follows. □

We now see that there cannot be an orthogonal set of more than n vectors in $\mathbb{R}^n$ or $\mathbb{C}^n$, answering a question from Chapter 1.

Example 6.4. In $\mathbb{R}^n$ or $\mathbb{C}^n$, the standard basis

$$\mathbf{e}_1 = (1, 0, \ldots, 0), \ldots, \mathbf{e}_n = (0, \ldots, 0, 1)$$

is an orthonormal set. Since span $\{\mathbf{e}_1, \cdots, \mathbf{e}_n\} = F^n$ for $F = \mathbb{C}$ or $\mathbb{R}$, $\{\mathbf{e}_1, \cdots, \mathbf{e}_n\}$ is called an orthonormal basis for F^n.

Example 6.5. Let $\mathbf{a}, \mathbf{b}$ be in $\mathbb{R}^3$, $\mathbf{a} \neq \mathbf{0}$, $\mathbf{b} \neq \mathbf{0}$, $\langle \mathbf{a}, \mathbf{b}\rangle = 0$. Then

$$\mathbf{a}, \mathbf{b}, \mathbf{a} \times \mathbf{b}$$

is an **orthogonal basis** of $\mathbb{R}^3$ (a basis that is an orthogonal set).

Let $v_1, \ldots, v_m$ be an orthogonal set in V. In particular $v_1, \ldots, v_m$ is a basis of

$$\text{span}\{v_1, \ldots, v_m\}. \tag{6.5}$$

Coordinate computations are easy in this basis.

Lemma 6.2. *Let $v_1, \ldots, v_m$ be an orthogonal set and suppose $w \in \text{span}\{v_1, \ldots, v_m\}$. Then*

$$w = \frac{\langle w, v_1\rangle}{\langle v_1, v_1\rangle} v_1 + \cdots + \frac{\langle w, v_m\rangle}{\langle v_m, v_m\rangle} v_m.$$

Proof. Certainly

$$w = x_1 v_1 + \cdots + x_m v_m$$

for scalars x_j. Take the inner product of both sides with v_j:

$$\begin{aligned}\langle w, v_j \rangle &= \langle x_1 v_1 + \cdots + x_m v_m, v_j \rangle \\ &= x_j \langle v_j, v_j \rangle. \ \square\end{aligned}$$

Recall that from Proposition 5.3,

$$e^{a+ib} = e^a (\cos(b) + i \sin(b))$$

and that

$$\frac{d}{dt} e^{(a+ib)t} = (a + ib)\, e^{(a+ib)t}.$$

Example 6.6. In Example 6.3, take $[a, b] = [0, 2\pi], F = \mathbb{C}$ and $h(x) = \frac{1}{2\pi}$. This inner product on $C_{\mathbb{C}}[0, 2\pi]$ is given by

$$\langle f, g \rangle = \frac{1}{2\pi} \int_0^{2\pi} f(x)\overline{g(x)} dx.$$

We claim that the set of functions $e^{-inx}, e^{-i(n-1)x}, \ldots, e^{-ix}, 1, e^{ix}, \ldots, e^{inx}$ is an orthonormal subset of $C_{\mathbb{C}}[0, 2\pi]$. To see this, recall $\overline{e^{i\theta}} = e^{-i\theta}$. Then

$$\begin{aligned}\langle e^{ijx}, e^{ikx} \rangle &= \frac{1}{2\pi} \int_0^{2\pi} e^{ijx} e^{-ikx} dx, \\ &= \frac{1}{2\pi} \int_0^{2\pi} e^{i(j-k)x} dx.\end{aligned}$$

If $j = k$, the integrand is 1, and

$$\langle e^{ijx}, e^{ijx} \rangle = 1.$$

If $j \neq k$, then

$$e^{i(j-k)x} = \frac{d}{dx} \left(\frac{e^{i(j-k)x}}{i(j-k)} \right) = f'(x),$$

say. Now

$$\int_0^{2\pi} e^{i(j-k)x} dx = f(2\pi) - f(0) = \frac{1}{i\,(j-k)} - \frac{1}{i\,(j-k)} = 0,$$

proving that our set is orthonormal.

This example lies at the foundation of the whole subject of Fourier series, which is important in boundary value problems in PDEs; see for example Polking, Boggess and Arnold (2002). We treat the 'best approximation' aspect of Fourier series below.

6.3 The Gram-Schmidt process, projections

It is natural to ask, given a finite dimensional subspace W of an inner product space, whether there is an orthogonal basis of W. The answer is yes, and there is a simple algorithm, the Gram-Schmidt process, to produce such a basis starting from any basis of W.

Lemma 6.3. *Let V be an inner product space and let W_j be a subspace of V having an orthogonal basis $\{v_1, \cdots, v_j\}$. If $w_{j+1} \notin \text{span}\{v_1, \cdots, v_j\}$, and v_{j+1} is given by*

$$v_{j+1} = w_{j+1} - \sum_{k=1}^{j} \frac{\langle w_{j+1}, v_k \rangle}{\langle v_k, v_k \rangle} v_k, \tag{6.6}$$

then $\{v_1, \cdots, v_j, v_{j+1}\}$ is an orthogonal set of vectors, and

$$\text{span}\{v_1, \cdots, v_j, v_{j+1}\} = \text{span}\{v_1, \cdots, v_j, w_{j+1}\}. \tag{6.7}$$

Proof. Note that (6.6) exhibits v_{j+1} as a linear combination of the vectors $v_1, \cdots, v_j, w_{j+1}$. Therefore,

$$\text{span}\{v_1, \cdots, v_j, v_{j+1}\} \subseteq \text{span}\{v_1, \cdots, v_j, w_{j+1}\}.$$

Also, (6.6) can be solved for w_{j+1}, obtaining w_{j+1} as a linear combination of the vectors $v_1, \cdots, v_j, v_{j+1}$. Therefore,

$$\text{span}\{v_1, \cdots, v_j, w_{j+1}\} \subseteq \text{span}\{v_1, \cdots, v_j, v_{j+1}\}.$$

Therefore, (6.7) follows.

It remains to verify that the vectors $v_1, \cdots, v_j, v_{j+1}$ are orthogonal. It is assumed that this is true of $v_1, \cdots, v_j$. Since $w_{j+1} \notin \text{span}\{v_1, \cdots, v_j\}$, it follows that $v_{j+1} \neq 0$. Therefore, it only remains to verify that $\langle v_{j+1}, v_m \rangle = 0$ for $m \leq j$.

$$\begin{aligned}
\langle v_{j+1}, v_m \rangle &= \left\langle w_{j+1} - \sum_{k=1}^{j} \frac{\langle w_{j+1}, v_k \rangle}{\langle v_k, v_k \rangle} v_k, v_m \right\rangle \\
&= \langle w_{j+1}, v_m \rangle - \sum_{k=1}^{j} \frac{\langle w_{j+1}, v_k \rangle}{\langle v_k, v_k \rangle} \langle v_k, v_m \rangle \\
&= \langle w_{j+1}, v_m \rangle - \langle w_{j+1}, v_m \rangle = 0. \quad \square
\end{aligned}$$

Proposition 6.1. Let W be a subspace of V, an inner product space, and suppose $\{w_1, \cdots, w_m\}$ is a basis for W. Then there exists an orthogonal basis for W

$$\{v_1, \cdots, v_m\}$$

such that for all $l \leq r$,

$$\text{span}\{v_1, \cdots, v_l\} = \text{span}\{w_1, \cdots, w_l\}.$$

Proof. Let $v_1 = w_1$. Now suppose $v_1, \cdots, v_j$ have been chosen such that

$$\text{span}\{v_1, \cdots, v_i\} = \text{span}\{w_1, \cdots, w_i\}$$

for all $i \leq j$ and $\{v_1, \cdots, v_j\}$ is an orthogonal set of vectors. Since $w_{j+1} \notin \text{span}\{v_1, \cdots, v_j\}$, it follows from Lemma 6.3 that there exists v_{j+1} such that

$$\begin{aligned} \text{span}\{v_1, \cdots, v_j, v_{j+1}\} &= \text{span}\{v_1, \cdots, v_j, w_{j+1}\} \\ &= \text{span}\{w_1, \cdots, w_j, w_{j+1}\} \end{aligned}$$

and $\{v_1, \cdots, v_j, v_{j+1}\}$ is an orthogonal set of vectors. Continuing this way, yields the existence of the orthogonal basis of the proposition. □

If each vector in the orthogonal basis is divided by its norm, the resulting set of vectors is still an orthogonal basis but has the additional property that each vector has norm 1. Thus the set of vectors is an orthonormal basis.

Example 6.7. In the inner product space $\mathbb{R}^4$, find an orthogonal basis of the column space of

$$A = \begin{pmatrix} -1 & 0 & 1 \\ 1 & 2 & 3 \\ 0 & 1 & 2 \\ 1 & 3 & 0 \end{pmatrix}.$$

Let $A = \begin{pmatrix} \mathbf{w}_1 & \mathbf{w}_2 & \mathbf{w}_3 \end{pmatrix}$,

$$\begin{aligned} \mathbf{v}_1 &= \mathbf{w}_1 = (-1, 1, 0, 1), \\ \mathbf{v}_2 &= \mathbf{w}_2 - \frac{\langle \mathbf{w}_2, \mathbf{v}_1 \rangle}{\langle \mathbf{v}_1, \mathbf{v}_1 \rangle} \mathbf{v}_1 \\ &= (0, 2, 1, 3) - \frac{5}{3}(-1, 1, 0, 1), \\ 3\mathbf{v}_2 &= (0, 6, 3, 9) - (-5, 5, 0, 5) = (5, 1, 3, 4). \end{aligned}$$

We observe as a check that $\langle \mathbf{v}_2, \mathbf{v}_1 \rangle = \mathbf{0}$. We replace $\mathbf{v}_2$ by $3\mathbf{v}_2$, which we rename $\mathbf{v}_2$. This simplification will not spoil our construction. Let

$$\begin{aligned} \mathbf{v}_3 &= \mathbf{w}_3 - \frac{\langle \mathbf{w}_3, \mathbf{v}_1 \rangle}{\langle \mathbf{v}_1, \mathbf{v}_1 \rangle} \mathbf{v}_1 - \frac{\langle \mathbf{w}_3, \mathbf{v}_2 \rangle}{\langle \mathbf{v}_2, \mathbf{v}_2 \rangle} \mathbf{v}_2 \\ &= (1, 3, 2, 0) - 2/3(-1, 1, 0, 1) - \frac{14}{51}(5, 1, 3, 4), \\ 51\mathbf{v}_3 &= (51, 153, 102, 0) - (-34, 34, 0, 34) - (70, 14, 42, 56) \\ &= (15, 105, 60, -90) = 15(1, 7, 4, -6). \end{aligned}$$

Replace $\mathbf{v}_3$ by $(1, 7, 4, -6)$ which we rename $\mathbf{v}_3$, and check $\langle \mathbf{v}_3, \mathbf{v}_1 \rangle = \langle \mathbf{v}_3, \mathbf{v}_2 \rangle = 0$. Our orthogonal basis is

$$\mathbf{v}_1 = (-1, 1, 0, 1), \mathbf{v}_2 = (5, 1, 3, 4), \mathbf{v}_3 = (1, 7, 4, -6).$$

It is often useful to convert an orthogonal basis $v_1, \ldots, v_m$ into an orthonormal basis

$$u_1 = \frac{v_1}{\|v_1\|}, \ldots, u_m = \frac{v_m}{\|v_m\|},$$

(this is called **normalizing**). In the above example,

$$\mathbf{u}_1 = \frac{1}{\sqrt{3}}(-1,1,0,1), \mathbf{u}_2 = \frac{1}{\sqrt{51}}(5,1,3,4), \mathbf{u}_3 = \frac{1}{\sqrt{102}}(1,7,4,-6).$$

The projection P in Example 5.2 of Chapter 5can be generalized to the setting of an inner product space V.

Let W be a finite-dimensional subspace of V and let $v \in V$. Then Pv (in this setting) is the element $w = Pv$ of W for which $d(v, w)$ is the smallest. (Recall $d(v, w)$ is notation which means distance from v to w.) It is easy to see that Example 5.2 of Chapter 5is a special case, with $V = \mathbb{R}^n, W = \{\mathbf{w} \in \mathbb{R}^n; w_n = 0\}$; the closest vector in W to $(x_1, \ldots, x_{n-1}, x_n)$ is $(x_1, \ldots, x_{n-1}, 0)$. In the general situation, it is natural to ask

(i) is there such a 'closest' point Pv?
(ii) is the mapping $P : V \to W$ a linear mapping?

Theorem 6.2. *Let W be a finite-dimensional subspace of the inner product space V. Let $v_1, \ldots, v_k$ be an orthogonal basis of W. Define $P : V \to W$ by*

$$Pv = \frac{\langle v, v_1 \rangle}{\langle v_1, v_1 \rangle} v_1 + \cdots + \frac{\langle v, v_k \rangle}{\langle v_k, v_k \rangle} v_k. \tag{6.8}$$

Then

$$\|v - w\| > \|v - Pv\| \qquad (w \in W, w \neq Pv) \tag{6.9}$$

For $z \in W$, $z = Pv$ if and only if

$$\langle v - z, w \rangle = 0 \text{ for all } w \in W. \tag{6.10}$$

It is obvious from (6.8) that the mapping $P : V \to W$ is linear. We call Pv the **orthogonal projection of v onto W**.

Proof. Let $j \in \{1, \ldots, k\}$. Using the definition of Pv in (6.8) and the properties of the inner product,

$$\begin{aligned}\langle v - Pv, v_j \rangle &= \langle v, v_j \rangle - \langle Pv, v_j \rangle \\ &= \langle v, v_j \rangle - \frac{\langle v, v_j \rangle}{\langle v_j, v_j \rangle} \langle v_j, v_j \rangle = 0.\end{aligned}$$

It follows that

$$\langle v - Pv, w \rangle = 0 \tag{6.11}$$

for every w in W because every w is a linear combination of the vectors $\{v_1, \ldots, v_k\}$.

We note that

$$\|z + w\|^2 = \|z\|^2 + \|w\|^2 + 2\Re \langle z, w \rangle$$

$$= \|z\|^2 + \|w\|^2 \tag{6.12}$$

whenever z, w are in V and $\langle z, w\rangle = 0$. Now let $w \in W, w \neq Pv$. To prove (6.9), we observe that

$$\langle v - Pv, Pv - w\rangle = 0 \tag{6.13}$$

as a particular case of (6.11). From (6.12), using the above orthogonality,

$$\begin{aligned}\|v - w\|^2 &= \|v - Pv + Pv - w\|^2 \\ &= \|v - Pv\|^2 + \|Pv - w\|^2 \\ &> \|v - Pv\|^2.\end{aligned}$$

Finally, take any z in W. Then using (6.13),

$$\langle v - z, Pv - z\rangle = \langle v - Pv + Pv - z, Pv - z\rangle$$

$$= \langle v - Pv, Pv - z\rangle + \langle Pv - z, Pv - z\rangle = \|Pv - z\|^2 .$$

If (6.10) holds, then the above equals 0, and so $Pv = z$. If $Pv = z$, then (6.10) follows from (6.11). □

The geometric significance of Pv being closest to v is illustrated in the following picture which captures the essence of the idea in every inner product space, but is drawn here to represent the situation in $\mathbb{R}^3$.

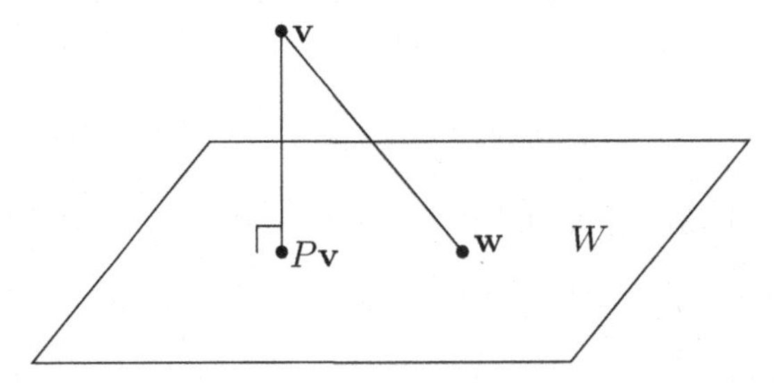

Example 6.8. Find the point in the plane W with equation

$$x_1 + x_2 - 2x_3 = 0$$

that is closest to $(1, 6, 1)$.

Solution. Choose two independent vectors in the plane,

$$\mathbf{w}_1 = (1, -1, 0), \mathbf{w}_2 = (2, 0, 1).$$

Construct an orthogonal basis $\mathbf{v}_1, \mathbf{v}_2$ of W:

$$\mathbf{v}_1 = \mathbf{w}_1,$$

$$\mathbf{v}_2 = \mathbf{w}_2 - \frac{\langle \mathbf{w}_2, \mathbf{w}_1\rangle}{\langle \mathbf{w}_1, \mathbf{w}_1\rangle}\mathbf{w}_1 = (2, 0, 1) - \frac{2}{2}(1, -1, 0)$$

$$= (1, 1, 1).$$

The closest point that we seek is

$$\begin{aligned} P\mathbf{v} &= \frac{\langle \mathbf{v}, \mathbf{v}_1 \rangle}{\langle \mathbf{v}_1, \mathbf{v}_1 \rangle}\mathbf{v}_1 + \frac{\langle \mathbf{v}, \mathbf{v}_2 \rangle}{\langle \mathbf{v}_2, \mathbf{v}_2 \rangle}\mathbf{v}_2 \\ &= \left(-\frac{5}{2}, \frac{5}{2}, 0\right) + \left(\frac{8}{3}, \frac{8}{3}, \frac{8}{3}\right) \\ &= \left(\frac{1}{6}, \frac{31}{6}, \frac{8}{3}\right) \end{aligned}$$

Note that, as expected,

$$\mathbf{v} - P\mathbf{v} = (1, 6, 1) - \left(\frac{1}{6}, \frac{31}{6}, \frac{8}{3}\right) = \left(\frac{5}{6}, \frac{5}{6}, -\frac{5}{3}\right)$$

which is a normal to W.

Example 6.9. In the inner product space $C[-1, 1]$ with

$$\langle f, g \rangle = \int_{-1}^{1} f(t)g(t)dt,$$

find the closest polynomial of degree 2 to $f(x) = x^4$.

Solution. We construct an orthogonal basis of $\text{span}\{1, x, x^2\}$. Let

$$\begin{aligned} v_1 &= 1, \\ v_2 &= x - \frac{\langle x, 1 \rangle}{\langle 1, 1 \rangle} 1 = x \end{aligned}$$

(note that $\langle x^m, x^n \rangle = 0$ for $m + n$ odd);

$$\begin{aligned} v_3 &= x^2 - \frac{\langle x^2, 1 \rangle}{\langle 1, 1 \rangle} 1 - \frac{\langle x^2, x \rangle}{\langle x, x \rangle} x \\ &= x^2 - 1/3. \end{aligned}$$

Now the closest polynomial of degree 2 to x^4 is

$$f(x) = \frac{\langle x^4, 1 \rangle}{\langle 1, 1 \rangle} 1 + \frac{\langle x^4, x \rangle}{\langle x, x \rangle} x + \frac{\langle x^4, x^2 - 1/3 \rangle}{\langle x^2 - 1/3, x^2 - 1/3 \rangle} (x^2 - 1/3).$$

Repeatedly using $\langle x^m, x^n \rangle = 0$ for $m + n$ odd and

$$\langle x^m, x^n \rangle = \frac{2}{m + n + 1} \qquad (m + n \text{ even}),$$

we are led to

$$\begin{aligned} \langle x^4, x^2 - 1/3 \rangle = 16/105, \langle x^2 - 1/3, x^2 - 1/3 \rangle &= 8/45, \\ f(x) &= 1/5 \cdot 1 + 0 \cdot x + 6/7(x^2 - 1/3), \\ &= \frac{6x^2}{7} - \frac{3}{35}. \end{aligned}$$

The following is the graph of the function $f(x) = x^4$ and the second degree polynomial just found, which is closest to this function using the definition of distance which comes from the above inner product.

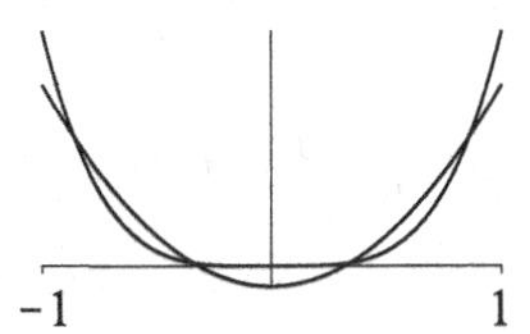

Example 6.10. Find the point in the **hyperplane** W in $\mathbb{R}^n$ with equation

$$a_1x_1 + \cdots + a_nx_n = 0$$

that is closest to $\mathbf{b} = (b_1, \ldots, b_n)$. Here $\mathbf{a} = (a_1, \ldots, a_n) \neq \mathbf{0}$.

Note that $\dim W = n-1$, the nullity of $(a_1 \cdots a_n)$. Our solution is more efficient than in Example 6.8. For any orthogonal basis $\{\mathbf{v}_1, \ldots, \mathbf{v}_{n-1}\}$, for W, it follows that $\{\mathbf{v}_1, \ldots, \mathbf{v}_{n-1}, \mathbf{a}\}$ is an orthogonal set, indeed an orthogonal basis of $\mathbb{R}^n$. So

$$\mathbf{b} = \overbrace{\frac{\langle \mathbf{b}, \mathbf{v}_1 \rangle}{\langle \mathbf{v}_1, \mathbf{v}_1 \rangle}\mathbf{v}_1 + \cdots + \frac{\langle \mathbf{b}, \mathbf{v}_{n-1} \rangle}{\langle \mathbf{v}_{n-1}, \mathbf{v}_{n-1} \rangle}\mathbf{v}_{n-1}}^{P\mathbf{b}} + \frac{\langle \mathbf{b}, \mathbf{a} \rangle}{\langle \mathbf{a}, \mathbf{a} \rangle}\mathbf{a},$$

and it follows that

$$P\mathbf{b} = \frac{\langle \mathbf{b}, \mathbf{v}_1 \rangle}{\langle \mathbf{v}_1, \mathbf{v}_1 \rangle}\mathbf{v}_1 + \cdots + \frac{\langle \mathbf{b}, \mathbf{v}_{n-1} \rangle}{\langle \mathbf{v}_{n-1}, \mathbf{v}_{n-1} \rangle}\mathbf{v}_{n-1} = \mathbf{b} - \frac{\langle \mathbf{b}, \mathbf{a} \rangle}{\langle \mathbf{a}, \mathbf{a} \rangle}\mathbf{a}.$$

(Lemma 6.2).

In Example 6.8,

$$\begin{aligned} P(1,6,1) &= (1,6,1) - \frac{\langle (1,1,-2),(1,6,1) \rangle}{6}(1,1,-2) \\ &= (1,6,1) - \frac{5}{6}(1,1,-2) = \left(\frac{1}{6}, \frac{31}{6}, \frac{8}{3}\right). \end{aligned}$$

Given a function $f \in C_{\mathbb{C}}[0, 2\pi]$, how closely can we approximate f in mean square by a **trigonometric polynomial** of **degree** at most n,

$$s(x) = \sum_{k=-n}^{n} a_k e^{ikx}?$$

That is, we want to minimize

$$\int_0^{2\pi} |f(x) - s(x)|^2 dx$$

over all possible choices of a_j $(j = -n, \ldots, n)$ in $\mathbb{C}$.

The **Fourier series** of f is the formal series

$$\sum_{k=-\infty}^{\infty} \hat{f}(k)e^{ikx}$$

where

$$\hat{f}(k) = \frac{1}{2\pi}\int_0^{2\pi} f(x)e^{-ikx}dx.$$

It is a 'formal series' because it may actually diverge for some x (but not for $f \in S_{\mathbb{C}}[0, 2\pi]$). See Katznelson (2004) or Stein and Shakarchi (2003). The nth **partial sum** of the Fourier series is

$$s_n(x) = \sum_{k=-n}^{n} \hat{f}(k)e^{ikx}.$$

Theorem 6.3. *Let $f \in C_{\mathbb{C}}[0, 2\pi]$. The minimum value of*

$$\int_0^{2\pi} |f(x) - s(x)|^2 dx,$$

over trigonometric polynomials s of degree at most n, is attained by $s(x) = s_n(x)$, the nth partial sum of the Fourier series of f.

Proof. With the inner product as in Example 6.6, we have to show that $\|f - s_n\| \le \|f - s\|$ for all s in $W = \text{span}\{e^{inx}, \ldots, e^{-ix}, 1, \ldots, e^{inx}\}$. This follows from Theorem 6.2 with $k = 2n+1, w_1 = e^{-inx}, \ldots, w_{2n+1} = e^{inx}$ (which we recall is an orthonormal set). In this particular case,

$$Pf = \langle f, w_1\rangle w_1 + \cdots + \langle f, w_{2n+1}\rangle w_{2n+1}.$$

That is,

$$(Pf)(x) = \sum_{j=-n}^{n} \hat{f}(j)e^{ijx} = s_n(x). \ \square$$

Example 6.11. Let $f(x) = x$ $(0 \le x \le 2\pi)$. Find the partial sum $s_n(x)$ of the Fourier series of f.

We first evaluate

$$\hat{f}(k) = \frac{1}{2\pi}\int_0^{2\pi} xe^{ikx}dx.$$

Of course $\hat{f}(0) = \frac{1}{2\pi}\frac{1}{2}(2\pi)^2 = \pi$. For $k \neq 0$, we use integration by parts:

$$\int_0^{2\pi} xe^{-ikx} = \frac{xe^{-ikx}}{-ik}\Big|_0^{2\pi} - \int_0^{2\pi} \frac{e^{-ikx}}{-ik}dx = \frac{2\pi i}{k}.$$

So $\hat{f}(k) = ik^{-1}$,

$$s_n(x) = \sum_{k=-n,\ k\neq 0}^{n} \frac{ie^{ikx}}{k} + \pi.$$

Grouping $-k$ and k together, $k = 1, \ldots, n$,

$$\frac{ie^{ikx}}{k} + \frac{ie^{-ikx}}{-k} = i\frac{(2i \sin kx)}{k} = -\frac{2\sin kx}{k}.$$

Hence

$$s_n(x) = \pi - 2\sum_{k=1}^{n} \frac{\sin kx}{k}.$$

The graph of $s_6(x)$ is sketched below along with the function $f(x) = x$.

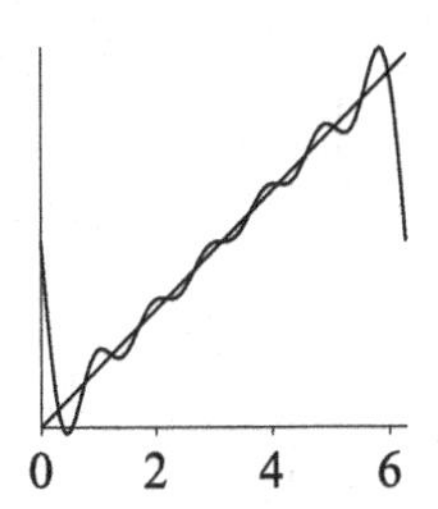

Example 6.12. Show that for u, v in the inner product space V over F,

$$\left\| u - \frac{\langle u, v \rangle}{\langle v, v \rangle} v \right\| \leq \|u - tv\| \tag{6.14}$$

for all $t \in F$. This shows that we made the best choice of t in the proof of Theorem 6.1 (ii).

The claim is a restatement of Theorem 6.2 with u, v in place of v, v_1 and $W = \text{span}\{v\}$. Here

$$Pu = \frac{\langle u, v \rangle v}{\langle v, v \rangle},$$

and (6.9) obviously reduces to (6.10).

6.4 The method of least squares

We shall only give a simple account of this method, which is tremendously important in applications. See Golub and Van Loan (1996) and Hogben (2007) for substantial information about the method. In this section, the field of scalars is $\mathbb{C}$. Suppose we have a possibly inconsistent linear system

$$A\mathbf{x} = \mathbf{b}, \tag{6.15}$$

where $A = (\mathbf{a}_1 \ldots \mathbf{a}_n)$ is $m \times n$ and $\mathbf{b} \in \mathbb{C}^m$. To get an 'approximate' solution, we find an $\widehat{\mathbf{x}}$ in $\mathbb{C}^n$ such that

$$\|\mathbf{b} - A\widehat{\mathbf{x}}\| \leq \|\mathbf{b} - A\mathbf{x}\| \tag{6.16}$$

for all $\mathbf{x}$ in $\mathbb{C}^n$. Any such $\widehat{\mathbf{x}}$ is called a **least-squares solution** of (6.15).

Definition 6.3. Let A be an $m \times n$ matrix. Then we define A^*as

$$A^* = \overline{A^t}.$$

In words: Take the transpose of A and then replace every entry by the complex conjugate. This matrix is called the **adjoint**.

Example 6.13.

$$\begin{pmatrix} 1-i & 2i & 4 & 3 \\ 2 & 3i & 2+i & 3 \\ 2 & 1 & 2 & 0 \end{pmatrix}^* = \overline{\begin{pmatrix} 1-i & 2i & 4 & 3 \\ 2 & 3i & 2+i & 3 \\ 2 & 1 & 2 & 0 \end{pmatrix}^t}$$

$$= \overline{\begin{pmatrix} 1-i & 2 & 2 \\ 2i & 3i & 1 \\ 4 & 2+i & 2 \\ 3 & 3 & 0 \end{pmatrix}} = \begin{pmatrix} 1+i & 2 & 2 \\ -2i & -3i & 1 \\ 4 & 2-i & 2 \\ 3 & 3 & 0 \end{pmatrix}$$

For A a real $m \times n$ matrix, $A^* = A^t$. The importance of the adjoint of a matrix is the first result in the following lemma.

Lemma 6.4. *Let A be an $m \times n$ matrix. Then denoting by $\langle \cdot, \cdot \rangle_{\mathbb{C}^n}, \langle \cdot, \cdot \rangle_{\mathbb{C}^m}$ the inner products on $\mathbb{C}^n$ and $\mathbb{C}^m$ respectively,*

$$\langle A\mathbf{x}, \mathbf{y} \rangle_{\mathbb{C}^m} = \langle \mathbf{x}, A^* \mathbf{y} \rangle_{\mathbb{C}^n}$$

for all $\mathbf{x} \in \mathbb{C}^n, \mathbf{y} \in \mathbb{C}^m$. Also, if the product AB exists, then

$$(AB)^* = B^* A^*.$$

Furthermore,

$$(A+B)^* = A^* + B^*, \ (A^*)^* = A.$$

Proof. By the definition of the inner product in $\mathbb{C}^n, \mathbb{C}^m$, it follows that for

$$\mathbf{x} = (x_1, \cdots, x_n), \ \mathbf{y} = (y_1, \cdots, y_m),$$

$$\begin{aligned} \langle A\mathbf{x}, \mathbf{y} \rangle_{\mathbb{C}^m} &= \sum_{i=1}^m \left(\sum_{j=1}^n A_{ij} x_j \right) \overline{y_i} \\ &= \sum_{j=1}^n \sum_{i=1}^m A_{ij} x_j \overline{y_i} = \sum_{j=1}^n \sum_{i=1}^m x_j A^t_{ji} \overline{y_i} \\ &= \sum_{j=1}^n x_j \sum_{i=1}^m \overline{\overline{A^t_{ji}} y_i} = \sum_{j=1}^n x_j \overline{\sum_{i=1}^m \overline{A^t_{ji}} y_i} \\ &= \langle \mathbf{x}, A^* \mathbf{y} \rangle_{\mathbb{C}^n}. \end{aligned}$$

Next, $(AB)^* = \overline{(AB)^t} = \overline{B^t A^t} = \overline{B^t}\ \overline{A^t} = B^* A^*$. The last two assertions are obvious. This proves the lemma. □

The following theorem is the main result. It equates solutions to the least squares problem with solutions to a simple system of equations involving the adjoint of the matrix.

Theorem 6.4. *Let A be an $m \times n$ matrix. For each $\mathbf{y} \in \mathbb{C}^m$, there exists $\mathbf{x} \in \mathbb{C}^n$ such that*

$$||A\mathbf{x} - \mathbf{y}|| \leq ||A\mathbf{x}_1 - \mathbf{y}||$$

for all $\mathbf{x}_1 \in \mathbb{C}^n$. Also, $\mathbf{x}$ is a solution to this minimization problem if and only if $\mathbf{x}$ is a solution to the equation, $A^* A\mathbf{x} = A^*\mathbf{y}$.

Proof. By Theorem 6.2 on Page 162, there exists a point, $A\mathbf{x}_0$, in the finite dimensional subspace $\mathrm{Im}(A)$ of $\mathbb{C}^m$ such that for all $\mathbf{x} \in \mathbb{C}^n$,

$$||A\mathbf{x} - \mathbf{y}||^2 \geq ||A\mathbf{x}_0 - \mathbf{y}||^2 .$$

Also, from this theorem, this inequality holds for all $\mathbf{x}$ if and only if $A\mathbf{x}_0 - \mathbf{y}$ is perpendicular to every $A\mathbf{x} \in \mathrm{Im}(A)$. Therefore, the solution is characterized by

$$\langle A\mathbf{x}_0 - \mathbf{y}, A\mathbf{x}\rangle = 0$$

for all $\mathbf{x} \in \mathbb{C}^n$. By Lemma 6.4, this is equivalent to the statement that

$$\langle A^* A\mathbf{x}_0 - A^*\mathbf{y}, \mathbf{x}\rangle = 0.$$

for all $\mathbf{x} \in \mathbb{C}^n$, in particular for $\mathbf{x} = A^* A\mathbf{x}_0 - A^*\mathbf{y}$. In other words, a solution is obtained by solving $A^* A\mathbf{x}_0 = A^*\mathbf{y}$ for $\mathbf{x}_0$. This proves the lemma. □

Example 6.14. Find a least-squares solution $\widehat{\mathbf{x}}$ of $A\mathbf{x} = \mathbf{b}$ for

$$A = \begin{pmatrix} 2 & 1 \\ 1 & 1 \\ 3 & -1 \\ -1 & 0 \end{pmatrix}, \mathbf{b} = \begin{pmatrix} 1 \\ 1 \\ 4 \\ 1 \end{pmatrix}.$$

By Theorem 6.4, this is equivalent to finding a solution $\widehat{\mathbf{x}}$ to the system

$$A^* A\widehat{\mathbf{x}} = A^*\mathbf{b}.$$

Thus we need a solution $\widehat{\mathbf{x}}$ to

$$\begin{pmatrix} 15 & 0 \\ 0 & 3 \end{pmatrix}\widehat{\mathbf{x}} = \begin{pmatrix} 14 \\ -2 \end{pmatrix}.$$

It follows right away that for $\widehat{\mathbf{x}} = (x_1, x_2)$,

$$x_1 = \frac{14}{15},\ x_2 = -\frac{2}{3}.$$

Example 6.15. Find a least-squares solution of $A\mathbf{x} = \mathbf{b}$ for

$$A = \begin{pmatrix} 1 & 1 & 0 & 0 \\ 1 & 1 & 0 & 0 \\ 1 & 0 & 1 & 0 \\ 1 & 0 & 1 & 0 \\ 1 & 0 & 0 & 1 \\ 1 & 0 & 0 & 1 \end{pmatrix}, \mathbf{b} = \begin{pmatrix} 2 \\ 1 \\ 0 \\ 3 \\ 4 \\ 1 \end{pmatrix}.$$

Solution. In this example, $A^* = A^t$. Maple gives

$$A^t A = \begin{pmatrix} 6 & 2 & 2 & 2 \\ 2 & 2 & 0 & 0 \\ 2 & 0 & 2 & 0 \\ 2 & 0 & 0 & 2 \end{pmatrix}, \quad A^t\mathbf{b} = \begin{pmatrix} 11 \\ 3 \\ 3 \\ 5 \end{pmatrix}.$$

The augmented matrix $[A^t A | A^t\mathbf{b}]$ has row reduced form

$$\begin{pmatrix} 1 & 0 & 0 & 1 & \frac{5}{2} \\ 0 & 1 & 0 & -1 & -1 \\ 0 & 0 & 1 & -1 & -1 \\ 0 & 0 & 0 & 0 & 0 \end{pmatrix}.$$

The general solution is

$$x_1 = 5/2 - x_4, x_2 = -1 + x_4, x_3 = x_4 - 1,$$

and we can choose any x_4 to provide a least squares solution of $A\mathbf{x} = \mathbf{b}$.

A nice application of least squares is finding a straight line in $\mathbb{R}^2$ that is a good fit to n given points $(x_1, y_1), \ldots, (x_n, y_n)$. The **least-squares line** (or line of regression of y on x) is the line

$$y = px + q$$

that minimizes the sum of the squares of the **residuals**

$$px_1 + q - y_1, \ldots, px_n + q - y_n.$$

In other words we are minimizing (over all p, q)

$$(px_1 + q - y_1)^2 + \cdots + (px_n + q - y_n)^2. \tag{6.17}$$

This is exactly the same as finding a least squares solution of

$$A\hat{\mathbf{x}} = \mathbf{y}$$

where

$$A = \begin{pmatrix} 1 & x_1 \\ 1 & x_2 \\ \vdots & \vdots \\ 1 & x_n \end{pmatrix}, \quad \widehat{\mathbf{x}} = \begin{pmatrix} q \\ p \end{pmatrix}, \quad \mathbf{y} = \begin{pmatrix} y_1 \\ \vdots \\ y_n \end{pmatrix}.$$

For the square of the distance $\|A\widehat{\mathbf{x}} - \mathbf{y}\|$ is the expression in (6.17).

For $n \geq 2$ there is a unique solution to this particular least squares problem involving the regression line and in fact, A^tA will be invertible. (See Problem 6 below.) However, in general least squares problems, the solution $\widehat{\mathbf{x}}$ is not unique. All that is known is that for $\widehat{\mathbf{x}}$ a least squares solution to $A\mathbf{x} = \mathbf{b}$, the $A\widehat{\mathbf{x}}$ is unique, being the closest point in $\text{Im}\,(A)$ to $\mathbf{b}$. When A is not one-to-one, there may be many $\mathbf{x}$ for which $A\mathbf{x} = A\widehat{\mathbf{x}}$. Therefore, an interesting problem involves picking the "best" least squares solution. This is discussed in the exercises. See Problem 26 and the exercises which follow for an introduction to the **Moore Penrose inverse**, which is taken up in the text in Chapter 11. In short, the best least squares solution is the one which is closest to $\mathbf{0}$.

Example 6.16. Find the least-squares line for the data points $(1, 7.1)$, $(2, 9.8)$, $(3, 12.9)$, $(4, 15.2)$, $(5, 18.1)$.

Solution. Here $A^t = A^*$.

$$A = \begin{pmatrix} 1 & 1 \\ 1 & 2 \\ 1 & 3 \\ 1 & 4 \\ 1 & 5 \end{pmatrix},$$

$$(A^tA)^{-1} = \begin{pmatrix} 5 & 15 \\ 15 & 55 \end{pmatrix}^{-1} = \frac{1}{10}\begin{pmatrix} 11 & -3 \\ -3 & 1 \end{pmatrix},$$

while

$$A^t\mathbf{y} = \begin{pmatrix} 1 & 1 & 1 & 1 & 1 \\ 1 & 2 & 3 & 4 & 5 \end{pmatrix}\begin{pmatrix} 7.1 \\ 9.8 \\ 12.9 \\ 15.2 \\ 18.1 \end{pmatrix} = \begin{pmatrix} 63.1 \\ 216.7 \end{pmatrix}.$$

Thus

$$\begin{pmatrix} q \\ p \end{pmatrix} = \frac{1}{10}\begin{pmatrix} 11 & -3 \\ -3 & 1 \end{pmatrix}\begin{pmatrix} 63.1 \\ 216.7 \end{pmatrix} = \begin{pmatrix} 4.4 \\ 2.74 \end{pmatrix}.$$

The approximating line

$$y = 2.74x + 4.4$$

is sketched along with the data points below.

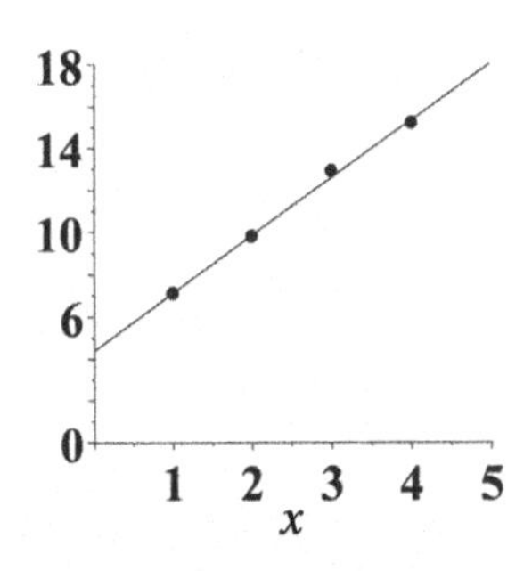

6.5 Exercises

(1) Suppose you define the dot product on $\mathbb{C}^2$ by $(x, y) \cdot (a, b) = ax + by$. As mentioned, this is **not** right, but suppose you did it this way. Give an example of a nonzero vector $\mathbf{x} \in \mathbb{C}^2$ such that $\mathbf{x} \cdot \mathbf{x} = 0$.

(2) Examine the Gram-Schmidt process for vectors and show that for $\{\mathbf{v}_1, \cdots, \mathbf{v}_n\}$ an independent set of vectors in $\mathbb{C}^n$ and $\{\mathbf{u}_1, \cdots, \mathbf{u}_n\}$ the orthonormal basis from the Gram-Schmidt process with the property that $\text{span}\{\mathbf{u}_1, \cdots, \mathbf{u}_k\} = \text{span}\{\mathbf{v}_1, \cdots, \mathbf{v}_k\}$ for each $k \leq n$, the process is of the form $\mathbf{u}_k = l_k \mathbf{v}_k + \mathbf{z}_k$ where $\mathbf{z}_k$ is a vector in $\text{span}\{\mathbf{v}_1, \cdots, \mathbf{v}_{k-1}\}$, and $l_k > 0$. Explain why, in terms of matrices, this implies

$$\begin{pmatrix} \mathbf{u}_1 & \cdots & \mathbf{u}_n \end{pmatrix} = \begin{pmatrix} \mathbf{v}_1 & \cdots & \mathbf{v}_n \end{pmatrix} \begin{pmatrix} l_1 & \cdots & * \\ & \ddots & \vdots \\ 0 & & l_n \end{pmatrix}$$

Explain why the inverse of an upper triangular matrix is also upper triangular. Multiplying on the right by the inverse of the upper triangular matrix, show that this yields

$$\begin{pmatrix} \mathbf{u}_1 & \cdots & \mathbf{u}_n \end{pmatrix} R = \begin{pmatrix} \mathbf{v}_1 & \cdots & \mathbf{v}_n \end{pmatrix}$$

where R is upper triangular having all positive entries down the main diagonal. Explain why this also shows that for all $k \leq n$,

$$\begin{pmatrix} \mathbf{u}_1 & \cdots & \mathbf{u}_n \end{pmatrix} R_k = \begin{pmatrix} \mathbf{v}_1 & \cdots & \mathbf{v}_k \end{pmatrix}$$

where R_k is the $n \times k$ matrix which comes from keeping only the first k columns of R. Denote the matrix $\begin{pmatrix} \mathbf{u}_1 & \cdots & \mathbf{u}_n \end{pmatrix}$ as Q. Explain why $Q^*Q = QQ^* = I$. Also explain why this shows that if A is an $n \times k$ matrix having rank k, then there exists an $n \times k$ matrix R with the property that $r_{ij} = 0$ whenever $j > i$ and a matrix Q satisfying $Q^*Q = QQ^* = I$ such that

$$A = QR$$

This is called the QR factorization of A. Note that the first k columns of Q yield an orthonormal basis for $\text{span}\{\mathbf{v}_1, \cdots, \mathbf{v}_k\}$.

(3) Find the best solution to the system

$$\begin{aligned} 2 + 2y &= 6 \\ 2x + 3y &= 5 \\ 3x + 2y &= 0 \end{aligned}$$

(4) Find the best solution to

$$\begin{aligned} x + y &= 5 \\ x + 2y &= 7 \\ 3x + 5y &= 19 \end{aligned}$$

(5) Find an orthonormal basis for $\mathbb{R}^3$, $\{\mathbf{w}_1, \mathbf{w}_2, \mathbf{w}_3\}$ given that $\mathbf{w}_1$ is a multiple of the vector $(1, 1, 2)$. Recall this is an orthogonal basis in which every vector has unit length.

(6) In the Example 6.15 for the linear regression line, you looked for a least squares solution to the system of equations

$$A\begin{pmatrix} p \\ q \end{pmatrix} = \mathbf{y}$$

which is of the form

$$\begin{pmatrix} 1 & x_1 \\ \vdots & \vdots \\ 1 & x_n \end{pmatrix}\begin{pmatrix} p \\ q \end{pmatrix} = \begin{pmatrix} y_1 \\ \vdots \\ y_n \end{pmatrix} \tag{6.18}$$

where the $x_1, x_2, \cdots$ are distinct numbers.

(a) Show the $n \times 2$ matrix on the left in (6.18) is one-to-one.

(b) Show that for any $m \times n$ matrix A,

$$\ker(A^* A) = \ker(A).$$

(c) Explain why the matrix $A^* A$, where A is as above, is always invertible if $n \geq 2$.

(d) Explain why there exists a unique p, q solving the problem for the least squares regression line.

(7) Relax the assumptions in the axioms for the inner product. Change the axiom about $\langle x, x \rangle \geq 0$ and equals 0 if and only if $x = 0$ to simply read $\langle x, x \rangle \geq 0$. Show that the Cauchy Schwarz inequality still holds in the following form.

$$|\langle x, y \rangle| \leq \langle x, x \rangle^{1/2} \langle y, y \rangle^{1/2}. \tag{6.19}$$

Hint: First let ω be a complex number having absolute value 1 such that

$$\omega \langle x, y \rangle = |\langle x, y \rangle|$$

Letting

$$p(t) = \langle \omega x + t y, \omega x + t y \rangle,$$

explain why $p(t) \geq 0$ for all $t \in \mathbb{R}$. Next explain why the remaining axioms of the inner product imply

$$p(t) = \langle x, x \rangle + 2t |\langle x, y \rangle| + t^2 \langle y, y \rangle \geq 0.$$

If $\langle y, y \rangle = 0$, explain why $|\langle x, y \rangle| = 0$, thus implying (6.19). If $\langle y, y \rangle > 0$, explain, using the quadratic formula, completing the square, or taking the derivative and setting equal to zero to find an optimal value for t, why the inequality of (6.19) is still valid.

(8) Let V be a complex inner product space and let $\{u_k\}_{k=1}^n$ be an orthonormal basis for V. Show that

$$\langle x, y\rangle = \sum_{k=1}^{n} \langle x, u_k\rangle \overline{\langle y, u_k\rangle}.$$

(9) Let the vector space V consist of real polynomials of degree no larger than 3. Thus a typical vector is a polynomial of the form

$$a + bx + cx^2 + dx^3.$$

For $p, q \in V$ define the inner product,

$$\langle p, q\rangle \equiv \int_0^1 p(x)\, q(x)\, dx.$$

Show that this is indeed an inner product. Then state the Cauchy Schwarz inequality in terms of this inner product. Show that $\{1, x, x^2, x^3\}$ is a basis for V. Finally, find an orthonormal basis for V. This is an example of orthonormal polynomials.

(10) Let P_n denote the polynomials of degree no larger than $n-1$, which are defined on an interval (a, b). Let $\{x_1, \cdots, x_n\}$ be n distinct points in (a, b). Now define for $p, q \in P_n$,

$$\langle p, q\rangle \equiv \sum_{j=1}^{n} p(x_j)\, \overline{q(x_j)}.$$

Show that this yields an inner product on P_n. **Hint:** Most of the axioms are obvious. The one which says $(p, p) = 0$ if and only if $p = 0$ is the only interesting one. To verify this one, note that a nonzero polynomial of degree no more than $n-1$ has at most $n-1$ zeros.

(11) Let $C([0,1])$ denote the vector space of continuous real valued functions defined on $[0,1]$. Let the inner product be given as

$$\langle f, g\rangle \equiv \int_0^1 f(x)\, g(x)\, dx.$$

Show that this is an inner product. Also let V be the subspace described in Problem 9. Using the result of this problem, find the vector in V which is closest to x^4.

(12) Consider the following system of equations

$$\alpha C_1 + \beta C_2 = 0$$
$$\gamma C_1 + \delta C_2 = 0$$

in which it is known that $C_1^2 + C_2^2 \neq 0$. Show that this requires $\alpha\delta - \beta\gamma = 0$.

(13) ↑A **regular Sturm Liouville problem** involves the differential equation, for an unknown function of x which is denoted here by y,

$$(p(x)\, y')' + (\lambda q(x) + r(x))\, y = 0,\ x \in [a, b].$$

It is assumed that $p(t), q(t) > 0$ for any $t \in [a,b]$ and also there are boundary conditions,

$$\begin{aligned} C_1 y(a) + C_2 y'(a) &= 0 \\ C_3 y(b) + C_4 y'(b) &= 0 \end{aligned}$$

where

$$C_1^2 + C_2^2 > 0, \text{ and } C_3^2 + C_4^2 > 0.$$

There is an immense theory connected to these important problems. The constant λ is called an eigenvalue. Show that if y is a solution to the above problem corresponding to $\lambda = \lambda_1$ and if z is a solution corresponding to $\lambda = \lambda_2 \neq \lambda_1$, then

$$\int_a^b q(x) y(x) z(x) \, dx = 0. \tag{6.20}$$

Show (6.20) defines an inner product. **Hint:** Do something like this:

$$\begin{aligned} (p(x) y')' z + (\lambda_1 q(x) + r(x)) yz &= 0, \\ (p(x) z')' y + (\lambda_2 q(x) + r(x)) zy &= 0. \end{aligned}$$

Now subtract and either use integration by parts or show that

$$(p(x) y')' z - (p(x) z')' y = ((p(x) y') z - (p(x) z') y)'$$

and then integrate. Use the boundary conditions to show that $y'(a) z(a) - z'(a) y(a) = 0$ and $y'(b) z(b) - z'(b) y(b) = 0$. The formula, (6.20) is called an orthogonality relation. It turns out that there are typically infinitely many eigenvalues. It is interesting to write given functions as an infinite series of these "eigenfunctions", solutions to the Sturm Liouville problem for a given λ.

(14) Consider the subspace $V \equiv \ker(A)$ where

$$A = \begin{pmatrix} 1 & 4 & -1 & -1 \\ 2 & 1 & 2 & 3 \\ 4 & 9 & 0 & 1 \\ 5 & 6 & 3 & 4 \end{pmatrix}.$$

Find an orthonormal basis for V. **Hint:** You might first find a basis and then use the Gram-Schmidt procedure.

(15) Let W be a subspace of a finite dimensional inner product space V. Also let P denote the map which sends every vector of V to its closest point in W. Show using Theorem 6.2 that P is linear and also that

$$||Px - Py|| \leq ||x - y||.$$

(16) Verify the parallelogram identity for any inner product space,

$$||x + y||^2 + ||x - y||^2 = 2\,||x||^2 + 2\,||y||^2.$$

Why is it called the parallelogram identity?

(17) Let H be an inner product space and let $K \subseteq H$ be a nonempty convex subset. This means that if $k_1, k_2 \in K$, then the line segment consisting of points of the form

$$tk_1 + (1-t)\,k_2 \text{ for } t \in (0,1)$$

is also contained in K. Suppose, for each $x \in H$, there exists Px defined to be a point of K closest to x. Show that Px is unique, so that P actually is a map. **Hint:** Suppose z_1 and z_2 both work as closest points. Consider the midpoint, $(z_1 + z_2)/2$ and use the parallelogram identity of Problem 16 in an auspicious manner.

(18) Give an example of two vectors $\mathbf{x}, \mathbf{y}$ in $\mathbb{R}^4$ and a subspace V such that $\langle \mathbf{x}, \mathbf{y} \rangle = 0$ but $\langle P\mathbf{x}, P\mathbf{y} \rangle \neq 0$ where P denotes the projection map which sends $\mathbf{x}$ to its closest point on V.

(19) Suppose you are given the data, $(1,2), (2,4), (3,8), (0,0)$. Find the linear regression line using the formulas derived above. Then graph the given data along with your regression line.

(20) Generalize the least squares procedure to the situation in which data is given and you desire to fit it with an expression of the form $y = af(x) + bg(x) + c$ where the problem would be to find a, b and c in order to minimize the sum of the squares of the error. Could this be generalized to higher dimensions? How about more functions?

(21) Let A be an $m \times n$ matrix of complex entries. Thus $A : \mathbb{C}^n \to \mathbb{C}^m$. Show that

$$\langle A\mathbf{x}, \mathbf{y} \rangle = \langle \mathbf{x}, A^*\mathbf{y} \rangle \tag{6.21}$$

if and only if $A^* = \overline{A^t}$. This shows that A^* can be defined in terms of the above equation. Note that if A has all real entries, A^* reduces to A^t.

(22) ↑For A an $m \times n$ matrix of complex entries, show $\ker(A^*A) = \ker(A)$.

(23) ↑Show using only the equation (6.21) that $(AB)^* = B^*A^*$ without making any reference to the entries of A.

(24) ↑Let A be an $m \times n$ matrix having complex entries. Show that the rank of A equals the rank of A^*A. Next justify the following inequality to conclude that the rank of A equals the rank of A^*.

$$\operatorname{rank}(A) = \operatorname{rank}(A^*A) \leq \operatorname{rank}(A^*)$$

$$= \operatorname{rank}(AA^*) \leq \operatorname{rank}(A).$$

Hint: Start with an orthonormal basis, $\{A\mathbf{x}_j\}_{j=1}^r$ of $A(\mathbb{C}^n)$ and verify that $\{A^*A\mathbf{x}_j\}_{j=1}^r$ is a basis for $A^*A(\mathbb{C}^n)$. You might also want to use the result of Problem 22 and Theorem 5.1.

(25) Let V be a vector space and let L be a subspace. Sketch a representative picture of L. Now sketch a representative picture of $L + x$ defined as

$$L + x = \{l + x : l \in L\}.$$

Why is $x \in L + x$? If $x_1 \in L + x$, is $L + x_1 = L + x$?

(26) Suppose V is an inner product space and L is a nonzero subspace. Define

$$M = x + L.$$

Show that there exists a point m of M which is closest to 0 and that this m is given by the following formula.

$$x - \sum_j \langle x, u_j \rangle u_j$$

where $\{u_1, \cdots, u_r\}$ is an orthonormal basis for L. **Hint:** First explain why a typical point $m \in M$ is of the form $x - \sum_j a_j u_j$. Thus it is desired to find the a_j in order to minimize $\left\| x - \sum_j a_j u_j \right\|$. Now review Theorem 6.2.

(27) ↑Let A be an $m \times n$ matrix and consider the set

$$M_y = \{\mathbf{x} \in \mathbb{C}^n : A^* A\mathbf{x} = A^*\mathbf{y}\}.$$

Thus M_y is the set of least squares solutions to the equation

$$A\mathbf{x} = \mathbf{y}.$$

Show $M_y = \mathbf{x}_0 + \ker(A^*A)$ where $\mathbf{x}_0$ is any vector in M_y.

(28) ↑Using Problem 26 explain why, for A an $m \times n$ matrix, there exists a unique point $A^+\mathbf{y}$ of M_y, the set of least squares solutions to

$$A\mathbf{x} = \mathbf{y}$$

which is closest to $\mathbf{0}$. This point is defined by the formula

$$A^+\mathbf{y} = \mathbf{x} - \sum_{j=1}^{r} \langle \mathbf{x}, \mathbf{u}_j \rangle \mathbf{u}_j.$$

where $\mathbf{x}$ is **any** point of M_y and $\{\mathbf{u}_1, \cdots, \mathbf{u}_r\}$ is an orthonormal basis for $\ker(A^*A)$.

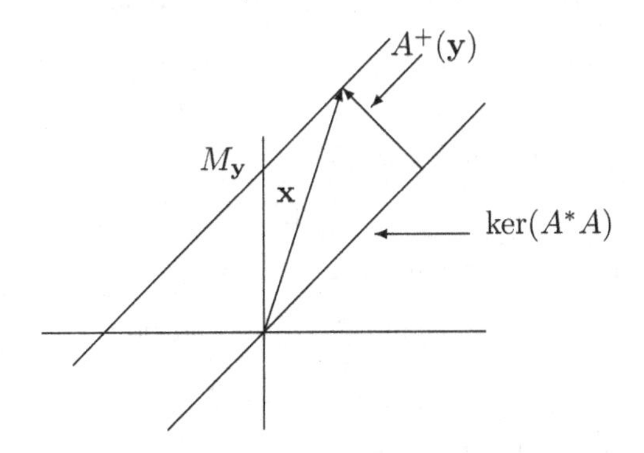

In a sense, this point is the 'best' least squares solution. The mapping A^+ just described is called the Moore Penrose inverse. Some of its properties are described in the following problems.

(29) ↑In the above problem, show that $A^+ : \mathbb{R}^m \to \mathbb{R}^n$ is a linear mapping and therefore, can be considered as a matrix which will also be denoted by A^+. **Hint:** First show that if $\mathbf{x}_1, \mathbf{x}_2$ are least squares solutions to $A\mathbf{x} = \mathbf{y}_1$ and $A\mathbf{x} = \mathbf{y}_2$ respectively, and if a, b are scalars, then $a\mathbf{x}_1 + b\mathbf{x}_2$ is a least squares solution to $A = a\mathbf{y}_1 + b\mathbf{y}_2$. Now use the above formula for $A^+\mathbf{y}$.

(30) ↑The Penrose conditions for the Moore Penrose inverse are the following.

$$AA^+A = A,\ A^+AA^+ = A^+,$$

and A^+A and AA^+ are **Hermitian**, meaning

$$(A^+A)^* = A^+A,\ (AA^+)^* = AA^+.$$

It can be shown that these conditions characterize the Moore Penrose inverse in the sense that the Moore Penrose inverse is the unique matrix which satisfies these conditions. This is discussed more conveniently later in the book in the context of the singular value decomposition. Show that if A is an $n \times n$ invertible matrix, the Penrose conditions hold for A^{-1} in place of A^+.

(31) ↑Recall that

$$A^+\mathbf{y} = \mathbf{x} - \sum_{j=1}^{r} \langle \mathbf{x}, \mathbf{u}_j \rangle \mathbf{u}_j$$

where the $\mathbf{u}_j$ are an orthonormal basis for $\ker(A^*A)$ and $\mathbf{x}$ is any least squares solution to $A\mathbf{x} = \mathbf{y}$. In particular, $A^+\mathbf{y} = A^+\mathbf{y} - \sum_{j=1}^{r} \langle A^+\mathbf{y}, \mathbf{u}_j \rangle \mathbf{u}_j$. Using this, explain why $A^+\mathbf{y}$ is the least squares solution to $A\mathbf{x} = \mathbf{y}$ which satisfies

$$A^*A(A^+\mathbf{y}) = A^*\mathbf{y},\ \langle A^+\mathbf{y}, \mathbf{z} \rangle = 0 \text{ for all } \mathbf{z} \in \ker(A^*A). \tag{6.22}$$

(32) ↑Establish the first Penrose condition,

$$AA^+A = A.$$

Hint: Follow the steps below.

(a) Using the description of the Moore Penrose inverse in Problem 31, show that

$$A^*AA^+A\mathbf{x} = A^*A\mathbf{x}.$$

(b) Show next that for any $\mathbf{y} \in \mathbb{C}^n$,

$$\langle (AA^+A - A)\mathbf{x}, A\mathbf{y} \rangle = 0.$$

(c) In the above equality, let $\mathbf{y} = (A^+A - I)\mathbf{x}$.

(33) ↑Now establish the second Penrose condition

$$A^+AA^+ = A^+.$$

Hint: Follow the steps below.

(a) From the first Penrose condition, explain why

$$A^*AA^+AA^+\mathbf{y} = A^*AA^+\mathbf{y}$$

(b) Now explain why $A^+ (AA^+\mathbf{y} - \mathbf{y}) \in \ker(A^*A)$.

(c) Using the second condition in (6.22), explain why

$$\langle A^+AA^+\mathbf{y} - A^+\mathbf{y}, A^+AA^+\mathbf{y} - A^+\mathbf{y}\rangle = 0.$$

(34) Find the least squares solution to

$$\begin{pmatrix} 1 & 1 \\ 1 & 1 \\ 1 & 1+\varepsilon \end{pmatrix} \begin{pmatrix} x \\ y \end{pmatrix} = \begin{pmatrix} a \\ b \\ c \end{pmatrix}.$$

Next suppose ε is so small that all ε^2 terms are ignored by the computer but the terms of order ε are not ignored. Show the least squares equations in this case reduce to

$$\begin{pmatrix} 3 & 3+\varepsilon \\ 3+\varepsilon & 3+2\varepsilon \end{pmatrix} \begin{pmatrix} x \\ y \end{pmatrix} = \begin{pmatrix} a+b+c \\ a+b+(1+\varepsilon)c \end{pmatrix}.$$

Find the solution to this and compare the y values of the two solutions. Show that one of these is -2 times the other. This illustrates a problem with the technique for finding least squares solutions presented in this chapter. One way of dealing with this problem is to use the QR factorization which is presented later in the book but also in the first problem above.

(35) For A an $m \times n$ matrix with $m \geq n$ and the rank of A equaling n, Problem 2 above shows that there exists a QR factorization for A. Show that the equations $A^*A\mathbf{x} = A^*\mathbf{y}$ can be written as $R^*R\mathbf{x} = R^*Q^*\mathbf{y}$ where R is upper triangular and R^* is lower triangular. Explain how to solve this system efficiently. **Hint:** You first find $R\mathbf{x}$ and then you find $\mathbf{x}$.

(36) Let $\{b_1, \cdots, b_n\}$ be a basis in an inner product space. Thus any vector v can be written in the form $v = \sum_{i=1}^n v_i b_i$ where x_i is in the complex numbers. Write a formula for $\langle v, w\rangle$ in terms of the components taken with respect the given basis. In particular, show that

$$\langle v, w\rangle = \begin{pmatrix} v_1 & \cdots & v_n \end{pmatrix} G \begin{pmatrix} \overline{w_1} \\ \vdots \\ \overline{w_n} \end{pmatrix}$$

where G is the **Gramian matrix** whose ij^{th} entry is $\langle b_i, b_j\rangle$. This is also called the **metric tensor.**

(37) ↑Let $\{b_1, \cdots, b_n\}$ be vectors in an inner product space. Show that the set of vectors is linearly independent if and only if the Gramian has rank n.

(38) ↑Let $\{b_1, \cdots, b_n\}$ be a basis for an inner product space. Show that there exists a unique dual basis $\{b^1, \cdots, b^n\}$ satisfying $\langle b^i, b_j\rangle = \delta_{ij}$. (In this context, δ_{ij} is usually written as δ^i_j. It equals 1 if $i = j$ and 0 if $i \neq j$.) In fact, $b^i = \sum_k G^{ik} b_k$ where G^{ik} is the ik^{th} entry of G^{-1}.

(39) ↑Let $\{b_1, \cdots, b_n\}$ be a basis for an inner product space and let $\{b^1, \cdots, b^n\}$ be the dual basis. Show that if v is any vector, $v = \sum_{i=1}^n \langle v, b^i\rangle b_i = \sum_{i=1}^n \langle v, b_i\rangle b^i$.

(40) Let $\{b_1, \cdots, b_n\}$ be linearly independent vectors in a real inner product space and let y be a vector in this inner product space. Show that $||y - \sum_{k=1}^{n} a_k b_k||$ is as small as possible when $\mathbf{a} \equiv \begin{pmatrix} a_1 \\ \vdots \\ a_n \end{pmatrix}$ is given by $\mathbf{a} = G^{-1} \begin{pmatrix} \langle y, b_1 \rangle \\ \vdots \\ \langle y, b_n \rangle \end{pmatrix}$ for G the Gramian.

(41) Let $\{b_1, \cdots, b_n\}$ be a basis for an inner product space H. We say $L \in H'$ if L is a linear transformation which maps H to the complex numbers. Show that there exists a unique $z \in H$ such that $Lv = \langle v, z \rangle$ for all $v \in H$. This is called the **Riesz representation theorem.**

(42) ↑With the Riesz representation theorem, it becomes possible to discuss the adjoint of a linear transformation in the correct way. Let $A : H \to F$ where H and F are finite dimensional inner product spaces and A is linear. Show that there exists a unique $A^* : F \to H$ such that $\langle Ax, y \rangle_F = \langle x, A^* y \rangle_H$ and A^* is linear. Also show that $(AB)^* = B^* A^*$ directly from this definition.

(43) Let S be a nonempty set in an inner product space. Let $S^\perp$ denote the set $S^\perp = \{x : \langle x, s \rangle = 0 \text{ for all } s \in S\}$. Show that $S^\perp$ is always a subspace.

(44) Let H, F be two finite dimensional inner product spaces and let A be a linear transformation which maps H to F. The **Fredholm alternative** says that there exists a solution to $A\mathbf{x} = \mathbf{y}$ if and only if $\mathbf{y} \in \ker(A^*)^\perp$. Here $S^\perp$ is defined as the collection of all vectors which have the property that the inner product of the vector with any vector of S equals 0. Prove the Fredholm alternative. Also show that A is onto, if and only if A^* is one-to-one.

(45) You can consider **any** complex finite dimensional vector space as an inner product space. Here is how you can do this. Start with a basis $\{b_1, \cdots, b_n\}$. Then **DECREE** that $\langle b_i, b_j \rangle = \delta_{ij}$. Then if u, w are any two vectors, it follows that $\langle u, w \rangle$ **must** equal $\sum_i u_i \overline{w_i}$ where $u = \sum_i u_i b_i$, similar for w. Show that this satisfies all the axioms of an inner product space. When you do this, you are bestowing a geometric structure on something which is only algebraic.

Chapter 7

Similarity and determinants

Chapter summary

We discuss the matrix of a linear transformation relative to a given arbitrary basis and the concept of similar matrices. This is used in an application to the problem of revolving about a given vector not necessarily parallel to any of the coordinate vectors. We discuss diagonalizable matrices. After this, the theory of determinants is presented, including all the major theorems about products, transposes, expansion along a row or column, and inverses of matrices. An application is given to Vandermonde determinants. Next is a discussion of Wronskians and independence. Then we use the theory of determinants to give a short proof of the important Cayley-Hamilton theorem. Finally there is a careful treatment of block multiplication of matrices.

7.1 Change of basis

Let V be a vector space over F and consider two bases $\beta = \{v_1, \ldots, v_n\}, \gamma = \{w_1, \ldots, w_n\}$ of V. Recall that for v a vector, the coordinate vector is given by

$$[v]_\beta = \begin{pmatrix} x_1 & \cdots & x_n \end{pmatrix}^t,$$

where

$$v = \sum_{i=1}^{n} x_i v_i.$$

Lemma 7.1. *The mapping $\theta_\beta(v)$ defined as $[v]_\beta$ is a well defined linear bijection from V to F^n.*

Proof. First observe that this mapping is well defined. This is because

$$\{v_1, \ldots, v_n\}$$

is a basis. If

$$0 = \sum_{j=1}^{n} x_j v_j - \sum_{j=1}^{n} y_j v_j = \sum_{j=1}^{n} (x_j - y_j) v_j,$$

so that $\mathbf{x}$ and $\mathbf{y}$ are both coordinate vectors for v, then by linear independence, it follows that $x_i - y_i = 0$ for all i.

Next consider the claim that this mapping is linear. Letting a, b be scalars, and v, w be vectors such that $\mathbf{x} = [v]_\beta$ and $\mathbf{y} = [v]_\beta$,

$$\begin{aligned}\theta_\beta (av + bw) &= \left[a\sum_{j=1}^{n} x_j v_j + b\sum_{j=1}^{n} y_j v_j \right]_\beta = \left[\sum_{j=1}^{n} (ax_j + by_j) v_j \right]_\beta \\ &= a\mathbf{x} + b\mathbf{y} = a\theta_\beta (v) + b\theta_\beta (v)\end{aligned}$$

Now consider the claim that the mapping is one-to-one. If

$$\mathbf{x} = [v]_\beta = \theta_\beta (v) = 0,$$

then from the definition,

$$v = \sum_{i=1}^{n} x_i v_i = \sum_{i=1}^{n} 0 v_i = 0,$$

and so $v = 0$. It follows that the mapping is one-to-one as claimed.

If $\mathbf{x} \in F^n$ is arbitrary, consider

$$v = \sum_{i=1}^{n} x_i v_i.$$

Then by definition,

$$\theta_\beta (v) = [v]_\beta = \mathbf{x},$$

and so the mapping θ_β is onto. □

Let

$$[v]_\beta, [v]_\gamma$$

be two coordinate vectors for the same vector v in V corresponding to two different bases β and γ. What is the relation between the coordinate vectors $[v]_\beta, [v]_\gamma$?

Lemma 7.2. *Let $\beta = \{v_1, \ldots, v_n\}, \gamma = \{w_1, \ldots, w_n\}$ be two bases for V. There exists an invertible matrix P such that*

$$[v]_\beta = P [v]_\gamma .$$

This matrix is given by $P = (p_{ji})$ where

$$w_i = \sum_{j=1}^{n} p_{ji} v_j. \tag{7.1}$$

Furthermore,

$$[v]_\gamma = P^{-1} [v]_\beta .$$

Proof. $[v]_\beta = \theta_\beta v = \theta_\beta \theta_\gamma^{-1} [v]_\gamma$. This shows that the mapping $[v]_\beta \to [v]_\gamma$ is a linear bijection and hence has an invertible matrix P associated to it. By Lemma 5.1, column i of P equals $\theta_\beta \theta_\gamma^{-1} \mathbf{e}_i$, that is, $\theta_\beta w_i$. This vector is $\begin{pmatrix} p_{1i} & \cdots & p_{ni} \end{pmatrix}^t$ by (7.1). Thus $[v]_\beta = P [v]_\gamma$ and so, multiplying both sides by P^{-1}, it follows that $[v]_\gamma = P^{-1} [v]_\beta$. □

Next we consider the matrix of a linear transformation.

Theorem 7.1. *Let T be a linear transformation mapping V to V where V is an n dimensional vector space. Then letting $\beta = \{v_1, \cdots, v_n\}$ be a basis for V, there exists a unique $n \times n$ matrix $[T]_\beta = (a_{ij})$ which satisfies*

$$[Tv]_\beta = [T]_\beta [v]_\beta .$$

This matrix is defined by

$$Tv_i = \sum_{k=1}^{n} a_{ki} v_k .$$

In terms of the coordinate map for the basis β,

$$[T]_\beta \mathbf{x} = \theta_\beta T \theta_\beta^{-1} (\mathbf{x}) .$$

Proof. The existence of the matrix follows from the following computation which comes directly from the definitions. Let $v \in V$ be arbitrary.

$$[Tv]_\beta = \left[T \left(\sum_{i=1}^{n} \left([v]_\beta \right)_i v_i \right) \right]_\beta = \left[\sum_{i=1}^{n} \left([v]_\beta \right)_i Tv_i \right]_\beta$$

$$= \left[\sum_{i=1}^{n} \left([v]_\beta \right)_i \sum_{k=1}^{n} a_{ki} v_k \right]_\beta = \left[\sum_{k=1}^{n} \left(\sum_{i=1}^{n} a_{ki} \left([v]_\beta \right)_i \right) v_k \right]_\beta .$$

It follows from the definition of the coordinate vector that

$$\left([Tv]_\beta \right)_k = \sum_{i=1}^{n} a_{ki} \left([v]_\beta \right)_i ,$$

and this implies that for all $v \in V$,

$$[Tv]_\beta = A [v]_\beta ,$$

where A is the matrix (a_{ij}).

Consider the claim of uniqueness. Let B be another $n \times n$ matrix which satisfies the above equation. By Lemma 7.1, the map $v \to [v]_\beta$ is onto, it follows that for all $\mathbf{x} \in F^n$,

$$A\mathbf{x} = B\mathbf{x},$$

which implies $A = B$. This proves the first part with $[T]_\beta$ defined as A.

Finally, the last claim follows from

$$\left[T\theta_\beta^{-1} \mathbf{x} \right]_\beta = [T]_\beta \left[\theta_\beta^{-1} \mathbf{x} \right]_\beta = [T]_\beta \theta_\beta \theta_\beta^{-1} \mathbf{x} = [T]_\beta \mathbf{x}. \ \square$$

We say that T **has matrix** $[T]_\beta$ **for the basis** β.

Example 7.1. Let $F = \mathbb{R}, \mathbf{w}_1 = (1,2), \mathbf{w}_2 = (2,3)$. Let T be defined by

$$T(x_1\mathbf{w}_1 + x_2\mathbf{w}_2) = 6x_1\mathbf{w}_1 + 5x_2\mathbf{w}_2. \tag{7.2}$$

Then, writing $\beta = \{\mathbf{w}_1, \mathbf{w}_2\}$,

$$[T]_\beta = \begin{pmatrix} 6 & 0 \\ 0 & 5 \end{pmatrix}.$$

This is because

$$\begin{pmatrix} 6 & 0 \\ 0 & 5 \end{pmatrix}\begin{pmatrix} x_1 \\ x_2 \end{pmatrix} = \begin{pmatrix} 6x_1 \\ 5x_2 \end{pmatrix} = [T(x_1\mathbf{w}_1 + x_2\mathbf{w}_2)]_\beta .$$

A natural question is: what is the relation between $[T]_\beta$ and $[T]_\gamma$ for distinct bases β and γ of V?

Lemma 7.3. *Let β, γ be the bases $\{v_1, \ldots, v_n\}, \{w_1, \ldots, w_n\}$ of V, related by*

$$[w]_\beta = P\,[w]_\gamma .$$

Then

$$[T]_\gamma = P^{-1}\,[T]_\beta\,P.$$

Proof. This follows from the following computation.

$$P^{-1}\,[T]_\beta\,P\,[v]_\gamma = P^{-1}\,[T]_\beta\,[v]_\beta = P^{-1}\,[Tv]_\beta = P^{-1}P\,[Tv]_\gamma = [Tv]_\gamma = [T]_\gamma\,[v]_\gamma$$

By Lemma 7.1, which says θ_γ is onto, it follows that

$$P^{-1}\,[T]_\beta\,P = [T]_\gamma . \ \square$$

We also have the following simple lemma.

Lemma 7.4. *Let V be a finite dimensional vector space with basis β and let A, B be two linear mappings from V to V. Then for a, b scalars,*

$$[aA + bB]_\beta = a\,[A]_\beta + b\,[B]_\beta .$$

Also, if id *is the identity linear transformation, then for any basis,*

$$[\mathrm{id}]_\beta = I,$$

the identity matrix.

Proof. This follows right away from the observation that the coordinate maps and their inverses are linear. Thus

$$[aA + bB]_\beta = \theta_\beta\,(aA + bB)\,\theta_\beta^{-1} = a\theta_\beta A\theta_\beta^{-1} + b\theta_\beta B\theta_\beta^{-1} = a\,[A]_\beta + b\,[B]_\beta .$$

As for the last claim,

$$[\mathrm{id}]_\beta = \theta_\beta\,\mathrm{id}\,\theta_\beta^{-1} = I. \ \square$$

Example 7.2. With $\beta = \{\mathbf{w}_1, \mathbf{w}_2\}$ and T as in Example 7.1, let $\mathbf{v}_1 = \mathbf{e}_1, \mathbf{v}_2 = \mathbf{e}_2, \gamma = \{\mathbf{e}_1, \mathbf{e}_2\}$. Thus

$$P = \begin{pmatrix} 1 & 2 \\ 2 & 3 \end{pmatrix}.$$

We obtain $P^{-1} = \begin{pmatrix} -3 & 2 \\ 2 & -1 \end{pmatrix}$ after a simple calculation. So

$$P^{-1}AP = \begin{pmatrix} 6 & 0 \\ 0 & 5 \end{pmatrix},$$

$$A = P \begin{pmatrix} 6 & 0 \\ 0 & 5 \end{pmatrix} P^{-1} = \begin{pmatrix} 2 & 2 \\ -6 & 9 \end{pmatrix}.$$

Here A is the matrix of T with respect to the standard basis (which we usually call 'the matrix of T', of course).

Check

$$\begin{aligned} T\mathbf{w}_1 &= \begin{pmatrix} 2 & 2 \\ -6 & 9 \end{pmatrix} \begin{pmatrix} 1 \\ 2 \end{pmatrix} = \begin{pmatrix} 6 \\ 12 \end{pmatrix} = 6\mathbf{w}_1, \\ T\mathbf{w}_2 &= \begin{pmatrix} 2 & 2 \\ -6 & 9 \end{pmatrix} \begin{pmatrix} 2 \\ 3 \end{pmatrix} = \begin{pmatrix} 10 \\ 15 \end{pmatrix} = 5\mathbf{w}_2, \end{aligned}$$

(using A), which conforms with the definition of T.

A very non - trivial example which is of interest in computer graphics is the problem of rotating all vectors about a given unit vector.

Consider three unit vectors $\mathbf{u}_1, \mathbf{u}_2, \mathbf{u}_3$ which form a right handed system and are mutually orthogonal. If T is the transformation which rotates all vectors about $\mathbf{u}_3$ through an angle θ, it follows from analogy to the case of rotating about the vector $\mathbf{k}$ that

$$\begin{aligned} T\mathbf{u}_1 &= \mathbf{u}_1 \cos\theta + \mathbf{u}_2 \sin\theta \\ T\mathbf{u}_2 &= -\mathbf{u}_1 \sin\theta + \mathbf{u}_2 \cos\theta \\ T\mathbf{u}_3 &= \mathbf{u}_3 \end{aligned}$$

Thus the matrix of this rotation with respect to the basis $\gamma = \{\mathbf{u}_1, \mathbf{u}_2, \mathbf{u}_3\}$ is

$$[T]_\gamma = \begin{pmatrix} \cos\theta & -\sin\theta & 0 \\ \sin\theta & \cos\theta & 0 \\ 0 & 0 & 1 \end{pmatrix}.$$

With $\beta = \{\mathbf{e}_1, \mathbf{e}_2, \mathbf{e}_3\}$, the problem is to find $[T]_\beta$. By Theorem 7.1, this matrix is given by

$$[T]_\beta = \theta_\beta T \theta_\beta^{-1} = \theta_\beta \theta_\gamma^{-1} [T]_\gamma \theta_\gamma \theta_\beta^{-1}$$

$$= \begin{pmatrix} \mathbf{u}_1 & \mathbf{u}_2 & \mathbf{u}_3 \end{pmatrix} [T]_\gamma \begin{pmatrix} \mathbf{u}_1 & \mathbf{u}_2 & \mathbf{u}_3 \end{pmatrix}^{-1}.$$

It follows that $[T]_\beta$ is

$$\begin{pmatrix} \mathbf{u}_1 & \mathbf{u}_2 & \mathbf{u}_3 \end{pmatrix} \begin{pmatrix} \cos\theta & -\sin\theta & 0 \\ \sin\theta & \cos\theta & 0 \\ 0 & 0 & 1 \end{pmatrix} \begin{pmatrix} \mathbf{u}_1 & \mathbf{u}_2 & \mathbf{u}_3 \end{pmatrix}^{-1}. \tag{7.3}$$

Note that, since the vectors are mutually orthogonal, the inverse on the right may be computed by taking the transpose. This gives the matrix which rotates about a given vector $\mathbf{u}_3$, provided $\mathbf{u}_1$ and $\mathbf{u}_2$ can be constructed such that $\mathbf{u}_1, \mathbf{u}_2, \mathbf{u}_3$ form a right handed system. We consider this now.

Recall the geometric description of the cross product in Problem 16 on Page 10. Given the vectors $\mathbf{x}, \mathbf{y}$, the cross product $\mathbf{x} \times \mathbf{y}$ satisfies the following conditions

$$|\mathbf{x} \times \mathbf{y}| = |\mathbf{x}| \, |\mathbf{y}| \sin\theta$$

where θ is the included angle, and $\mathbf{x}, \mathbf{y}, \mathbf{x} \times \mathbf{y}$ forms a right handed system with $\mathbf{x} \times \mathbf{y}$ being perpendicular to both $\mathbf{x}$ and $\mathbf{y}$.

Let $\mathbf{u}_3 = (a, b, c)$ a given unit vector. We will assume for the sake of convenience that $|c| < 1$. A unit vector perpendicular to the given vector is $\mathbf{u}_1 = \frac{(-b,a,0)}{\sqrt{a^2+b^2}}$. Now it remains to choose $\mathbf{u}_2$, a unit vector, such that $\mathbf{u}_1, \mathbf{u}_2, \mathbf{u}_3$ is a right hand system and the vectors are mutually orthogonal.

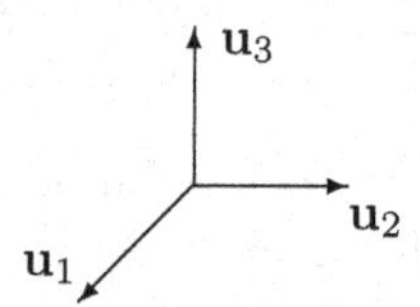

Using the right hand rule for the cross product, such a vector is $\mathbf{u}_3 \times \mathbf{u}_1 =$

$$\mathbf{u}_2 = (a, b, c) \times \frac{(-b, a, 0)}{\sqrt{a^2 + b^2}} = \frac{-1}{\sqrt{a^2 + b^2}} \left(ac, bc, c^2 - 1\right),$$

where we used the fact that $a^2 + b^2 + c^2 = 1$. It follows that the matrix of interest is

$$\begin{pmatrix} \frac{-b}{\sqrt{a^2+b^2}} & \frac{-ac}{\sqrt{a^2+b^2}} & a \\ \frac{a}{\sqrt{a^2+b^2}} & \frac{-bc}{\sqrt{a^2+b^2}} & b \\ 0 & \frac{1-c^2}{\sqrt{a^2+b^2}} & c \end{pmatrix} \begin{pmatrix} \cos\theta & -\sin\theta & 0 \\ \sin\theta & \cos\theta & 0 \\ 0 & 0 & 1 \end{pmatrix} \begin{pmatrix} \frac{-b}{\sqrt{a^2+b^2}} & \frac{-ac}{\sqrt{a^2+b^2}} & a \\ \frac{a}{\sqrt{a^2+b^2}} & \frac{-bc}{\sqrt{a^2+b^2}} & b \\ 0 & \frac{1-c^2}{\sqrt{a^2+b^2}} & c \end{pmatrix}^t.$$

Doing the routine computations and simplifications, it follows that $[T]_\beta$ equals

$$= \begin{pmatrix} a^2 + \left(1 - a^2\right)\cos\theta & ab\,(1 - \cos\theta) - c\sin\theta & ac\,(1 - \cos\theta) + b\sin\theta \\ ab\,(1 - \cos\theta) + c\sin\theta & b^2 + \left(1 - b^2\right)\cos\theta & bc\,(1 - \cos\theta) - a\sin\theta \\ ac\,(1 - \cos\theta) - b\sin\theta & bc\,(1 - \cos\theta) + a\sin\theta & c^2 + \left(1 - c^2\right)\cos\theta \end{pmatrix}.$$

This also gives the correct answer if $|c| = 1$ as can be verified directly. In this case, $c = \pm 1$ and both $a, b = 0$. Summarizing, if $\mathbf{u} = (a, b, c)$ is a unit vector in $\mathbb{R}^3$, rotation about $\mathbf{u}$ may be accomplished by multiplying the vector of $\mathbb{R}^3$ on the left by the above matrix.

Definition 7.1. We say that $n \times n$ matrices A and B over F are **similar over** F (or, A is similar to B over F) if $A = PBP^{-1}$ for an invertible $n \times n$ matrix P over F.

Note that $B = P^{-1}AP = QAQ^{-1}$ with $Q = P^{-1}$, so the order of A and B in defining similarity is of no consequence. Lemma 7.3 says that the change of basis from β to γ has the effect of replacing $[T]\beta$ by a similar matrix $[T]\gamma$.

Definition 7.2. An $n \times n$ matrix of the form

$$D = \begin{pmatrix} d_1 & & \\ & \ddots & \\ & & d_n \end{pmatrix}$$

(that is, $D = [d_{ij}]$ with $d_{ij} = 0$ for $i \neq j, d_{ii} = d_i$) is said to be **diagonal.**

Be careful in using the convention, adopted from here on, that 'blank' entries in a matrix are 0.

Definition 7.3. If the matrix A over F is similar over F to a diagonal matrix D over F, we say that A is **diagonalizable** over F.

Thus in Examples 1, 2, $\begin{pmatrix} 2 & 2 \\ -6 & 9 \end{pmatrix}$ is diagonalizable. The point of the definition is that the linear mapping $T\mathbf{x} = A\mathbf{x}$ 'looks much simpler after a change of basis', as in (7.2). But how are we to tell which matrices are diagonalizable? The answer (find the eigenvalues and eigenvectors) must be postponed until Chapter 8.

The repetition of the phrase 'over F' in the above definition is necessary. We shall see examples where a real matrix A is diagonalizable over $\mathbb{C}$ but not $\mathbb{R}$. That is, we can write

$$A = PDP^{-1}.$$

only by allowing complex entries on the right-hand side.

7.2 Determinants

Every square matrix A over a field F has a *determinant*, an element $\det A$ of F. It turns out that $\det A \neq 0$ if A is invertible and $\det A = 0$ if A is singular. To take a simple example,

$$\det \begin{pmatrix} a_{11} & a_{12} \\ a_{21} & a_{22} \end{pmatrix} = a_{11}a_{22} - a_{12}a_{21}.$$

The simplest way to describe determinants in the $n \times n$ case is to use permutations. A **permutation** of $1, \ldots, n$ is a rearrangement $\{p(1), p(2), \ldots, p(n)\}$ of

$J_n = \{1, \ldots, n\}$. Formally, we have $p : J_n \to J_n$and p is a bijection. For example, when $n = 3$ the notation

$$p : 2, 1, 3$$

means that $p(1) = 2, p(2) = 1, p(3) = 3$.

A permutation contains a number (possibly 0) of **disordered pairs**, that is pairs of integers listed in reverse order. Such a pair need not be adjacent. In

$$p : 2, 1, 3, 4 \tag{7.4}$$

there is one disordered pair, $(2, 1)$. In

$$p' : 4, 3, 2, 1$$

there are six: $(4, 3), (4, 2), (4, 1), (3, 2), (3, 1), (2, 1)$.

Definition 7.4. The permutation p is **even** if the number of disordered pairs in p is even. Otherwise, p is **odd**.

A useful fact is that the transposition of two integers alters the **parity** (changes an even permutation to odd and vice versa). This with p as in (7.4),

$$p'' : 4, 1, 3, 2$$

is obtained from p by a transposition, and has 4 disordered pairs. So p is odd, p'' is even.

Lemma 7.5. *Suppose that p' is obtained from*

$$p : p(1), p(2), \ldots, p(n)$$

by a transposition. Then p and p' have opposite parity.

Proof. Say $p(i), p(j)$ are interchanged to produce p'. Pairs which change places are $(p(i), p(j))$ and $(p(i), p(k)), (p(j), p(k))$ for all $k, i < k < j$. That is an odd number of place changes, so an odd number of ± 1's is added to the disordered pair count. □

For example, consider $J_4 = \{1, 2, 3, 4\}$, and switch the 3 and the 1. This yields

$$p : 3\ 2\ 1\ 4$$

There are now three disordered pairs, $(3, 2), (3, 1), (2, 1)$ where initially there were none.

Lemma 7.6. *There are $n!$ permutations of $1, \ldots, n$, where $n! = n \cdot (n-1) \cdots 2 \cdot 1$.*

Proof. This is obvious if $n = 1$. Suppose it is so for $k \leq n - 1$ and consider the permutations of $\{1, 2, \cdots, n\}$. There are n choices for the last number in the list and for each of these choices, induction implies there are $(n-1)!$ ways to order the remaining numbers. Thus there are $n!$ permutations in all. □

Definition 7.5. The **determinant** of

$$A = \begin{pmatrix} a_{11} & \cdots & a_{1n} \\ \vdots & & \\ a_{n1} & \cdots & a_{nn} \end{pmatrix}$$

is

$$\det A = \sum_p (-1)^p a_{ip(1)} \ldots a_{np(n)}. \tag{7.5}$$

Here

(i) the sum is over all $n!$ permutations p of J_n;
(ii) $(-1)^p$ means 1 if p is even and -1 if p is odd.

Note that the products which appear in (7.5) contain one factor from each row of A, and also contain one factor from each column of A.

Example 7.3. Let $n = 1$. The only permutation p has $p(1) = 1$ and p is even. So

$$\det (a_{11}) = a_{11}.$$

Example 7.4. Let $n = 2$. The permutations are

$$p : 1, 2 \text{ (even)}$$
$$p' : 2, 1 \text{ (odd)}.$$

So

$$\det \begin{pmatrix} a_{11} & a_{12} \\ a_{21} & a_{22} \end{pmatrix} = a_{1p(1)}a_{2p(2)} - a_{1p'(1)}a_{2p'(2)} = a_{11}a_{22} - a_{12}a_{21}.$$

Example 7.5. Let $n = 3$. The permutations are:

$$\begin{array}{llll} p_1 : 1, 2, 3 & p_2 : 2, 3, 1 & p_3 : 3, 1, 2 & \text{(even)}, \\ p_4 : 1, 3, 2 & p_5 : 2, 1, 3 & p_6 : 3, 2, 1 & \text{(odd)}. \end{array}$$

Listing terms in (7.5) in the order $p_1, \ldots, p_6$,

$$\det \begin{pmatrix} a_{11} & a_{12} & a_{13} \\ a_{21} & a_{22} & a_{23} \\ a_{31} & a_{32} & a_{33} \end{pmatrix} = a_{11}a_{22}a_{33} + a_{12}a_{23}a_{31} + a_{13}a_{21}a_{32}$$
$$- a_{11}a_{23}a_{32} - a_{12}a_{21}a_{33} - a_{13}a_{22}a_{31}.$$

Evaluating $\det A$ from the definition becomes unworkable as n increases. Determinants (whether computed by hand or via Maple) are actually found using row reduction.

Lemma 7.7. *Let B be obtained from the $n \times n$ matrix A by interchanging row i and row j. Then*

$$\det B = -\det A.$$

Proof. A product of the form

$$b_{1p(1)} \cdots b_{np(n)} \tag{7.6}$$

is equal to

$$a_{1p'(1)} \cdots a_{np'(n)}, \tag{7.7}$$

where p' is obtained from p by transposing $p(i), p(j)$. To see this, $p(k) = p'(k)$ for $k \notin \{i, j\}, p(i) = p'(j)$ and $p(j) = p'(i)$,

$$b_{jp(j)} = a_{ip(j)} = a_{ip'(i)},$$

and similarly

$$b_{ip(i)} = a_{jp'(j)}.$$

The other factors are the same in (7.6), (7.7).

As p runs over all permutations, so does p'. Further, because of Lemma 7.5,

$$(-1)^p b_{1p(1)} \cdots b_{np(n)} = -(-1)^{p'} a_{1p'(1)} \cdots a_{np'(n)}. \tag{7.8}$$

Adding all $n!$ terms in equation (7.8) gives $\det B = -\det A$. □

Lemma 7.8. *Let A be an $n \times n$ matrix with two identical rows. Then* $\det A = 0$.

Proof. Suppose rows i, j are identical. Associate p' with p as in the last proof. Then

$$\det A = \sum_{p \text{ even}} \left(a_{1p(1)} \cdots a_{np(n)} - a_{1p'(1)} \cdots a_{np'(n)} \right)$$

because p' runs over the odd permutations when p runs over the even ones. The subtracted pair inside the brackets are identical. (They are given by (7.6), (7.7) with $A = B$.) So

$$\det A = \sum_{p \text{ even}} 0. \ \square$$

Alternatively, letting B equal the matrix which comes from A by switching the two identical rows, it follows that $A = B$ and so, from Lemma 7.7 that $-\det(A) = \det(B) = \det(A)$. Thus $\det(A) = 0$.

A determinant can be regarded as a fixed linear function of a given row. Thus

$$L(\mathbf{x}) = \det \begin{pmatrix} x_1 & x_2 & x_3 \\ a_{21} & a_{22} & a_{23} \\ a_{31} & a_{32} & a_{33} \end{pmatrix} = u_1 x_1 + u_2 x_2 + u_3 x_3 \tag{7.9}$$

where, after a short calculation, we obtain

$$u_1 = \det \begin{pmatrix} a_{22} & a_{23} \\ a_{32} & a_{33} \end{pmatrix}, u_2 = -\det \begin{pmatrix} a_{21} & a_{23} \\ a_{31} & a_{33} \end{pmatrix}, u_3 = \det \begin{pmatrix} a_{21} & a_{22} \\ a_{31} & a_{32} \end{pmatrix}. \tag{7.10}$$

We have

$$L(c\mathbf{x} + d\mathbf{y}) = cL(\mathbf{x}) + dL(\mathbf{y}) \tag{7.11}$$

(without needing to specify u_1, u_2, u_3). More generally, if the ith row is $\mathbf{r}_i = (a_{i1}, \ldots, a_{in})$, and if this row is replaced by a vector $\mathbf{x} = \begin{pmatrix} x_1 & \cdots & x_n \end{pmatrix}$, then if $L(\mathbf{x})$ is the function of $\mathbf{x}$ which is the determinant of the resulting matrix,

$$L(\mathbf{x}) = u_1 x_1 + \cdots + u_n x_n. \tag{7.12}$$

This holds simply because each product in the determinant contains exactly one of $x_1, \ldots, x_n$. Again, (7.11) holds.

Lemma 7.9. *Let B be obtained from the $n \times n$ matrix A by multiplying row i by c. Then*

$$\det B = c \det A.$$

Proof. In (7.12),

$$\det B = L(c\mathbf{r}_i) = cL(\mathbf{r}_i) = c \det A. \ \square$$

Lemma 7.10. *Let B be obtained from the $n \times n$ matrix A by adding c times row i to row j, where $j \neq i$. Then*

$$\det B = \det A.$$

Proof. In (7.12)

$$\det B = L(\mathbf{r}_j + c\mathbf{r}_i) = L(\mathbf{r}_j) + cL(\mathbf{r}_i) = \det A + cL(\mathbf{r}_i).$$

But $L(\mathbf{r}_i)$ is the determinant of a matrix with two identical rows. So $L(\mathbf{r}_i) = 0$ by Lemma 7.8. $\square$

Certain determinants are rather easy to evaluate.

Lemma 7.11. *Let A be* ***upper triangular,***

$$A = \begin{pmatrix} a_{11} & \cdots & a_{1n} \\ & \ddots & \vdots \\ & & a_{nn} \end{pmatrix}.$$

That is, $a_{ij} = 0$ for $i > j$. Then

$$\det A = a_{11} a_{22} \cdots a_{nn}.$$

Proof. Consider a nonzero product $a_{1p(1)} \cdots a_{np(n)}$. Since the product is nonzero, this requires $p(n) = n$. This forces $p(n-1) = n-1$. We can continue the process and show that $p(j) = j$ for $j = n, \ldots, 1$. Hence the only nonzero term in the sum $\det A$ is $a_{11} \cdots a_{nn}$. $\square$

Lemma 7.12. *Let A be $n \times n$. Then*

$$\det A^t = \det A.$$

Proof. The $n!$ products appearing in the sum $\det A^t$ are of the form

$$a_{p(1)1} \cdots a_{p(n)n}. \tag{7.13}$$

These are all products which contain one term from each row, and one term from each column, of A. They are the products for $\det A$, but we have to check that the $\pm$ signs are correct.

Let us rewrite (7.13) by rearranging the product as

$$a_{1q(1)} \cdots a_{nq(n)}, \tag{7.14}$$

where q is a permutation that depends on p. Now p, q have the same number of disordered pairs. In (7.14) this is the number of 'subproducts' $a_{iu}a_{jv}$ with $i < j$ and $u > v$. But $i = p(u)$ and $j = p(v)$, so this is equal to the number of pairs u, v with $v < u$ and $p(v) > p(u)$. Thus $(-1)^p = (-1)^q$ as required. □

Example 7.6. Let $F = \mathbb{R}$. Evaluate $\det A$,

$$A = \begin{pmatrix} 1 & 3 & 5 & 6 \\ 2 & 1 & 1 & 1 \\ 3 & 1 & 6 & 2 \\ 0 & 1 & 1 & 1 \end{pmatrix}.$$

We note that, doing appropriate row operations,

$$A \to \begin{pmatrix} 1 & 3 & 5 & 6 \\ 0 & -5 & -9 & -11 \\ 3 & 1 & 6 & 2 \\ 0 & 1 & 1 & 1 \end{pmatrix} \to \begin{pmatrix} 1 & 3 & 5 & 6 \\ 0 & -5 & -9 & -11 \\ 0 & -8 & -9 & -16 \\ 0 & 1 & 1 & 1 \end{pmatrix} \overset{\text{switch 2 and 3}}{\to}$$

$$\begin{pmatrix} 1 & 3 & 5 & 6 \\ 0 & 1 & 1 & 1 \\ 0 & -8 & -9 & -16 \\ 0 & -5 & -9 & -11 \end{pmatrix} \to \begin{pmatrix} 1 & 3 & 5 & 6 \\ 0 & 1 & 1 & 1 \\ 0 & 0 & -1 & -8 \\ 0 & 0 & -4 & -6 \end{pmatrix} \to \begin{pmatrix} 1 & 3 & 5 & 6 \\ 0 & 1 & 1 & 1 \\ 0 & 0 & -1 & -8 \\ 0 & 0 & 0 & 26 \end{pmatrix}$$

Write this chain as $A \to B_1 \to B_2 \to B_3 \to B_4 \to B_5$. We have (using Lemmas 7.7 and 7.10)

$$\det B_5 = \det B_4 = \det B_3 = -\det B_2 = -\det B_1 = \det A.$$

Since $\det B_5 = -26$, we get $\det A = 26$.

It is of course quicker to use Maple. But Example 7.6 illustrates the fact that in any chain obtained from row operations

$$A \to B_1 \to \cdots \to B_k. \tag{7.15}$$

starting with an $n \times n$ matrix and ending in a row reduced matrix B_k, each determinant is a nonzero multiple of the preceding one (by Lemmas 7.7, 7.9, 7.10). So

$$\det A = 0 \text{ if } \det B_k = 0$$
$$\det A \neq 0 \text{ if } \det B_k \neq 0.$$

Note that if A is singular then B_k, which has the same rank as A, has a zero row; so $\det B_k = 0$ and $\det A = 0$.

Lemma 7.13. *Let A and B be $n \times n$. Then*

$$\det(AB) = \det A \det B.$$

If A is an $n \times n$ matrix, $\det(A) \neq 0$ *if and only if* $R(A) = n$.

Proof. If A is singular, $(R(A) < n)$, then so is AB, and

$$\det AB = 0 = \det A \det B.$$

In the other case that A is invertible, $(R(A) = n)$ we note the particular case

$$\det(EB) = \det E \det B, \tag{7.16}$$

where E is an elementary matrix. This is a restatement of Lemmas 7.7, 7.9 and 7.10.

Now if A is invertible, we recall that

$$A = E_1 \cdots E_k$$

with each E_j elementary. See Problem 34 on Page 91. Using (7.16) repeatedly,

$$\det AB = \det(E_1 \ldots E_k B) = \det E_1 \det E_2 \cdots \det E_k \det B.$$

Since $\det A = \det E_1 \cdots \det E_k$ from the particular case $B = I$, the first claim of the lemma follows.

Now consider the claim about rank. If $R(A) = n$, then the columns of A are linearly independent, and so they are each pivot columns. Therefore, the row reduced echelon form of A is I. It follows that a product of elementary matrices corresponding to various row operations, when applied to the left of A, yields I,

$$E_1 E_2 \cdots E_p A = I.$$

Now take the determinant of both sides of the above expression to conclude that $\det(A)$ cannot equal 0. Next suppose $\det(A) \neq 0$. Then the row reduced echelon form for A also has nonzero determinant. It follows that this row reduced echelon form must equal I since otherwise the row reduced echelon form of A would have a row of zeros and would have zero determinant. Hence, by Lemma 3.5 the rank of A must equal n. $\square$

Example 7.7. (The Vandermonde determinant.) Show that

$$\det \begin{pmatrix} 1 & z_1 & \cdots & z_1^{n-1} \\ 1 & z_2 & \cdots & z_2^{n-1} \\ \vdots & \vdots & & \vdots \\ 1 & z_n & \cdots & z_n^{n-1} \end{pmatrix} = \prod_{j>i} (z_j - z_i). \tag{7.17}$$

Notation 7.1. The large $\prod$ indicates a product, and the '$j > i$' underneath indicates that we multiply all $z_j - z_i$ for pairs i, j with $1 \leq i < j \leq n$. In the 3×3 case, the value of the determinant is

$$(z_3 - z_1)(z_3 - z_2)(z_2 - z_1).$$

To get (7.17) we use polynomials in several variables. A **polynomial in** $z_1, \ldots, z_n$ is a sum of *monomials*

$$a(i_1, \ldots, i_n) z_1^{i_1} \cdots z_n^{i_n}, \; a(i_1, \ldots, i_n) \neq 0.$$

(We can work over any field F, so that $a(i_1, \ldots, i_n) \in F$.) The **degree** of this monomial is $i_1 + \cdots + i_n$, and the degree of a polynomial P is the largest degree among the monomials whose sum is P.

Both the left-hand and right-hand sides of (7.17) are polynomials in $z_1, \ldots, z_n$. A short calculation shows that both have degree $0 + 1 + 2 + \cdots + n - 1$. This is obvious for P the polynomial on the left, given by the determinant. It is also true for Q, the polynomial on the right.

Lemma 7.14. *The degree of Q is $1 + 2 + \cdots + n - 1$.*

Proof. This is obvious if $n = 2$. Suppose the statement is true for $n - 1$ where $n \geq 3$. Then

$$\prod_{n \geq j > i} (z_j - z_i) = \prod_{n > i} (z_n - z_i) \prod_{(n-1) \geq j > i} (z_j - z_i).$$

Now the first of the two factors clearly has degree $n - 1$ and the second has degree

$$1 + 2 + \cdots + n - 2$$

by induction. Hence the degree of their product is $1 + 2 + \cdots + n - 1$. □

Lemma 7.15. *$P(z_1, \ldots, z_n)$ is divisible by $z_j - z_i$ if $j > i$. That is,*

$$\frac{P(\mathbf{z})}{z_j - z_i}$$

is a polynomial.

Proof. Fixing each z_j for $j \neq i$, the function $z_i \to P(z_1, \ldots, z_n) = q(z_i)$ is an ordinary polynomial which is a function of the single variable z_i. Then when $z_i = z_j$, $q(z_i)$ becomes the determinant of a matrix which has two equal rows and is therefore, equal to 0. Now from the Euclidean algorithm for polynomials Theorem 1.2, there exists a polynomial $k(z_i)$ such that

$$q(z_i) = k(z_i)(z_i - z_j) + r(z_i)$$

where the degree of $r(x_i) < 1$ or else $r(z_i)$ equals 0. However, since $q(z_j) = 0$, it follows that $r(z_j) = 0$, and so $r(z_i) = 0$. This proves the lemma. □

Thus

$$P(\mathbf{z}) = (z_j - z_i) P'(\mathbf{z}).$$

The same reasoning applies to $P'(\mathbf{z})$. It equals 0 when $z_i = z_k$ for another $k > i$. Thus, repeating this argument a finite number of times, the quotient

$$R(\mathbf{z}) = \frac{P(\mathbf{z})}{Q(\mathbf{z})}$$

is a polynomial of degree 0. That is, $R(\mathbf{z})$ is a constant c. In fact, $c = 1$, since P, Q both contain the term $z_2 z_3^2 \ldots z_n^{n-1}$. This establishes (7.17).

Determinants can be used to give an algebraic formula for A^{-1}, e.g.

$$\det A = \frac{1}{\det A}\begin{pmatrix} a_{22} & -a_{12} \\ -a_{21} & a_{11} \end{pmatrix}$$

when $n = 2, \det A \neq 0$. To obtain the general formula, we begin with

$$\det A = \sum_{j=1}^{n} (-1)^{i+j} a_{ij} \det A_{ij}, \tag{7.18}$$

where A_{ij} is the matrix obtained by deleting row i and column j. This is an **expansion by row** i. Expansion by column j is the similar formula

$$\det A = \sum_{i=1}^{n} (-1)^{i+j} a_{ij} \det A_{ij}. \tag{7.19}$$

See the exercises for a proof of (7.19).

Definition 7.6. It is customary to refer to the quantity $(-1)^{j+k} A_{jk}$ described above as the jk^{th} cofactor. It is often written as

$$(-1)^{j+k} A_{jk} = \operatorname{cof}(A)_{jk}.$$

The determinant A_{jk} is called the jk^{th} minor.

Example 7.8. To evaluate $\det \begin{pmatrix} 0 & 3 & 1 \\ 2 & 4 & 6 \\ 5 & 2 & 1 \end{pmatrix}$, the natural choice is 'expand by row 1' or 'expand by column 1'. We get by the first expansion the value

$$\det A = -3(2 - 30) + 1(4 - 20) = 68.$$

The second expansion yields

$$\det A = -2(3 - 2) + 5(18 - 4) = 68.$$

Theorem 7.2. *The formula (7.18) is valid.*

Proof. To begin with, it is clear that (7.18) is correct apart perhaps from the $\pm$ signs $(-1)^{i+j}$. Every j on the right-hand side corresponds to $(n-1)!$ distinct products that appear in $\det A$, namely all

$$a_{1p(1)} \cdots a_{np(n)}$$

for which the given integer i has $p(i) = j$. Adding these up, we get all the terms of $\det A$, but it is not clear that the $\pm$ signs are correct.

Consider the matrix B obtained from A, by interchanging adjacent rows $i-1$ times, then interchanging adjacent columns $j-1$ times, to bring a_{ij} to top left position ($b_{11} = a_{ij}$). Clearly $B_{11} = A_{ij}$. Now

$$\det B = (-1)^{i+j-2} \det A; \ \det A = (-1)^{i+j} \det B. \tag{7.20}$$

The sign of a term $\pm b_{2p(2)} \dots b_{np(n)}$ of $\det B_{11}$ is evidently the same as the sign of $\pm b_{11} b_{2p(2)} \dots b_{np(n)}$ in $\det B$. Hence $\det B$ includes the terms $b_{11} \det B_{11} = a_{ij} \det A_{ij}$, and recalling (7.20), $\det A$ includes the terms $(-1)^{i+j} a_{ij} \det A_{ij}$. □

Definition 7.7. The **adjugate** of A is the matrix $C = (c_{ij})$ over F with entries

$$c_{ij} = (-1)^{i+j} \det A_{ji} \qquad (i, j = 1, \dots, n).$$

We write $\operatorname{adj} A$ for C.

It is a simple matter to deduce from (7.18) that

$$A \operatorname{adj} A = (\det A) I. \tag{7.21}$$

The ij^{th} entry of $A \operatorname{adj} A$ is

$$\sum_{k=1}^{n} a_{ik} (-1)^{j+k} \det A_{jk}. \tag{7.22}$$

This is of course just $\det A$ if $i = j$, since we can refer to (7.18). If $i \neq j$, then in (7.22) we have the expansion by row j of the matrix obtained from A **by replacing row j with a copy of row i.** It follows that the ij^{th} entry of $A \operatorname{adj} A$ is 0 if $i \neq j$, and we have proved (7.21). In terms of indices,

$$\sum_{k=1}^{n} a_{ik} (-1)^{j+k} \det A_{jk} = \det (A) \, \delta_{ij}.$$

As a corollary, if $\det A \neq 0$,

$$A^{-1} = (\det A)^{-1} \operatorname{adj} \ A. \tag{7.23}$$

In terms of the ij^{th} entries, letting a_{ij}^{-1} denote the ij^{th} entry of the inverse matrix,

$$a_{ij}^{-1} = \frac{1}{\det (A)} \operatorname{cof} (A)_{ji} \, .$$

This formula for the inverse also implies a famous procedure known as Cramer's rule. Cramer's rule gives a formula for the solutions $\mathbf{x}$, to a system of equations $A\mathbf{x} = \mathbf{y}$.

In case you are solving such a system of equations $A\mathbf{x} = \mathbf{y}$ for $\mathbf{x}$, it follows that if A^{-1} exists,

$$\mathbf{x} = \left(A^{-1} A\right) \mathbf{x} = A^{-1} \left(A\mathbf{x}\right) = A^{-1} \mathbf{y}$$

thus solving the system. Now in the case that A^{-1} exists, there is a formula for A^{-1} given above. Using this formula,

$$x_i = \sum_{j=1}^{n} a_{ij}^{-1} y_j = \sum_{j=1}^{n} \frac{1}{\det (A)} \operatorname{cof} (A)_{ji} \, y_j.$$

By the formula for the expansion of a determinant along a column,

$$x_i = \frac{1}{\det(A)} \det \begin{pmatrix} * & \cdots & y_1 & \cdots & * \\ \vdots & & \vdots & & \vdots \\ * & \cdots & y_n & \cdots & * \end{pmatrix},$$

where here the ith column of A is replaced with the column vector, $(y_1 \cdots, y_n)^t$, and the determinant of this modified matrix is taken and divided by $\det(A)$. This formula is known as Cramer's rule.

Corollary 7.1. *Suppose A is an $n \times n$ matrix and it is desired to solve the system $A\mathbf{x} = \mathbf{y}, \mathbf{y} = (y_1, \cdots, y_n)^t$ for $\mathbf{x} = (x_1, \cdots, x_n)^t$. Then Cramer's rule says*

$$x_i = \frac{\det A_i}{\det A},$$

where A_i is obtained from A by replacing the ith column of A with the column $(y_1, \cdots, y_n)^t$.

7.3 Wronskians

So far we have only proved linear independence of a set of functions $f_1, \ldots, f_n$ in some rather simple cases (including orthogonality). A useful general approach is based on the **Wronskian determinant.** The Wronskian determinant of $f_1, \ldots, f_n$, functions in $S[a, b]$, is the $n \times n$ determinant

$$W(x) = \det \begin{pmatrix} f_1(x) & \cdots & f_n(x) \\ f_1'(x) & \cdots & f_n'(x) \\ \vdots & & \\ f_1^{(n-1)}(x) & \cdots & f_n^{(n-1)}(x) \end{pmatrix}.$$

Example 7.9. For $f_1(x) = \cos x, f_2(x) = \sin x$,

$$W(x) = \det \begin{pmatrix} \cos x & \sin x \\ -\sin x & \cos x \end{pmatrix} = 1.$$

Lemma 7.16. *Let $f_1, \ldots, f_n$ be as above. If there is some $y \in [a, b]$ with $W(y) \neq 0$, then $f_1, \ldots, f_n$ is a linearly independent set.*

Proof. Suppose that $c_1, \ldots, c_n$ are scalars with

$$c_1 f_1 + \cdots + c_n f_n = 0.$$

Differentiating j times,

$$c_1 f_1^{(j)} + \cdots + c_n f_n^{(j)} = 0.$$

Fix y with $W(y) \neq 0$. The linear system

$$\begin{aligned} 2x_1 f_1(y) + \cdots + x_n f_n(y) &= 0 \\ x_1 f_1'(y) + \cdots + x_n f_n'(y) &= 0 \\ &\vdots \\ x_1 f_1^{(n-1)}(y) + \cdots + x_n f_n^{(n-1)}(y) &= 0 \end{aligned}$$

is satisfied by $x_1 = c_1, \ldots, x_n = c_n$. Since the coefficient matrix of this system has determinant $W(y)$, the matrix is invertible. We infer that $c_1 = \cdots = c_n = 0$. □

Example 7.10. Show that

$$\frac{1}{x}, \sin x, e^x$$

is a linearly independent set in $S[1,2]$.

Solution. The Wronskian in this case is

$$W(x) = \det \begin{pmatrix} \frac{1}{x} & \sin x & e^x \\ -\frac{1}{x^2} & \cos x & e^x \\ \frac{1}{x^3} & -\sin x & e^x \end{pmatrix}.$$

Try an 'easy' value, $x = \frac{\pi}{2}$.

$$W\left(\frac{\pi}{2}\right) = e^{\pi/2} \det \begin{pmatrix} \frac{2}{\pi} & 1 & 1 \\ \frac{-4}{\pi^2} & 0 & 1 \\ \frac{16}{\pi^3} & -1 & 1 \end{pmatrix} = e^{\pi/2} 2 \frac{\pi^2 + 4\pi + 8}{\pi^3} \neq 0.$$

Lemma 7.16 gives the linear independence. (The choice of $\pi/2$ is not particularly lucky, since there are rather few zeros of W to avoid.)

7.4 The Cayley-Hamilton theorem

Definition 7.8. Let A be an $n \times n$ matrix. The characteristic polynomial is defined as

$$p_A(t) \equiv \det(tI - A)$$

and the solutions to $p_A(t) = 0$ are called eigenvalues. For A a matrix and $p(t) = t^n + a_{n-1}t^{n-1} + \cdots + a_1 t + a_0$, denote by $p(A)$ the matrix defined by

$$p(A) \equiv A^n + a_{n-1}A^{n-1} + \cdots + a_1 A + a_0 I.$$

The explanation for the last term is that A^0 is interpreted as I, the identity matrix.

The Cayley-Hamilton theorem states that every matrix satisfies its characteristic equation, that equation defined by $p_A(t) = 0$. It is one of the most important theorems in linear algebra[1]. The proof in this section is not the most general proof, but works well when the field of scalars is the real or complex numbers. This will be assumed in the remainder of this section. The proof of the Cayley-Hamilton theorem is based on the following lemma.

Lemma 7.17. *Suppose for all $|\lambda|$ large enough,*

$$A_0 + A_1\lambda + \cdots + A_m\lambda^m = 0,$$

where the A_i are $n \times n$ matrices. Then each $A_i = 0$.

Proof. Multiply by λ^{-m} to obtain

$$A_0\lambda^{-m} + A_1\lambda^{-m+1} + \cdots + A_{m-1}\lambda^{-1} + A_m = 0.$$

Now let $|\lambda| \to \infty$ to obtain $A_m = 0$. With this, multiply by λ to obtain

$$A_0\lambda^{-m+1} + A_1\lambda^{-m+2} + \cdots + A_{m-1} = 0.$$

Now let $|\lambda| \to \infty$ to obtain $A_{m-1} = 0$. Continue multiplying by λ and letting $\lambda \to \infty$ to obtain that all the $A_i = 0$. $\square$

With the lemma, here is a simple corollary.

Corollary 7.2. *Let A_i and B_i be $n \times n$ matrices and suppose*

$$A_0 + A_1\lambda + \cdots + A_m\lambda^m = B_0 + B_1\lambda + \cdots + B_m\lambda^m$$

for all $|\lambda|$ large enough. Then $A_i = B_i$ for all i. If $A_i = B_i$ for each A_i, B_i then one can substitute an $n \times n$ matrix M for λ and the identity will continue to hold.

Proof. Subtract and use the result of the lemma. The last claim is obvious by matching terms. $\square$

With this preparation, here is a relatively easy proof of the Cayley-Hamilton theorem.

Theorem 7.3. *Let A be an $n \times n$ matrix and let $p(\lambda) \equiv \det(\lambda I - A)$ be the characteristic polynomial. Then $p(A) = 0$.*

Proof. Let $C(\lambda)$ equal the transpose of the cofactor matrix of $(\lambda I - A)$ for $|\lambda|$ large. (If $|\lambda|$ is large enough, then λ cannot be in the finite list of eigenvalues of A and so for such λ, $(\lambda I - A)^{-1}$ exists.) Therefore, by the description of the inverse in terms of the cofactor matrix,

$$C(\lambda) = p(\lambda)(\lambda I - A)^{-1}.$$

Say

$$p(\lambda) = a_0 + a_1\lambda + \cdots + \lambda^n$$

[1] A special case was first proved by Hamilton in 1853. The general case was announced by Cayley some time later and a proof was given by Frobenius in 1878.

Note that each entry in $C(\lambda)$ is a polynomial in λ having degree no more than $n-1$. Therefore, collecting the terms,

$$C(\lambda) = C_0 + C_1\lambda + \cdots + C_{n-1}\lambda^{n-1}$$

for C_j some $n \times n$ matrix. Then

$$C(\lambda)(\lambda I - A) = \left(C_0 + C_1\lambda + \cdots + C_{n-1}\lambda^{n-1}\right)(\lambda I - A) = p(\lambda) I$$

Then multiplying out the middle term, it follows that for all $|\lambda|$ sufficiently large,

$$a_0 I + a_1 I\lambda + \cdots + I\lambda^n = C_0\lambda + C_1\lambda^2 + \cdots + C_{n-1}\lambda^n$$

$$- \left[C_0 A + C_1 A\lambda + \cdots + C_{n-1} A\lambda^{n-1}\right]$$

$$= -C_0 A + (C_0 - C_1 A)\lambda + (C_1 - C_2 A)\lambda^2 + \cdots + (C_{n-2} - C_{n-1}A)\lambda^{n-1} + C_{n-1}\lambda^n$$

Then, using Corollary 7.2, one can replace λ on both sides with A. Then the right side is seen to equal 0. Hence the left side, $p(A) I$ is also equal to 0. $\square$

7.5 Exercises

(1) In the following examples, a linear transformation T is given by specifying its action on a basis β. Find its matrix with respect to this basis. Then find its matrix with respect to the usual basis.

(a) $T\begin{pmatrix} 1 \\ 2 \end{pmatrix} = 2\begin{pmatrix} 1 \\ 2 \end{pmatrix} + 1\begin{pmatrix} -1 \\ 1 \end{pmatrix}, T\begin{pmatrix} -1 \\ 1 \end{pmatrix} = \begin{pmatrix} -1 \\ 1 \end{pmatrix}$

(b) $T\begin{pmatrix} 0 \\ 1 \end{pmatrix} = 2\begin{pmatrix} 0 \\ 1 \end{pmatrix} + 1\begin{pmatrix} -1 \\ 1 \end{pmatrix}, T\begin{pmatrix} -1 \\ 1 \end{pmatrix} = \begin{pmatrix} 0 \\ 1 \end{pmatrix}$

(c) $T\begin{pmatrix} 1 \\ 0 \end{pmatrix} = 2\begin{pmatrix} 1 \\ 2 \end{pmatrix} + 1\begin{pmatrix} 1 \\ 0 \end{pmatrix}, T\begin{pmatrix} 1 \\ 2 \end{pmatrix} = 1\begin{pmatrix} 1 \\ 0 \end{pmatrix} - \begin{pmatrix} 1 \\ 2 \end{pmatrix}$

(2) Let $\beta = \{\mathbf{u}_1, \cdots, \mathbf{u}_n\}$ be a basis for F^n and let $T : F^n \to F^n$ be defined as follows

$$T\left(\sum_{k=1}^{n} a_k \mathbf{u}_k\right) = \sum_{k=1}^{n} a_k b_k \mathbf{u}_k.$$

First show that T is a linear transformation. Next show that the matrix of T with respect to this basis is $[T]_\beta =$

$$\begin{pmatrix} b_1 & & \\ & \ddots & \\ & & b_n \end{pmatrix}.$$

Show that the above definition is equivalent to simply specifying T on the basis vectors of β by

$$T(\mathbf{u}_k) = b_k \mathbf{u}_k.$$

(3) ↑In the situation of the above problem, let $\gamma = \{\mathbf{e}_1, \cdots, \mathbf{e}_n\}$ be the standard basis for F^n where $\mathbf{e}_k$ is the vector which has 1 in the kth entry and zeros elsewhere. Show that $[T]_\gamma$ is

$$\begin{pmatrix} \mathbf{u}_1 & \cdots & \mathbf{u}_n \end{pmatrix}^{-1} [T]_\beta \begin{pmatrix} \mathbf{u}_1 & \cdots & \mathbf{u}_n \end{pmatrix}. \tag{7.24}$$

(4) ↑Generalize the above problem to the situation where T is given by specifying its action on the vectors of a basis $\beta = \{\mathbf{u}_1, \cdots, \mathbf{u}_n\}$ as follows

$$T\mathbf{u}_k = \sum_{j=1}^{n} a_{jk}\mathbf{u}_j.$$

Letting $A = (a_{ij})$, verify that for $\gamma = \{\mathbf{e}_1, \cdots, \mathbf{e}_n\}$, (7.24) still holds and that $[T]_\beta = A$.

(5) Let V, W be real inner product spaces. For $w \in W$ and $v \in V$, the tensor product, denoted as $w \otimes v$ is defined as a mapping from V to W as follows

$$w \otimes v\,(u) = \langle u, v \rangle_V\, w.$$

Show that $w \otimes v$ is a linear mapping from V to W. If $\{w_1, \cdots, w_m\}$ is a basis for W and $\{v_1, \cdots, v_n\}$ is a basis for V, show that $\{w_i \otimes v_j\}_{i,j}$ is a basis for the vector space of linear transformations mapping V to W. In the case where $V = W$ and $\{v_1, \cdots, v_n\}$ a basis for V, let L be a linear transformation, and let a_{ij} denote the unique scalars such that

$$L = \sum_{i,j} a_{ij} v_i \otimes v_j\,.$$

Let $[L]$ denote the matrix of the linear transformation with respect to this basis. Show that $[L]_{ij} = a_{ij}$ if and only if $\{v_1, \cdots, v_n\}$ is an orthonormal basis. In general, show that $[L] = AG$ where A is the $n \times n$ matrix having components a_{ij} and G is the metric tensor, $G_{ij} = \langle v_i, v_j \rangle$.

(6) Let P_3 denote the set of real polynomials of degree no more than 3, defined on an interval $[a, b]$. Show that P_3 is a subspace of the vector space of all functions defined on this interval. Show that a basis for P_3 is $\{1, x, x^2, x^3\}$. Now let D denote the differentiation operator which sends a function to its derivative. Show D is a linear transformation which sends P_3 to P_3. Find the matrix of this linear transformation with respect to the given basis.

(7) Generalize the above problem to P_n, the space of polynomials of degree no more than n with basis $\{1, x, \cdots, x^n\}$.

(8) In the situation of the above problem, let the linear transformation be $T = D^2 + 1$, defined as $Tf = f'' + f$. Find the matrix of this linear transformation with respect to the given basis $\{1, x, \cdots, x^n\}$.

(9) Consider the block matrix, $M = \begin{pmatrix} A & B \\ 0 & C \end{pmatrix}$ where A is $m \times m$ and C is $k \times k$. Thus the matrix is $(m + k) \times (m + k)$. Show that $\det(M) = \det(A)\det(C)$.

(10) *Suppose $\{\mathbf{v}_1, \cdots, \mathbf{v}_k\}$ are vectors in $\mathbb{R}^n$. Then the k dimensional volume of the parallelepiped determined by these vectors is defined as

$$\left(\det\left(\begin{pmatrix} \mathbf{v}_1^t \\ \vdots \\ \mathbf{v}_k^t \end{pmatrix} \begin{pmatrix} \mathbf{v}_1 & \cdots & \mathbf{v}_k \end{pmatrix}\right)\right)^{1/2}.$$

Verify that this makes sense by showing that it gives the right answer if there is only one vector, and also gives the right answer if the vectors are orthogonal. Next verify that for any list of vectors, the above determinant is nonnegative so one can at least take the square root as required. What is the k dimensional volume if $k > n$? If $k = n$, show that the expression reduces to

$$\left|\det\begin{pmatrix} \mathbf{v}_1 & \cdots & \mathbf{v}_k \end{pmatrix}\right|.$$

(11) *Show that the definition of volume given in the above is the only one which is geometrically reasonable. See the picture. You know what the volume of the base is. The volume of the $p+1$ dimensional parallelpiped needs to equal the height times the volume of the base to make sense geometrically.

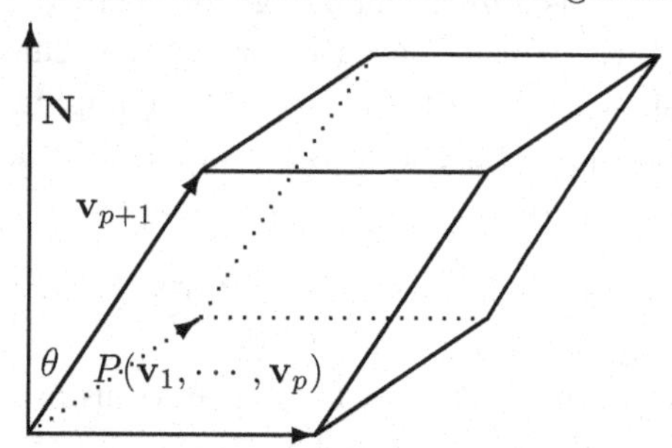

(12) In calculus, the following situation is encountered. Let $\mathbf{f} : U \to \mathbb{R}^m$ where U is an open subset of $\mathbb{R}^n$. Such a function is said to have a **derivative** or to be **differentiable** at $\mathbf{x} \in U$ if there exists a linear transformation $T : \mathbb{R}^n \to \mathbb{R}^m$ such that

$$\lim_{\mathbf{v}\to\mathbf{0}} \frac{|\mathbf{f}(\mathbf{x}+\mathbf{v}) - \mathbf{f}(\mathbf{x}) - T\mathbf{v}|}{|\mathbf{v}|} = 0.$$

First show that this linear transformation, if it exists, must be unique. Next show that for $\beta = \{\mathbf{e}_1, \cdots, \mathbf{e}_n\}$, the kth column of $[T]_\beta$ is

$$\frac{\partial \mathbf{f}}{\partial x_k}(\mathbf{x}).$$

Actually, the result of this problem is a well kept secret. People typically don't see this in calculus. It is seen for the first time in advanced calculus.

(13) Recall that A is similar to B if there exists a matrix P such that $A = P^{-1}BP$. Show that if A and B are similar, then they have the same determinant. Give an example of two matrices which are not similar but have the same determinant.

(14) It was shown in Lemma 7.3 that when two matrices come from the same linear mapping but with respect to two different bases, then they are similar. Prove the converse of this fact, that if two $n \times n$ matrices are similar, then there exists a vector space V, a linear transformation T, and two different bases, β, γ on V such that one matrix is $[T]_\beta$ and the other is $[T]_\gamma$.

(15) Prove the formula (7.19) for expansion of a determinant along a column .

(16) Let A be an $n \times n$ matrix, $A = (a_{ij})$. The **trace** of A is defined as

$$\operatorname{trace}(A) = \sum_i a_{ii}.$$

Show that for A, B two $n \times n$ matrices, $\operatorname{trace}(AB) = \operatorname{trace}(BA)$. Now show that if A and B are similar $n \times n$ matrices, then $\operatorname{trace}(A) = \operatorname{trace}(B)$.

(17) Find the inverse, if it exists, of the matrix

$$\begin{pmatrix} e^t & \cos t & \sin t \\ e^t & -\sin t & \cos t \\ e^t & -\cos t & -\sin t \end{pmatrix}.$$

(18) Consider the following two functions which are defined on $(-1, 1)$.

$$f_1(x) = x^2,\ f_2(x) = \begin{cases} x^2 \text{ if } x \geq 0 \\ -x^2 \text{ if } x < 0 \end{cases}$$

Show that these functions are linearly independent but that their Wronskian is identically equal to 0.

(19) Let $Ly = y^{(n)} + a_{n-1}(x)\, y^{(n-1)} + \cdots + a_1(x)\, y' + a_0(x)\, y$ where the a_i are given continuous functions defined on a closed interval, (a, b) and y is some function which has n derivatives, so it makes sense to write Ly. Suppose $Ly_k = 0$ for $k = 1, 2, \cdots, n$. The Wronskian of these functions, y_i is defined as

$$W(y_1, \cdots, y_n)(x) \equiv \det \begin{pmatrix} y_1(x) & \cdots & y_n(x) \\ y_1'(x) & \cdots & y_n'(x) \\ \vdots & & \vdots \\ y_1^{(n-1)}(x) & \cdots & y_n^{(n-1)}(x) \end{pmatrix}.$$

Show that for $W(x) = W(y_1, \cdots, y_n)(x)$ to save space,

$$W'(x) = \det \begin{pmatrix} y_1(x) & \cdots & y_n(x) \\ y_1'(x) & \cdots & y_n'(x) \\ \vdots & & \vdots \\ y_1^{(n-2)}(x) & \cdots & y_n^{(n-2)}(x) \\ y_1^{(n)}(x) & \cdots & y_n^{(n)}(x) \end{pmatrix}.$$

Now use the differential equation, $Ly = 0$ which is satisfied by each of these functions, y_i and properties of determinants presented above to verify that $W' + a_{n-1}(x)\, W = 0$. Give an explicit solution of this linear differential equation, Abel's formula, and use your answer to verify that the Wronskian of these solutions to the equation, $Ly = 0$ either vanishes identically on (a, b) or never. **Hint:** Multiply both sides of the differential equation by $\exp(A(x))$ where $A'(x) = a_{n-1}(x)$. Then show that what results is the derivative of something.

(20) Recall that you can find the determinant $A = (a_{ij})$ from expanding along the jth column.

$$\det(A) = \sum_i (-1)^{i+j} a_{ij} A_{ij}$$

where A_{ij} is the determinant obtained by deleting the ith row and the jth column. It is customary to write

$$(-1)^{i+j} A_{ij} = \operatorname{cof}(A)_{ij}.$$

Think of $\det(A)$ as a function of the entries, a_{ij}. Explain why the ij^{th} cofactor is really just

$$\frac{\partial \det(A)}{\partial a_{ij}}.$$

(21) Let U be an open set in $\mathbb{R}^n$ and let $\mathbf{g} : U \to \mathbb{R}^n$ be such that all the first partial derivatives of all components of $\mathbf{g}$ exist and are continuous. Under these conditions form the matrix $D\mathbf{g}(\mathbf{x})$ given by

$$D\mathbf{g}(\mathbf{x})_{ij} \equiv \frac{\partial g_i(\mathbf{x})}{\partial x_j} \equiv g_{i,j}(\mathbf{x}).$$

The best kept secret in calculus courses is that the linear transformation determined by this matrix $D\mathbf{g}(\mathbf{x})$ is called the derivative of $\mathbf{g}$ and is the correct generalization of the concept of derivative of a function of one variable. Suppose the second partial derivatives also exist and are continuous. Then show that

$$\sum_j (\operatorname{cof}(D\mathbf{g}))_{ij,j} = 0.$$

Hint: First explain why

$$\sum_i g_{i,k} \operatorname{cof}(D\mathbf{g})_{ij} = \delta_{jk} \det(D\mathbf{g}).$$

Next differentiate with respect to x_j and sum on j using the equality of mixed partial derivatives. Assume $\det(D\mathbf{g}) \neq 0$ to prove the identity in this special case. Then explain why there exists a sequence $\varepsilon_k \to 0$ such that for $\mathbf{g}_{\varepsilon_k}(\mathbf{x}) \equiv \mathbf{g}(\mathbf{x}) + \varepsilon_k \mathbf{x}$, $\det(D\mathbf{g}_{\varepsilon_k}) \neq 0$, and so the identity holds for $\mathbf{g}_{\varepsilon_k}$. Then take a limit to get the desired result in general. This is an extremely important identity which has surprising implications.

(22) Recall that a determinant of the form

$$\begin{vmatrix} 1 & 1 & \cdots & 1 \\ a_0 & a_1 & \cdots & a_n \\ a_0^2 & a_1^2 & \cdots & a_n^2 \\ \vdots & \vdots & & \vdots \\ a_0^{n-1} & a_1^{n-1} & \cdots & a_n^{n-1} \\ a_0^n & a_1^n & \cdots & a_n^n \end{vmatrix}$$

is called a Vandermonde determinant. Give a new proof that this determinant equals

$$\prod_{0\leq i<j\leq n}(a_j - a_i)$$

Hint: Show it works if $n = 1$, so you are looking at

$$\begin{vmatrix} 1 & 1 \\ a_0 & a_1 \end{vmatrix}.$$

Then suppose it holds for $n-1$ and consider the case n. Consider the following polynomial.

$$p(t) \equiv \begin{vmatrix} 1 & 1 & \cdots & 1 \\ a_0 & a_1 & \cdots & t \\ a_0^2 & a_1^2 & \cdots & t^2 \\ \vdots & \vdots & & \vdots \\ a_0^{n-1} & a_1^{n-1} & \cdots & t^{n-1} \\ a_0^n & a_1^n & \cdots & t^n \end{vmatrix}.$$

Explain why $p(a_j) = 0$ for $i = 0, \cdots, n-1$. Thus

$$p(t) = c\prod_{i=0}^{n-1}(t - a_i).$$

Of course c is the coefficient of t^n. Find this coefficient from the above description of $p(t)$ and the induction hypothesis. Then plug in $t = a_n$ and observe that you have the formula valid for n.

(23) Let $T : V \to V$ be linear where V is a vector space of dimension n. Letting β be a basis for V define $\det(T) = \det[T]_\beta$. Show that this definition is well defined. That is, show that you get the same number if you use another definition for the basis.

(24) ↑Using the above problem, show that there exists a unique polynomial $p(\lambda)$ which is of the form

$$p(\lambda) = \lambda^m + a_{m-1}\lambda^{m-1} + \cdots + a_1\lambda + a_0$$

such that

$$p(T) = T^m + a_{m-1}T^{m-1} + \cdots + a_1T + a_0I = 0.$$

and m is as small as possible. This is called the minimal polynomial. **Hint:** You can show this easily by using the result of Problem 21 on Page 151 to get the existence of a polynomial which sends T to the zero linear transformation. Now let m be the smallest integer such that for some polynomial $p(\lambda)$ of degree m, it follows that $p(T) = 0$. Pick such a polynomial and divide by its leading coefficient to obtain a monic polynomial. If you have two such polynomials, $p(\lambda), p'(\lambda)$, then from the Euclidean algorithm, Theorem 1.2,

$$p'(\lambda) = p(\lambda)\,l(\lambda) + r(\lambda)$$

where the degree of $r(\lambda)$ is less than the degree of $p(\lambda)$ or else $r(\lambda) = 0$. Explain why this requires $r(\lambda) = 0$. Now note that since $p'(\lambda)$ and $p(\lambda)$ have the same degree, the degree of $l(\lambda)$ must be 0 and hence, since they are both monic, they must coincide.

(25) The example in this exercise was contributed by Marc van Leeuwen. It is an excellent example to keep in mind when considering the proof of the Cayley-Hamilton theorem presented in this chapter. If $p(\lambda) = q(\lambda)$ for all λ or for all λ large enough where $p(\lambda), q(\lambda)$ are polynomials having matrix coefficients, then it is not necessarily the case that $p(A) = q(A)$ for A a matrix of an appropriate size. Let

$$E_1 = \begin{pmatrix} 1 & 0 \\ 0 & 0 \end{pmatrix}, E_2 = \begin{pmatrix} 0 & 0 \\ 0 & 1 \end{pmatrix}, N = \begin{pmatrix} 0 & 1 \\ 0 & 0 \end{pmatrix}$$

Show that for all λ,

$$(\lambda I + E_1)(\lambda I + E_2) = (\lambda^2 + \lambda) I = (\lambda I + E_2)(\lambda I + E_1)$$

However,

$$(NI + E_1)(NI + E_2) \neq (NI + E_2)(NI + E_1)$$

Explain why this can happen. **Hint:** Multiply both sides out with N in place of λ. Does N commute with E_i? In the above proof of the Cayley-Hamilton theorem, this issue was avoided by considering only polynomials which are of the form $C_0 + C_1\lambda + \cdots$ in which the polynomial identity held because the corresponding matrix coefficients were equal. However, in the proof of the Cayley-Hamilton theorem given in the chapter, one can show that the matrix A does commute with the matrices C_i in that argument.

(26) One often has to raise a matrix to a high power. That is, the matrix will be multiplied by itself k times where k is fairly large. Computer algebra systems are very good at doing this. How does this relate to the Cayley-Hamilton theorem and the minimal polynomial? **Hint**: Use what you know about division of polynomials, Theorem 1.2, to write $A^n = p(A)q(A) + r(A)$ where $p(A)$ is either the minimal polynomial or the characteristic polynomial so that $p(A) = 0$ and $r(\lambda)$ has degree less than p where A is a $p \times p$ matrix.

(27) Show that the minimal polynomial divides the characteristic polynomial. Recall that the minimal polynomial $p(\lambda)$ is the polynomial which satisfies $p(A) = 0$ has leading coefficient equal to 1 and has the smallest degree of all polynomials $q(\lambda)$ which satisfy $q(A) = 0$. **Hint:**Use Theorem 1.2 and the Cayley-Hamilton theorem. This theorem **always** holds even for fields not equal to the real or complex numbers, but we have only proved the Cayley-Hamilton theorem for fields which are such that it makes sense to take $\lim_{|\lambda|\to\infty}$. Later, more general proofs of this important theorem are outlined.

Chapter 8

Characteristic polynomial and eigenvalues of a matrix

Chapter summary

The concept of eigenvectors and eigenvalues is presented and the characteristic polynomial is defined. A theorem about when a matrix is similar to a diagonal matrix is presented. After this, there is an exercise set containing many applications of this idea. Finally, in an optional section, a proof is given of the Jordan Canonical form. Some major applications of this theorem are presented in the exercises.

8.1 The characteristic polynomial

Let A be an $n \times n$ matrix over F. The **characteristic polynomial** is the expression

$$P_A(\lambda) = \det(\lambda I - A).$$

It is a polynomial of degree n in the variable λ of the form

$$P_A(\lambda) = \lambda^n + b_1\lambda^{n-1} + \cdots + b_{n-1}\lambda + b_n.$$

Example 8.1. The characteristic polynomial of

$$A = \begin{pmatrix} a_{11} & a_{12} \\ a_{21} & a_{22} \end{pmatrix}$$

is

$$\begin{aligned} P_A(\lambda) &= \det \begin{pmatrix} \lambda - a_{11} & -a_{12} \\ -a_{21} & \lambda - a_{22} \end{pmatrix} \\ &= \lambda^2 - (a_{11} + a_{22})\lambda + \det A. \end{aligned}$$

Example 8.2. Let $F = \mathbb{R}$. The characteristic polynomial of

$$A = \begin{pmatrix} 1 & 3 & 0 \\ 3 & 1 & 0 \\ 3 & 4 & 6 \end{pmatrix}$$

is

$$P_A(\lambda) = \det \begin{pmatrix} \lambda - 1 & -3 & 0 \\ -3 & \lambda - 1 & 0 \\ -3 & -4 & \lambda - 6 \end{pmatrix}.$$

Expanding by column 3,

$$P_A(\lambda) = (\lambda - 6) \det \begin{pmatrix} \lambda - 1 & -3 \\ -3 & \lambda - 1 \end{pmatrix}$$

$$= (\lambda - 6)((\lambda - 1)^2 - 9)$$
$$= (\lambda - 6)(\lambda - 4)(\lambda + 2).$$

A Maple command will evaluate $P_A(\lambda)$ (and give its zeros) for any given $n \times n$ matrix A, $n < 5$ e.g.

$$A = \begin{pmatrix} 1 & 0 & 0 & 1 \\ 1 & 1 & 0 & 0 \\ 1 & 0 & 0 & 1 \\ 1 & 1 & 1 & 1 \end{pmatrix}$$

has characteristic polynomial $\lambda^4 - 3\lambda^3 + \lambda^2$. Its zeros (which are just the *eigenvalues* referred to in the definition below) are $0, 0, 3/2 \pm 1/2\sqrt{5}$ (shown with multiplicities). Try changing a_{12} to 2; you may be surprised by the horrific appearance of the eigenvalues.

Definition 8.1. Let $\lambda \in F$. We say that λ is an eigenvalue of A if there is a vector $\mathbf{x} \neq \mathbf{0}$ with

$$A\mathbf{x} = \lambda \mathbf{x}.$$

Such an $\mathbf{x}$ is called an **eigenvector** of A.

Example 8.3. Since

$$\begin{pmatrix} 1 & 3 & 0 \\ 3 & 1 & 0 \\ 3 & 4 & 6 \end{pmatrix} \begin{pmatrix} 0 \\ 0 \\ 1 \end{pmatrix} = \begin{pmatrix} 0 \\ 0 \\ 6 \end{pmatrix},$$

we conclude that $\mathbf{e}_3$ is an eigenvector of

$$\begin{pmatrix} 1 & 3 & 0 \\ 3 & 1 & 0 \\ 3 & 4 & 6 \end{pmatrix},$$

and the corresponding eigenvalue is 6.

The usefulness of the characteristic polynomial comes from the following result.

Lemma 8.1. *Let A be an $n \times n$ matrix over F. The eigenvalues of A are the zeros of $P_A(\lambda)$ in F.*

Proof. Clearly λ is an eigenvalue if, and only if, $\ker(A - \lambda I) \neq \{\mathbf{0}\}$. So λ is an eigenvalue if, and only if, $R(A - \lambda I) < n$. (Theorem 5.1 of Chapter 5. This is equivalent in turn to $\det(A - \lambda I) = 0$ (Lemma 7.13). □

The eigenvalues of the matrix A in Example 8.1 are now known to be 6, 4 and -2. A short calculation is needed to find the eigenvectors. These lie in what are called *eigenspaces*.

Definition 8.2. For an eigenvalue λ of A, the **eigenspace** corresponding to λ is $\ker(\lambda I - A)$, in other words the set of solutions of

$$A\mathbf{x} = \lambda\mathbf{x}.$$

Thus the eigenspace corresponding to λ contains all the eigenvectors with eigenvalue λ, plus the vector $\mathbf{0}$.

Example 8.4. Find the eigenspaces of

$$A = \begin{pmatrix} 1 & 3 & 0 \\ 3 & 1 & 0 \\ 3 & 4 & 6 \end{pmatrix}.$$

The eigenspace for $\lambda = 6$ is

$$\ker \begin{pmatrix} 5 & -3 & 0 \\ -3 & 5 & 0 \\ -3 & -4 & 0 \end{pmatrix} = \ker A_6, \text{ say.}$$

Here $A_6 \sim \begin{pmatrix} 1 & 0 & 0 \\ 0 & 1 & 0 \\ 0 & 0 & 0 \end{pmatrix} = B_6$, since the first two columns are a basis for $\mathrm{Col}\,(A_6)$. The corresponding set of equations $B_6\mathbf{x} = 0$ reads $x_1 = x_2 = 0$. The general solution is $(0, 0, x_3)$, so $\mathrm{span}\{\mathbf{e}_3\}$ is the eigenspace for $\lambda = 6$.

Similarly, after row reduction, the eigenspace for $\lambda = 4$ is

$$\ker \begin{pmatrix} 3 & -3 & 0 \\ -3 & 3 & 0 \\ -3 & -4 & -2 \end{pmatrix} = \ker \begin{pmatrix} 1 & 0 & 2/7 \\ 0 & 1 & 2/7 \\ 0 & 0 & 0 \end{pmatrix}.$$

The corresponding equations read $x_1 + 2/7x_3 = 0, x_2 + 2/7x_3 = 0$. The eigenspace for $\lambda = 4$ is seen to be $\mathrm{span}\{(-2, -2, 7)\}$.

Similarly the eigenspace for $\lambda = -2$ is

$$\ker \begin{pmatrix} -3 & -3 & 0 \\ -3 & -3 & 0 \\ -3 & -4 & -8 \end{pmatrix} = \ker \begin{pmatrix} 1 & 0 & -8 \\ 0 & 1 & 8 \\ 0 & 0 & 0 \end{pmatrix}$$

$$= \mathrm{span}\{(8, -8, 1)\}.$$

Example 8.5. (Maple) Find the eigenvalues and eigenspaces of

$$\begin{pmatrix} 5 & 12 & -18 & -3 \\ 2 & 15 & -18 & -3 \\ 2 & 10 & -13 & -3 \\ -4 & -8 & 18 & 8 \end{pmatrix}.$$

Solution. Right click and choose in turn 'Eigenvalues etc.', 'Eigenvectors'. The display is

$$\begin{pmatrix} 5 \\ 5 \\ 3 \\ 2 \end{pmatrix}, \begin{pmatrix} 1/4 & 3/2 & 0 & -1 \\ 1/4 & 3/2 & -1/2 & -1 \\ 0 & 1 & -1/2 & -1 \\ 1 & 0 & 1 & 1 \end{pmatrix},$$

meaning that 5 is an eigenvalue with multiplicity 2 and the corresponding eigenspace is

$$\text{span}\{(1/4, 1/4, 0, 1), (3/2, 3/2, 1, 0)\}$$

3 is an eigenvalue with corresponding eigenspace

$$\text{span}\{(0, -1/2, -1/2, 1)\}$$

and 2 is an eigenvalue with corresponding eigenspace

$$\text{span}\{(-1, -1, -1, 1)\}.$$

We now recall Definition 7.3. We say an $n \times n$ matrix A can be **diagonalized** over F if there exists an invertible $n \times n$ matrix P and a diagonal matrix

$$D = \begin{pmatrix} \lambda_1 & \cdots & 0 \\ \vdots & \ddots & \vdots \\ 0 & \cdots & \lambda_n \end{pmatrix}$$

where each λ_k is in F and

$$P^{-1}AP = D,$$

equivalently

$$A = PDP^{-1}.$$

We can now answer the question: Which matrices A can be diagonalized over F? We can even write down P and D in the equation

$$A = PDP^{-1},$$

(P invertible, D diagonal). The next theorem says that a matrix is diagonalizable if and only if there is a basis of F^n consisting of eigenvectors. Another common term for this situation is to say that when there is such a basis, the matrix is **non-defective**. When there is no such basis of eigenvectors, the matrix is called **defective**.

Theorem 8.1. *Let A be an $n \times n$ matrix over F. The following statements are equivalent.*

(i) A can be diagonalized over F,

(ii) there is a basis $\mathbf{w}_1, \ldots, \mathbf{w}_n$ *of* F^n *with each* $\mathbf{w}_j$ *an eigenvector of* A, $A\mathbf{w}_j = \lambda_j \mathbf{w}_j$.

Proof. If (i) holds, then $A = PDP^{-1}$ (D diagonal, P inverible), D having $\lambda_1 \ldots \lambda_n$ down the main diagonal. Let

$$P = \begin{pmatrix} \mathbf{w}_1 & \cdots & \mathbf{w}_n \end{pmatrix}. \tag{8.1}$$

Now

$$AP = PD. \tag{8.2}$$

This can be rewritten

$$\begin{pmatrix} A\mathbf{w}_1 & \cdots & A\mathbf{w}_n \end{pmatrix} = \begin{pmatrix} \lambda_1\mathbf{w}_1 & \cdots & \lambda_n\mathbf{w}_n \end{pmatrix} \tag{8.3}$$

which yields (ii).

If (ii) holds, then this implies (8.3) which we rewrite in the form (8.1), (8.2). Hence $P^{-1}AP = D$ and this implies (i). □

Example 8.6. We revisit Example 8.2.

Given the eigenvalues and eigenspaces that were found there,

$$\begin{pmatrix} 1 & 3 & 0 \\ 3 & 1 & 0 \\ 3 & 4 & 6 \end{pmatrix} = PDP^{-1},$$

with

$$P = \begin{pmatrix} 0 & -2 & 8 \\ 0 & -2 & -8 \\ 1 & 7 & 1 \end{pmatrix}, \; D = \begin{pmatrix} 6 & 0 & 0 \\ 0 & 4 & 0 \\ 0 & 0 & -2 \end{pmatrix}.$$

All we have done is place a vector from each eigenspace as a column of P (in a position corresponding to the eigenvalue in D). It is easily checked that P is invertible.

To use Maple to decide diagonalizability of A place the cursor over A, right-click on 'Solvers and forms' and select 'Jordan form'. The Jordan form J is a matrix similar to A which is diagonal except possibly for some $1's$ immediately above the diagonal. For instance,

$$\begin{pmatrix} 5 & 0 & 0 & 0 \\ 0 & 5 & 0 & 0 \\ 0 & 0 & 3 & 0 \\ 0 & 0 & 0 & 3 \end{pmatrix}, \begin{pmatrix} 5 & 1 & 0 & 0 \\ 0 & 5 & 0 & 0 \\ 0 & 0 & 3 & 0 \\ 0 & 0 & 0 & 3 \end{pmatrix}, \begin{pmatrix} 5 & 1 & 0 & 0 \\ 0 & 5 & 0 & 0 \\ 0 & 0 & 3 & 1 \\ 0 & 0 & 0 & 3 \end{pmatrix}$$

are examples of Jordan forms of 4×4 matrices of characteristic polynomial $(\lambda - 5)^2(\lambda - 3)^2$. The diagonalizability of A is equivalent to the Jordan form of A being a diagonal matrix. See Hogben (2007) for more detailed information. However, it is impossible to compute the Jordan form for a general matrix unless the eigenvalues can be found exactly, even for computer algebra systems, so this procedure will not always work.

The linear independence of the columns of P in Example 8.6 could have been predicted from the following result.

Lemma 8.2. *Let $\lambda_1, \ldots, \lambda_k$ be distinct eigenvalues of A with corresponding eigenvectors $\mathbf{v}_1, \ldots, \mathbf{v}_k$. Then $\mathbf{v}_1, \ldots, \mathbf{v}_k$ is a linearly independent set.*

Proof. Suppose the claim of the lemma is not true. Then there exists a subset of this set of vectors

$$\{\mathbf{w}_1, \cdots, \mathbf{w}_r\} \subseteq \{\mathbf{v}_1, \cdots, \mathbf{v}_k\}$$

such that

$$\sum_{j=1}^{r} c_j \mathbf{w}_j = \mathbf{0} \tag{8.4}$$

where each $c_j \neq 0$. Say $A\mathbf{w}_j = \mu_j \mathbf{w}_j$ where

$$\{\mu_1, \cdots, \mu_r\} \subseteq \{\lambda_1, \cdots, \lambda_k\},$$

the μ_j being distinct eigenvalues of A. Out of all such subsets, let this one be such that r is as small as possible. Then necessarily, $r > 1$ because otherwise, $c_1 \mathbf{w}_1 = \mathbf{0}$ which would imply $\mathbf{w}_1 = \mathbf{0}$, which is not allowed for eigenvectors.

Now apply A to both sides of (8.4).

$$\sum_{j=1}^{r} c_j \mu_j \mathbf{w}_j = \mathbf{0}. \tag{8.5}$$

Next pick $\mu_k \neq 0$ and multiply both sides of (8.4) by μ_k. Such a μ_k exists because $r > 1$. Thus

$$\sum_{j=1}^{r} c_j \mu_k \mathbf{w}_j = \mathbf{0} \tag{8.6}$$

Subtract the sum in (8.6) from the sum in (8.5) to obtain

$$\sum_{j=1}^{r} c_j \left(\mu_k - \mu_j\right) \mathbf{w}_j = \mathbf{0}$$

Now one of the constants $c_j \left(\mu_k - \mu_j\right)$ equals 0, when $j = k$. Therefore, r was not as small as possible after all. $\square$

Example 8.7. We can easily construct a matrix that cannot be diagonalized.

With any field F, let

$$A = \begin{pmatrix} 1 & 1 \\ 0 & 1 \end{pmatrix}.$$

Then A cannot be diagonalized over F. For obviously 1 is the only eigenvalue. The eigenspace is

$$\ker \begin{pmatrix} 0 & -1 \\ 0 & 0 \end{pmatrix} = \text{span}\{(1,0)\}.$$

There is no basis of F^2 consisting of eigenvectors of A.

Example 8.8. Let θ be an angle, $0 < \theta < \pi$, and $A = \begin{pmatrix} \cos\theta & -\sin\theta \\ \sin\theta & \cos\theta \end{pmatrix}$.

Then A can be diagonalized over $\mathbb{C}$ but cannot be diagonalized over $\mathbb{R}$.

The characteristic polynomial is

$$(\lambda - \cos\theta)^2 + \sin^2\theta = (\lambda - \cos\theta + i\sin\theta)(\lambda - \cos\theta - i\sin\theta).$$

So the eigenvalues are $\cos\theta \pm i\sin\theta$ (note that $\sin\theta > 0$). Lemma 8.2 guarantees two linearly independent eigenvectors (we need not compute them) and A can be diagonalized over $\mathbb{C}$. On the other hand, if we work over the field $\mathbb{R}$, there are no eigenvalues and no eigenvectors.

A simple geometrical argument also shows the absence of eigenvectors, since the rotated vector $A\mathbf{x}$ lies outside $\text{span}\{\mathbf{x}\}$ whenever $\mathbf{x} \neq \mathbf{0}, \mathbf{x} \in \mathbb{R}^2$.

It is important to note that transposing a square matrix does not affect the determinant or the eigenvalues. This follows from Lemma 7.12.

It is a simple consequence of Lemma 7.12 that **column operations** affect the determinant in just the same way as row operations, e.g.

$$\det \begin{pmatrix} \mathbf{a}_1 + k\mathbf{a}_2 & \mathbf{a}_2 & \mathbf{a}_3 \end{pmatrix} = \det \begin{pmatrix} \mathbf{a}_1 & \mathbf{a}_2 & \mathbf{a}_3 \end{pmatrix}.$$

In this particular case we would argue as follows that:

$$\begin{aligned} \det \begin{pmatrix} \mathbf{a}_1 + k\mathbf{a}_2 & \mathbf{a}_2 & \mathbf{a}_3 \end{pmatrix} &= \det \begin{pmatrix} \mathbf{a}_1^t + k\mathbf{a}_2^t \\ \mathbf{a}_2^t \\ \mathbf{a}_3^t \end{pmatrix} \\ &= \det \begin{pmatrix} \mathbf{a}_1^t \\ \mathbf{a}_2^t \\ \mathbf{a}_3^t \end{pmatrix} = \det \begin{pmatrix} \mathbf{a}_1 & \mathbf{a}_2 & \mathbf{a}_3 \end{pmatrix}. \end{aligned}$$

It can be shown that if λ has multiplicity h as a zero of the characteristic polynomial P_A, the corresponding eigenspace has dimension *at most* h. See Corollary 8.1 presented later in the section on the Jordan form. Also see Problem 23 on Page 218.

An application of eigenvalues and eigenvectors is to the computation of powers of a matrix. We illustrate with a representative example.

Example 8.9. Let $A = \begin{pmatrix} 3 & 2 & 2 \\ 3 & 5 & 3 \\ -7 & -8 & -6 \end{pmatrix}$. Find A^{35} in terms of powers of the matrix.

The eigenvalues and eigenvectors are

$$\begin{pmatrix} 0 \\ -1 \\ 1 \end{pmatrix} \leftrightarrow 2, \begin{pmatrix} 1 \\ 1 \\ -3 \end{pmatrix} \leftrightarrow -1, \begin{pmatrix} -1 \\ 0 \\ 1 \end{pmatrix} \leftrightarrow 1$$

Therefore, a matrix which will produce the desired diagonalization is

$$P = \begin{pmatrix} 0 & 1 & -1 \\ -1 & 1 & 0 \\ 1 & -3 & 1 \end{pmatrix}.$$

Thus

$$P^{-1} = \begin{pmatrix} -1 & -2 & -1 \\ -1 & -1 & -1 \\ -2 & -1 & -1 \end{pmatrix}$$

and $P^{-1}AP$ is a diagonal matrix D having the diagonal entries $2, -1, 1$ from upper left to lower right. Now

$$\begin{aligned} A^n &= PDP^{-1}PDP^{-1} \cdots PDP^{-1} \\ &= PD^nP^{-1}. \end{aligned}$$

It is very easy to raise a diagonal matrix to a power:

$$\begin{pmatrix} \lambda_1 & & & \\ & \lambda_2 & & \\ & & \ddots & \\ & & & \lambda_n \end{pmatrix}^r = \begin{pmatrix} \lambda_1^r & & & \\ & \lambda_2^r & & \\ & & \ddots & \\ & & & \lambda_n^r \end{pmatrix}$$

and so, continuing with our example,

$$\begin{aligned} A^{35} &= \begin{pmatrix} -1 & -2 & -1 \\ -1 & -1 & -1 \\ -2 & -1 & -1 \end{pmatrix} \begin{pmatrix} 2^{35} & 0 & 0 \\ 0 & -1 & 0 \\ 0 & 0 & 1 \end{pmatrix} \begin{pmatrix} 0 & 1 & -1 \\ -1 & 1 & 0 \\ 1 & -3 & 1 \end{pmatrix} \\ &= \begin{pmatrix} -3 & -2^{35}+5 & 2^{35}-1 \\ -2 & -2^{35}+4 & 2^{35}-1 \\ -2 & -2^{36}+4 & 2^{36}-1 \end{pmatrix}. \end{aligned}$$

8.2 Exercises

(1) Here are some matrices. The eigenvalues are 1,2, and 3. Find the eigenvectors which go with the given eigenvalues.

(a) $\begin{pmatrix} -11 & -3 & -7 \\ 4 & 3 & 2 \\ 22 & 5 & 14 \end{pmatrix}$

(b) $\begin{pmatrix} -7 & -2 & -5 \\ -4 & 1 & -2 \\ 18 & 4 & 12 \end{pmatrix}$

(c) $\begin{pmatrix} 1 & -2 & 2 \\ -1 & 0 & 2 \\ -1 & -3 & 5 \end{pmatrix}$

(d) $\begin{pmatrix} -17 & -4 & -11 \\ -14 & -1 & -8 \\ 38 & 8 & 24 \end{pmatrix}$

(2) The following matrix has eigenvalues $-3, 1$.

$$\begin{pmatrix} -25 & -11 & -78 \\ -4 & -5 & -12 \\ 8 & 4 & 25 \end{pmatrix}$$

Describe all eigenvectors which correspond to the given eigenvalues. Does there exist a basis of eigenvectors? When there is no basis of eigenvectors, the matrix is called **defective.** By Theorem 8.1 this implies that a defective matrix cannot be diagonalized.

(3) The following matrix has eigenvalues $-1, 1$.

$$\begin{pmatrix} -1 & 0 & 0 \\ 0 & -1 & 0 \\ 2 & 2 & 1 \end{pmatrix}$$

Find all eigenvectors which go with each of these eigenvalues. Does there exist a basis of eigenvectors?

(4) Write each of the above matrices in Problem 1 in the form $S^{-1}DS$ where

$$D = \begin{pmatrix} 1 & 0 & 0 \\ 0 & 2 & 0 \\ 0 & 0 & 3 \end{pmatrix}.$$

(5) For each matrix A in Problem 1, find A^{23} in terms of powers of the eigenvalues.

(6) Suppose A, B are two $n \times n$ matrices and that there exists a single matrix P such that $P^{-1}AP$ and $P^{-1}BP$ are both diagonal matrices. Show that $AB = BA$.

(7) Let T be an upper triangular matrix which can be diagonalized because all the entries on the main diagonal are distinct. Show that, corresponding to the ith diagonal entry from the top left corner, there exists an eigenvector of the form $\begin{pmatrix} \mathbf{a} \\ \mathbf{0} \end{pmatrix}$ where $\mathbf{a}$ is an $i \times 1$ matrix. Show that this implies that there exists an upper triangular S such that $S^{-1}TS$ is a diagonal matrix.

(8) A **discrete dynamical system** is of the form

$$\mathbf{x}_n = A\mathbf{x}_{n-1},\ \mathbf{x}_0 = \mathbf{a}$$

where A is an $p \times p$ matrix. Suppose that A can be diagonalized. Recall that this means that there exists a basis $\{\mathbf{y}_1, \ldots, \mathbf{y}_p\}$ of eigenvectors for F^p. Here we assume $F = \mathbb{R}$. Letting λ_k be an eigenvalue which corresponds to $\mathbf{y}_k$, show that the solution to the dynamical system is

$$\mathbf{x}_m = \sum_{k=1}^{p} a_k \lambda_k^m \mathbf{y}_k$$

where the a_k are defined by

$$\mathbf{a} = \sum_{k=1}^{p} a_k \mathbf{y}_k.$$

(9) ↑An important example of a dynamical system is the case where A is a **Markov matrix.**

Definition 8.3. A matrix $A = (a_{ij})$ is a Markov matrix if each $a_{ij} \geq 0$ and $\sum_i a_{ij} = 1$. That is, the sum along any column equals 1.

Show that 1 is always an eigenvalue of a Markov matrix and that if λ is an eigenvalue, then $|\lambda| \leq 1$. **Hint:** Recall that the eigenvalues of A and A^t are the same because these matrices have the same characteristic polynomials. Establish the result for A^t and then draw the desired conclusion for A. Consider the column vector consisting entirely of ones.

(10) ↑Suppose A is a Markov matrix $A = (a_{ij})$ where each $a_{ij} > 0$. Show that other than 1, all eigenvalues λ satisfy $|\lambda| < 1$. **Hint:** Let $B = A^t$, so that the sum along the rows equals 1. Then if $|\lambda| = 1$ where λ is an eigenvalue with eigenvector $\mathbf{x}$,

$$\sum_j b_{ij} x_j = \lambda x_i.$$

Take the product of both sides with $\overline{\lambda x_i}$. Explain why

$$|x_i|^2 \geq \sum_j b_{ij} |x_j| |x_i| \geq \sum_j b_{ij} x_j \overline{\lambda x_i} = |x_i|^2$$

and why this requires that for each j, $x_j \overline{\lambda x_i} \geq 0$. Hence λ is real and nonnegative.

(11) Suppose that A is a Markov matrix which can be diagonalized. Suppose also that 1 is an eigenvalue of algebraic multiplicity 1 and that all the other eigenvalues have absolute value less than 1. Show that for $A\mathbf{x} = \mathbf{x}$, and $\mathbf{x}_n$ the solution to the dynamical system

$$\mathbf{x}_n = A\mathbf{x}_{n-1}, \ \mathbf{x}_n = \mathbf{a},$$

the following limit is obtained.

$$\lim_{n \to \infty} \mathbf{x}_n = k\mathbf{x}$$

where k is chosen in such a way that $k \left(\sum_i x_i\right) = \sum_i a_i$. **Hint:** Show that for A a Markov matrix, $\sum_i (A\mathbf{x})_i = \sum_i x_i$.

(12) Using the method of the above problem, find $\lim_{n\to\infty} \mathbf{x}_n$ where $\mathbf{x}_n$ is the solution to the dynamical system

$$\mathbf{x}_n = A\mathbf{x}_{n-1},\ \mathbf{x}_0 = \mathbf{a}$$

where $\mathbf{a} = (1,1,1)^t$ and A is the Markov matrix

$$A = \begin{pmatrix} .5 & .4 & .8 \\ .3 & .1 & .1 \\ .2 & .5 & .1 \end{pmatrix}.$$

(13) Let A be an $n \times n$ matrix. Show that $(-1)^n \det(A)$ is the constant term of the characteristic polynomial of A, $\det(\lambda I - A)$.

(14) Let A be a complex $n \times n$ matrix. Show that the coefficient of λ^{n-1} in the characteristic polynomial is the trace of A, defined as $\sum_i a_{ii}$, the sum of the entries down the main diagonal. Also verify that the trace of A is equal to the sum of the eigenvalues.

(15) If $\det(A) \neq 0$, show that A^{-1} may be obtained as a polynomial in A. **Hint:** Use the Cayley-Hamilton theorem.

(16) Recall Problem 24 on Page 100 which shows that if V, W are two vector spaces over a field of scalars F then $\mathcal{L}(V, W)$ is also a vector space over the same field of scalars. Let $\{v_1, \ldots, v_n\}$ be a basis for V and suppose $Lv_i = w_i \in W$. Thus L is defined on the basis vectors of V. Now define L on all linear combinations of these basis vectors as follows.

$$L\left(\sum_{k=1}^{n} c_k v_k\right) = \sum_{k=1}^{n} c_k L v_k.$$

Verify that with this definition, $L \in \mathcal{L}(V, W)$ and is well defined and uniquely determined. Thus you can define a linear transformation by specifying what it does to a basis.

(17) ↑In the case that V and W are finite dimensional vector spaces, it is natural to ask for the dimension of $\mathcal{L}(V, W)$. Let $\{v_1, \ldots, v_n\}$ be a basis for V and let $\{w_1, \ldots, w_m\}$ be a basis for W. Define the linear transformation denoted by $w_i v_j$ as follows.

$$w_i v_j (v_k) = w_i \delta_{jk}$$

Thus $w_i v_j (v_k) = w_i$ if $k = j$ and $w_i v_j (v_k) = 0$ if $k \neq j$. Show that if $L \in \mathcal{L}(V, W)$, then

$$L = \sum_{i,j} L_{ij} w_i v_j$$

where the L_{ij} are defined by

$$L v_i = \sum_j L_{ji} w_j$$

Thus $\{w_i v_j\}_{i,j}$ is a spanning set of $\mathcal{L}(V, W)$. These special transformations are called **dyadics**.

(18) ↑In the situation of the above problem, show $\{w_i v_j\}_{i,j}$ are linearly independent in addition to spanning $\mathcal{L}(V,W)$. Therefore, these form a basis. Conclude that the dimension of $\mathcal{L}(V,W) = mn$. The matrix of $L \in \mathcal{L}(V,W)$ is defined as (L_{ij}) where

$$L = \sum_{i,j} L_{ij} w_i v_j.$$

(19) ↑Show that if $T \in \mathcal{L}(V,V)$ where $\dim(V) = n$, there exists a unique monic polynomial $p(\lambda)$ of minimal degree which satisfies $p(T) = 0$. This is a more elegant version of Problem 24 on Page 205 because it makes no reference to matrices or particular coordinate systems.

(20) For $T \in \mathcal{L}(V,V)$, give conditions on the minimal polynomial for T which will ensure that T^{-1} exists. Then find a formula for T^{-1} in terms of a polynomial in T.

(21) Let V, W be finite dimensional vector spaces having bases $\{v_1, \ldots, v_n\}$ and $\{w_1, \ldots, w_m\}$ respectively. Also let $\{w_i v_j\}_{i,j}$ be as defined in Problem 17 and the problems following that one. Show that $(w_i v_j)(v_k w_l) = \delta_{jk} w_i w_l$. Also show that $\mathrm{id}_W = \sum_{i,j} \delta_{ij} w_i w_j$, where id_W denotes the identity map on W. Thus the matrix of the identity map is just the usual identity matrix. Now consider the identity with respect to two different bases on W.

$$\underset{W}{\mathrm{id}} = \sum_{i,j} c_{ij} w_i' w_j, \ \underset{W}{\mathrm{id}} = \sum_{r,s} d_{rs} w_r w_s'.$$

Show that the matrices (c_{ij}) and (d_{rs}) are inverses of each other.

(22) ↑Now suppose $\{v_1, \ldots, v_n\}$ and $\{w_1, \ldots, w_m\}$ are two bases for V and W respectively and let $\{v_1', \ldots, v_n'\}$ and $\{w_1', \ldots, w_m'\}$ denote respectively two other bases for V and W and let $L \in \mathcal{L}(V,W)$. How do the matrices of L with respect to these bases compare? **Hint:**

$$\sum_{i,j} L_{ij} w_i v_j = L = \underset{W}{\mathrm{id}}\, L \,\underset{V}{\mathrm{id}} = \sum_{i,j,p,q,r,s} c_{ij} w_i' w_j \left(L_{pq} w_p v_q\right) d_{rs} v_r v_s'.$$

Now use the preceding problem to get

$$\sum_{i,j,r,s} w_i' c_{ij} L_{jr} d_{rs} v_s'.$$

In the case where $W = V$, establish the usual result that $[L]_\beta = P^{-1} [L]_{\beta'} P$ for a suitable matrix P as a special case.

(23) Let A be a complex $n \times n$ matrix with characteristic polynomial $P_A(\lambda)$. Suppose an eigenvalue μ has algebraic multiplicity h. This means that when the characteristic polynomial is factored, it is of the form

$$P_A(\lambda) = (\lambda - \mu)^h q(\lambda)$$

where $q(\mu) \neq 0$. Show that the eigenspace for μ has dimension at most h. **Hint:** Use the Cayley-Hamilton theorem and Problem 20 on Page 150 to show that

$$\mathbb{C}^n = \ker\left(\left(A - \mu I\right)^h\right) \oplus \ker\left(q(A)\right).$$

Now let $\beta_1 = \{\mathbf{v}_1, \ldots, \mathbf{v}_r\}$ be a basis for $\ker(A - \mu I)$ and extend this to a basis for $\ker\left((A - \mu I)^h\right)$ which is of the form $\{\mathbf{v}_1, \ldots, \mathbf{v}_r, \beta_2\}$. Now letting β_3 be a basis for $\ker(q(A))$ and $\beta = \{\beta_1, \beta_2, \beta_3\}$, explain why β is a basis and why the matrix $[A]_\beta$ is a block diagonal matrix of the form

$$\begin{pmatrix} D & & \\ & E & \\ & & F \end{pmatrix}$$

where D is a diagonal matrix which has r entries down the diagonal, each equal to μ. Show that $(\lambda - \mu)^r$ divides the characteristic polynomial, and so $r \leq h$.

(24) Let A be an $m \times n$ matrix and let B be an $n \times m$ matrix for $m \leq n$. Then denoting by $p_N(t)$ the characteristic polynomial of the matrix N, show that

$$p_{BA}(t) = t^{n-m} p_{AB}(t),$$

so the eigenvalues of BA and AB are the same including multiplicities except that BA has $n - m$ extra zero eigenvalues. **Hint:** Use block multiplication to write

$$\begin{pmatrix} AB & 0 \\ B & 0 \end{pmatrix} \begin{pmatrix} I & A \\ 0 & I \end{pmatrix} = \begin{pmatrix} AB & ABA \\ B & BA \end{pmatrix}$$

$$\begin{pmatrix} I & A \\ 0 & I \end{pmatrix} \begin{pmatrix} 0 & 0 \\ B & BA \end{pmatrix} = \begin{pmatrix} AB & ABA \\ B & BA \end{pmatrix}.$$

Explain why this implies that the following two matrices are similar and therefore have the same characteristic equation.

$$\begin{pmatrix} AB & 0 \\ B & 0 \end{pmatrix}, \begin{pmatrix} 0 & 0 \\ B & BA \end{pmatrix}$$

Now note that AB is $m \times m$ and BA is $n \times n$. Explain why

$$t^m \det(tI - BA) = t^n \det(tI - AB)$$

(25) In the case where A is an $m \times n$ matrix with $m \geq n$ and A has full rank, Problem 2 on Page 172 outlined why there exists an $n \times n$ matrix Q having the property that $Q^* Q = QQ^* = I$ obtained from the Gram-Schmidt procedure by having the columns of Q form an orthonormal set such that $A = QR$ where R is upper triangular in the sense that $r_{ij} = 0$ if $i > j$. Show, using block multiplication that there exists Q_1 an $m \times n$ matrix and R an upper triangular $n \times n$ matrix such that $A = Q_1 R$. This is called the thin QR factorization by Golub and Van Loan. [3]

(26) Suppose A is a complex $n \times n$ matrix which can be diagonalized. Consider the first order initial value problem

$$\mathbf{x}' = A\mathbf{x}, \ \mathbf{x}(0) = \mathbf{x}_0.$$

Letting $\{\mathbf{v}_1, \ldots, \mathbf{v}_n\}$ be a basis of eigenvectors, $A\mathbf{v}_k = \lambda_k \mathbf{v}_k$, show that there is a unique solution to the initial value problem and it is given by

$$\mathbf{x}(t) = \sum_{k=1}^{n} a_k \mathbf{v}_k e^{\lambda_k t}$$

where a_k is defined by

$$\mathbf{x}_0 = \sum_{k=1}^{n} a_k \mathbf{v}_k.$$

For the meaning of $e^{\lambda_k t}$, see Proposition 5.3. **Hint:** By assumption, $P = \begin{pmatrix} \mathbf{v}_1 & \cdots & \mathbf{v}_n \end{pmatrix}$ is such that

$$P^{-1}AP = D,\ PDP^{-1} = A$$

where D is a diagonal matrix having the eigenvalues of A down the main diagonal. Hence

$$\mathbf{x}' = PDP^{-1}\mathbf{x}.$$

Now let $\mathbf{y} = P^{-1}\mathbf{x}$ and argue that

$$y_i'(t) = \lambda_i y_i(t),\ y_i(0) = \left(P^{-1}\mathbf{x}_0\right)_i(0)$$

Use Proposition 5.3 to determine $\mathbf{y}$ and then use $\mathbf{x} = P\mathbf{y}$.

(27) Let $\Phi(t) = \begin{pmatrix} \mathbf{y}_1(t) & \cdots & \mathbf{y}_n(t) \end{pmatrix}$ where each $\mathbf{y}_k(t)$ is a differentiable vector valued function. Show that

$$(\det \Phi)'(t) = \sum_{k=1}^{n} \det \begin{pmatrix} \mathbf{y}_1(t) & \cdots & \mathbf{y}_k'(t) & \cdots & \mathbf{y}_n(t) \end{pmatrix}.$$

(28) Establish Abel's formula in the case where A can be diagonalized. This formula says that if $\Phi(t)$ is an $n \times n$ matrix whose columns are solutions of the differential equation

$$\mathbf{x}'(t) = A\mathbf{x}(t),$$

then there exists a constant C such that

$$\det \Phi(t) = Ce^{\text{trace}(A)t}$$

which shows that $\Phi(t)$ is either invertible for all t or for no t. This determinant is called the Wronskian. **Hint:** Reduce to the case where

$$\Psi^{t\prime} = \Psi^t D.$$

for D a diagonal matrix having the eigenvalues of A down the main diagonal and $\Psi = P^{-1}\Phi$ where $D = P^{-1}AP$. Use Problem 14 above or else note that, since D is similar to A, its trace equals the trace of A. (Problem 23 on Page 218.)

(29) Let A be a complex $n \times n$ matrix. The ith Gerschgorin disk is the set of all $z \in \mathbb{C}$ such that

$$|a_{ii} - z| \leq \sum_{j \neq i} |a_{ij}| .$$

Thus there are n **Gerschgorin disks**, one for each diagonal entry of A. Gerschgorin's theorem says that all eigenvalues of A are contained in the union of the Gerschgorin disks. Prove this important theorem.

(30) Explain why A is invertible if and only if it has no zero eigenvalues. A complex $n \times n$ matrix is called **diagonally dominant** if for each i,

$$|a_{ii}| > \sum_{j \neq i} |a_{ij}| .$$

Show that such a diagonally dominant matrix must be invertible.

(31) Explain carefully why every complex $n \times n$ matrix has an eigenvector.

(32) ↑Let A be a complex $n \times n$ matrix. Show that there exists an orthonormal basis $\{\mathbf{u}_1, \cdots, \mathbf{u}_n\}$ such that $A\mathbf{u}_1 = \lambda \mathbf{u}_1$.

(33) ↑Let A be a complex $n \times n$ matrix. Show that there exists a matrix U such that $U^* A U$ is of the form

$$\begin{pmatrix} \lambda & \mathbf{r} \\ \mathbf{0} & A_1 \end{pmatrix}$$

and $U^* U = I$ where $U^* = \left(\overline{U}\right)^t$, the Cayley-Hamilton of the conjugate of U. That is, you replace every entry of U with its complex conjugate and then take the transpose of the result. Thus the ij^{th} entry of $U^* U$ is $\overline{\mathbf{u}_i} \cdot \mathbf{u}_j = \langle \mathbf{u}_j, \mathbf{u}_i \rangle$. In the formula, A_1 is an $(n-1) \times (n-1)$ matrix, $\mathbf{0}$ is an $(n-1) \times 1$ matrix and $\mathbf{r}$ is a $1 \times (n-1)$ matrix.

(34) ↑Let A be a complex $n \times n$ matrix. Show that there exists a matrix U such that $U^* U = I$ and $U^* A U$ is an upper triangular matrix T. **Hint:** Use the above result and induction. This is Schur's theorem, possibly the most important theorem in spectral theory.

(35) ↑Let T be a complex upper triangular $n \times n$ matrix. Show by induction on the size of T that there exists a unique solution to the first order initial value problem given by

$$\begin{pmatrix} x_1'(t) \\ \vdots \\ x_n'(t) \end{pmatrix} = T \begin{pmatrix} x_1(t) \\ \vdots \\ x_n(t) \end{pmatrix} + \begin{pmatrix} f_1(t) \\ \vdots \\ f_n(t) \end{pmatrix}, \begin{pmatrix} x_1(0) \\ \vdots \\ x_n(0) \end{pmatrix} = \begin{pmatrix} x_{10} \\ \vdots \\ x_{n0} \end{pmatrix}.$$

This is written more compactly as $\mathbf{x}' = T\mathbf{x} + \mathbf{f}$, $\mathbf{x}(0) = \mathbf{x}_0$. Here you must assume only that each f_i is a continuous function so that it is possible to take an integral. **Hint:** To solve the scalar equation $y' = ay + f$, write as $y' - ay = f$ and multiply both sides by e^{-at}. Verify that this reduces the left side to $\frac{d}{dt}\left(e^{-at} y\right)$. Then integrate both sides to get rid of the derivative. Hence, in particular, there exists a unique solution in the case that T is 1×1.

(36) ↑Let A be an $n \times n$ complex matrix. Show that there exists a unique solution to the initial value problem $\mathbf{y}' = A\mathbf{y} + \mathbf{f}$, $\mathbf{y}(0) = \mathbf{y}_0$. **Hint:** Use the above problem and Schur's theorem. When you have done this problem, you have almost done all the mathematical content of a typical ordinary differential equations class.

(37) Let A be a complex $n \times n$ matrix. The trace of A denoted as trace(A) is defined as the sum of the entries on the main diagonal of A. Thus the trace equals $\sum_i A_{ii}$. Show that if A is simlar to B then they have the same trace. Then show using Schur's theorem that the trace is always equal to the sum of the eigenvalues.

(38) ↑Let A, B be complex $m \times n$ matrices. The Frobenius inner product is defined as $\langle A, B \rangle = \text{trace}(AB^*)$ where B^* is defined as $\left(\overline{B}\right)^t$. Show this is an inner product and that $||A||^2 = \sum_{i,j} |A_{ij}|^2$. Show that the matrices e_{ij} obtained by placing a 1 in the ij^{th} position and a 0 in every other position is an orthonormal basis for the space of $m \times n$ matrices. Thus the dimension of the vector space of $m \times n$ matrices is mn.

(39) ↑Let A be an $n \times n$ complex matrix. Show that there exists a unique monic polynomial $p(\lambda)$ having smallest possible degree such that $p(A) = 0$. Also describe an easy way to find this minimal polynomial. (Monic means the coefficient of the highest power of λ equals 1.) Using the Cayley-Hamilton theorem, explain why the minimal polynomial must divide the characteristic polynomial.

(40) Find the minimal polynomial for the matrix

$$\begin{pmatrix} -14 & -30 & 0 \\ 5 & 11 & 0 \\ -11 & -23 & 1 \end{pmatrix}$$

Is this the same as the characteristic polynomial?

(41) Find the minimal polynomial for the matrix

$$\begin{pmatrix} 5 & 12 & 16 \\ 0 & 1 & 0 \\ -1 & -3 & -3 \end{pmatrix}$$

Is this the same as the characteristic polynomial?

(42) If you have an $n \times n$ matrix A, then you can obtain its characteristic polynomial $\det(\lambda I - A)$. This is a monic polynomial of degree n. Suppose you start with a monic polynomial of degree n. Can you produce a matrix which has this polynomial as its characteristic polynomial? The answer is yes, and such a matrix is called a companion matrix. Show that the matrix

$$\begin{pmatrix} 0 & \cdots & 0 & -a_0 \\ 1 & 0 & \cdots & -a_1 \\ & \ddots & & \vdots \\ & & 1 & -a_{n-1} \end{pmatrix}$$

is a companion matrix for the polynomial $p(\lambda) = \lambda^n + a_{n-1}\lambda^{n-1} + \cdots + a_1\lambda + a_0$.

8.3 The Jordan canonical form*

Earlier we showed how to find the matrix of a linear transformation with respect to various bases. The Jordan form of a matrix is one of these matrices. It exists exactly when the minimal polynomial, described above in Problem 24 on Page 205, **splits.** This term means that there exist scalars $\lambda_i \in F$ such that

$$p(\lambda) = (\lambda - \lambda_1)^{r_1} (\lambda - \lambda_2)^{r_2} \cdots (\lambda - \lambda_p)^{r_p}.$$

One situation in which this always happens is when the field of scalars equals $\mathbb{C}$. The Jordan form is the next best thing to a diagonal matrix.

Recall the definition of direct sum of subspaces given in Problem 19 on Page 150. One writes

$$V = V_1 \oplus \cdots \oplus V_p \tag{8.7}$$

if every $v \in V$ can be written in a unique way in the form $v = v_1 + \cdots + v_p$ where each $v_k \in V_k$. Recall also from this problem that if β_j is a basis for V_j, then $\{\beta_1, \cdots, \beta_p\}$ is a basis for V.

Theorem 8.2. *Let T be a linear transformation mapping V to V, an n dimensional vector space, and suppose (8.7) holds for the subspaces V_i of V. Suppose also that $T : V_i \to V_i$. Letting T_i denote the restriction of T to V_i, suppose that the matrix of T_i with respect to V_i is M_i with respect to β_i, a basis for V_i. Then it follows that that the matrix of T with respect to $\{\beta_1, \cdots, \beta_p\}$ is the block diagonal matrix*

$$M = \begin{pmatrix} M_1 & & \\ & \ddots & \\ & & M_p \end{pmatrix}.$$

Proof. Let θ denote the coordinate map corresponding to $\beta = \{\beta_1, \cdots, \beta_p\}$. Also, letting β_i be the above basis for V_i, let θ_i denote the coordinate map from V_i to F^{d_i} where d_i is the dimension of V_i. It follows that for $v = v_1 + \cdots + v_p$ where each $v_i \in V_i$,

$$\theta v = (\theta_1 v_1, \cdots, \theta_p v_p).$$

Thus from the definition of the matrix of a linear transformation, and using block multiplication,

$$M\theta v = \begin{pmatrix} M_1\theta_1 v_1 \\ \vdots \\ M_p\theta_p v_p \end{pmatrix} = \begin{pmatrix} \theta_1 T_1 v_1 \\ \vdots \\ \theta_p T_p v_p \end{pmatrix} = \theta (T_1 v_1 + \cdots + T_p v_p) = \theta T v.$$

Therefore, M is the matrix of T with respect to β as claimed. □

Definition 8.4. Let W be a vector space and let $N : W \to W$ be linear and satisfy $N^r w = 0$ for all $w \in W$. Such a linear mapping is called **nilpotent.**

We also need the following definition.

Definition 8.5. $J_k(\alpha)$ is a **Jordan block** if it is a $k \times k$ matrix of the form

$$J_k(\alpha) = \begin{pmatrix} \alpha & 1 & & 0 \\ 0 & \ddots & \ddots & \\ \vdots & \ddots & \ddots & 1 \\ 0 & \cdots & 0 & \alpha \end{pmatrix}$$

In words, there is an unbroken string of ones down the super diagonal, and the number α filling every space on the main diagonal, with zeros everywhere else.

Proposition 8.1. Let $N \in \mathcal{L}(W, W)$ be nilpotent, $N^m = 0$ for some $m \in \mathbb{N}$. Here W is an p dimensional vector space with field of scalars $\mathbb{F}$. Then there exists a basis for W such that the matrix of N with respect to this basis is of the form

$$\begin{pmatrix} J_{r_1}(0) & & & \\ & J_{r_2}(0) & & \\ & & \ddots & \\ & & & J_{r_s}(0) \end{pmatrix}$$

where $r_1 \geq r_2 \geq \cdots \geq r_s \geq 1$ and $\sum_{i=1}^{s} r_i = p$. In the above, the $J_{r_j}(0)$ is a Jordan block of size $r_j \times r_j$ with 0 down the main diagonal.

Proof. First note the only eigenvalue of N is 0. This is because if $Nv = \lambda v$, $v \neq 0$, and $\lambda \neq 0$, then $0 = N^m v = \lambda^{m+1} v$ which contradicts the requirement that $v \neq 0$.

Let v_1 be an eigenvector. Then $\{v_1, v_2, \cdots, v_r\}$ is called a chain based on v_1 if $Nv_{k+1} = v_k$ for all $k = 0, 1, 2, \cdots, r$ where v_0 is defined as 0. It will be called a maximal chain if there is no solution v, to the equation $Nv = v_r$.

Claim 1. The vectors in any chain are linearly independent and for

$$\{v_1, v_2, \cdots, v_r\}$$

a chain based on v_1,

$$N : \text{span}\{v_1, v_2, \cdots, v_r\} \to \text{span}\{v_1, v_2, \cdots, v_r\}. \tag{8.8}$$

Also if $\{v_1, v_2, \cdots, v_r\}$ is a chain, then $r \leq p$.

Proof. First note that (8.8) is obvious because $N \sum_{i=1}^{r} c_i v_i = \sum_{i=2}^{r} c_i v_{i-1}$. It only remains to verify that the vectors of a chain are independent. This is true if $r = 1$. Suppose true for $r - 1 \geq 1$. Then if $\sum_{k=1}^{r} c_k v_k = 0$, an application of N to both sides yields $\sum_{k=2}^{r} c_k v_{k-1} = 0$. By induction, each $c_k, k > 1$ equals 0. Hence $c_1 = 0$ also and so the claim follows.

Consider the set of all chains based on eigenvectors. Since all have total length no larger than p, it follows that there exists one which has maximal length,

$$\{v_1^1, \cdots, v_{r_1}^1\} \equiv B_1.$$

If span $\{B_1\}$ contains all eigenvectors of N, then stop. Otherwise, consider all chains based on eigenvectors not in span $\{B_1\}$ and pick one, $B_2 \equiv \{v_1^2, \cdots, v_{r_2}^2\}$ which is as long as possible. Thus $r_2 \leq r_1$. If span $\{B_1, B_2\}$ contains all eigenvectors of N, stop. Otherwise, consider all chains based on eigenvectors not in span $\{B_1, B_2\}$ and pick one, $B_3 \equiv \{v_1^3, \cdots, v_{r_3}^3\}$ such that r_3 is as large as possible. Continue this way. Thus $r_k \geq r_{k+1}$.

Claim 2. The above process terminates with a finite list of chains

$$\{B_1, \cdots, B_s\}$$

because for any k, $\{B_1, \cdots, B_k\}$ is linearly independent.

Proof of Claim 2. The claim is true if $k = 1$. This follows from **Claim 1**. Suppose it is true for $k-1, k \geq 2$. Then $\{B_1, \cdots, B_{k-1}\}$ is linearly independent. Now consider the induction step going from $k-1$ to k. Suppose that there exists a linear combination of vectors from $\{B_1, \cdots, B_{k-1}, B_k\}$.

$$\sum_{q=1}^{p} c_q w_q = 0,\ c_q \neq 0$$

By induction, some of these w_q must come from B_k. Let v_i^k be the vector of B_k for which i is as large as possible. Then apply N^{i-1} to both sides of the above equation to obtain v_1^k, the eigenvector upon which the chain B_k is based, is a linear combination of $\{B_1, \cdots, B_{k-1}\}$ contrary to the construction. Since $\{B_1, \cdots, B_k\}$ is linearly independent, the process terminates. This proves the claim.

Claim 3. Suppose $Nw = 0$ (w is an eigenvector). Then there exist scalars c_i such that

$$w = \sum_{i=1}^{s} c_i v_1^i.$$

Recall that the ith chain is based on v_1^i.

Proof of Claim 3. From the construction, $w \in$ span $\{B_1, \cdots, B_s\}$ since otherwise, it could serve as a base for another chain. Therefore,

$$w = \sum_{i=1}^{s} \sum_{k=1}^{r_i} c_i^k v_k^i.$$

Now apply N to both sides. It follows that $0 = \sum_{i=1}^{s} \sum_{k=2}^{r_i} c_i^k v_{k-1}^i$, and so by **Claim 2**, $c_i^k = 0$ if $k \geq 2, i = 1, \cdots, s$. Therefore, $w = \sum_{i=1}^{s} c_i^1 v_1^i$ and this proves the claim.

In fact, there is a generalization of this claim.

Claim 4. Suppose $N^k w = 0, k$ a positive integer. Then $w \in$ span $\{B_1, \cdots, B_s\}$.

Proof of Claim 4. From **Claim 3**, this is true if $k = 1$. Suppose that it is true for $k-1 \geq 1$ and $N^k w = 0$. Then if $N^{k-1} w = 0$ the result follows by induction, so suppose $N^{k-1} w \neq 0$. Then $N^{k-1} w, \cdots, Nw, w$ is a chain based on $N^{k-1} w$ having length k. It follows that each B_i has length at least as long as k

because the construction chooses chains of maximal length. From **Claim 3** there exist scalars c_1^j such that

$$N^{k-1}w = \sum_{j=1}^{s} c_1^j v_1^j = \sum_{j=1}^{s} c_1^j N^{k-1} v_k^j.$$

Therefore, $0 = N^{k-1}\left(w - \sum_{j=1}^{s} c_1^j v_k^j\right)$. Now by induction

$$w - \sum_{j=1}^{s} c_1^j v_k^j \in \text{span}\,(B_1, \cdots, B_s)$$

which shows $w \in \text{span}\,\{B_1, \cdots, B_s\}$.

From this claim, it is clear that $\text{span}\,\{B_1, \cdots, B_s\} = W$ because $N^m = 0$ on W. Since $\{B_1, \cdots, B_s\}$ is linearly independent, this shows it is a basis for W.

Now consider the matrix of N with respect to this basis. Since $\{B_1, \cdots, B_s\}$ is a basis, it follows that

$$W = \text{span}\,(B_1) \oplus \cdots \oplus \text{span}\,(B_s).$$

Also each $\text{span}\,\{B_k\}$ is N invariant. It follows from Theorem 8.2 that the matrix of N with respect to this basis is block diagonal in which the kth block is the matrix of N_k, the restriction of N to $\text{span}\,(B_k)$. Denote this matrix as J^k. Then by the definition of the matrix of a linear transformation,

$$v_{j-1}^k = N v_j^k = \sum_i J_{ij}^k v_i^k,$$

and so

$$J_{ij}^k = \begin{cases} 1 \text{ if } i = j-1 \\ 0 \text{ otherwise} \end{cases}$$

Hence this block is of the form

$$\begin{pmatrix} 0 & 1 & & \\ & 0 & \ddots & \\ & & \ddots & 1 \\ & & & 0 \end{pmatrix} \equiv J_{r_k}(0)$$

where there are r_k vectors in B_k. This proves the proposition. $\square$

Let V be an n dimensional vector space and let T be a linear mapping which maps V to V. Also suppose the minimal polynomial (See Problem 19 on Page 218.) splits as described above. Then it follows from a repeat of the arguments of Problems 20–22 beginning on Page 150 that there exist scalars $\lambda_1, \cdots, \lambda_p$ such that

$$\begin{aligned} V &= \ker\left((A - \lambda_1 I)^{r_1}\right) \oplus \cdots \oplus \ker\left((A - \lambda_p)^{r_p}\right) \\ &= V_1 \oplus \cdots \oplus V_p. \end{aligned}$$

Just replace the term $n \times n$ matrix with the words "linear mapping" and the results of these problems give the above direct sum.

Each $(A - \lambda_i I)$ is nilpotent on V_i. Therefore, letting β_i denote the basis on V_i for which

$$[A - \lambda_i I]_{\beta_i}$$

is of the form given in the above proposition, it follows that from Lemma 7.4 that

$$[A]_{\beta_i} = [A - \lambda_i I]_{\beta_i} + \lambda_i I.$$

$$= \begin{pmatrix} J_{r_1}(\lambda_i) & & & \\ & J_{r_2}(\lambda_i) & & \\ & & \ddots & \\ & & & J_{r_s}(\lambda_i) \end{pmatrix} = J_i. \tag{8.9}$$

Then from Theorem 8.2, the matrix of A with respect to the basis

$$\{\beta_1, \beta_2, \cdots, \beta_p\} \tag{8.10}$$

is of the form

$$J = \begin{pmatrix} J_1 & & & \\ & J_2 & & \\ & & \ddots & \\ & & & J_p \end{pmatrix}. \tag{8.11}$$

This proves the following theorem.

Theorem 8.3. *Let $T : V \to V$ be a linear mapping for V a finite dimensional vector space. Suppose also that the minimal polynomial splits. Then there exists a basis β for V such that $[T]_\beta$ is of the form shown in (8.11) where each of the blocks J_i is of the form shown in (8.9) for $J_r(\lambda)$ described in Definition 8.5.*

Much more can be said relative to the uniqueness of the Jordan form, but we will only need its existence. Note that the Jordan form of the linear transformation, determined by multiplication by an $n \times n$ matrix A, is an upper triangular matrix. Therefore, the diagonal entries are the eigenvalues of A, listed according to multiplicity since the Jordan form of A and A are similar and therefore have the same characteristic polynomial.

Letting J be a Jordan form for T. Then both T and J have the same eigenvalues and the dimensions of the corresponding eigenspaces are the same. This follows easily from the observation that J is just the matrix of T with respect to a suitable basis. Then we can give the following interesting corollary as an application of the existence of the Jordan form.

Corollary 8.1. *Let λ be an eigenvalue of T. Then the dimension of the eigenspace equals the number of Jordan blocks in the Jordan canonical form which are associated with λ. In particular, if λ has algebraic multiplicity h, then the dimension of the eigenspace associated with λ is no larger than h.*

Proof. The Jordan form is as described in the above theorem. Without loss of generality, assume λ corresponds to J_1 in the above theorem. Otherwise permute the bases of (8.10) used to get the Jordan form. Then $J - \lambda I$ is a matrix of the form

$$\begin{pmatrix} J_{r_1}(0) & & & & \\ & J_{r_2}(0) & & & \\ & & \ddots & & \\ & & & J_{r_s}(0) & \\ & & & & \hat{J} \end{pmatrix}$$

where $\hat{J}$ is upper triangular and has all non zero entries down the diagonal. It is desired to find the dimension of $\ker(J - \lambda I)$. Recall that this is just the number of free variables in the system of equations which result from the augmented matrix obtained by adding a column of zeros to the above matrix on the right. Recall that $J_r(0)$ is an $r \times r$ matrix of the form

$$\begin{pmatrix} 0 & 1 & & 0 \\ & 0 & \ddots & \\ & & \ddots & 1 \\ 0 & & & 0 \end{pmatrix}$$

Thus exactly one free variable occurs from the first column, then the column separating $J_{r_1}(0)$ and $J_{r_2}(0)$, then the column separating $J_{r_2}(0)$ from $J_{r_3}(0)$ and so forth till the column separating $J_{r_{s-1}}(0)$ from $J_{r_s}(0)$. There are exactly r_s of these columns and so the dimension of the eigenspace for λ equals the number of Jordan blocks as claimed. $\square$

The existence of the Jordan form makes possible the proof of many significant results. A few of these are illustrated in the following exercises.

8.4 Exercises

(1) Here is a matrix. Find its Jordan canonical form by finding chains of generalized eigenvectors based on eigenvectors to obtain a basis which will yield the Jordan form. The eigenvalues are 1 and 2.

$$\begin{pmatrix} -3 & -2 & 5 & 3 \\ -1 & 0 & 1 & 2 \\ -4 & -3 & 6 & 4 \\ -1 & -1 & 1 & 3 \end{pmatrix}$$

Why is it typically impossible to find the Jordan canonical form?

(2) Here is a matrix in Jordan form.

$$\begin{pmatrix} 2 & 1 & 0 & 0 \\ 0 & 2 & 0 & 0 \\ 0 & 0 & 3 & 0 \\ 0 & 0 & 0 & 4 \end{pmatrix}.$$

What is the Jordan form of

$$\begin{pmatrix} 2 & 1 & 0 & 0 \\ 0 & 2.000001 & 0 & 0 \\ 0 & 0 & 3 & 0 \\ 0 & 0 & 0 & 4 \end{pmatrix}?$$

(3) Let A be a complex $n \times n$ matrix. Show that there exists a sequence of $n \times n$ matrices $\{A_k\}_{k=1}^{\infty}$ such that the ij^{th} entry of A_k converges to the ij^{th} entry of A and each A_k can be diagonalized. **Hint:** If A can be diagonalized, there is nothing to prove. Suppose then that $J = P^{-1}AP$ where J is in Jordan form, being a block diagonal matrix with each block having a single eigenvalue down the main diagonal. Describe a modified matrix J_k such that J_k has n distinct entries down the main diagonal and such that the entries of J_k all converge as $k \to \infty$ to the corresponding entries of J. Explain why J_k can be diagonalized. (See Lemma 8.2.) Then consider $A_k = PJ_kP^{-1}$.

(4) Let A be an $n \times n$ matrix over a field F and assume that A can be diagonalized. Give an exceedingly easy proof of the Cayley-Hamilton theorem for such a matrix.

(5) Now use the above Problem 3 to prove the Cayley-Hamilton theorem for any complex $n \times n$ matrix.

(6) Using the Jordan form, show that for any complex $n \times n$ matrix A and p a positive integer,

$$\sigma(A^p) = \sigma(A)^p.$$

Here $\sigma(B)$ denotes the set of eigenvalues of B and $\sigma(A)^p$ denotes the set $\{\lambda^p : \lambda \in \sigma(A)\}$. **Hint:** Use block multiplication in the Jordan form. Next give a very easy proof which does not require the Jordan form.

(7) Let f be an analytic function, one which is given correctly by a power series,

$$f(z) = \sum_{k=0}^{\infty} a_k z^k.$$

Define for A a complex $n \times n$ matrix

$$||A|| = \max\{|a_{ij}|, i, j \leq n\}.$$

Show first that $||AB|| \leq n\,||A||\,||B||$ and $||A + B|| \leq ||A|| + ||B||$. Next show that if $\sum_{k=0}^{\infty} a_k z^k$ converges for all complex z, then if one defines

$$B_m = \sum_{k=0}^{m} a_k A^k$$

it follows that for every $\varepsilon > 0$, there exists N_ε such that if $m, p \geq N_\varepsilon$, then $||B_m - B_p|| < \varepsilon$. Explain why this shows that for each i, j the ij^{th} entry of B_m is a Cauchy sequence in $\mathbb{C}$ and so, by completeness, it must converge. Denote by $f(A)$ the matrix whose ij^{th} entry is this limit. Using block multiplication and the Jordan form, show that $f(\sigma(A)) = \sigma(f(A))$ where

$$f(\sigma(A)) = \{f(\lambda) : \lambda \in \sigma(A)\}.$$

(8) ↑Let $||A||$ be as defined in the above problem for A a complex $n \times n$ matrix and suppose $|\lambda| > n\,||A||$. Show that then $\lambda I - A$ must be invertible so that $\lambda \notin \sigma(A)$. **Hint:** It suffices to verify that $\lambda I - A$ is one-to-one. If it is not, then there exists $\mathbf{x} \neq \mathbf{0}$ such that

$$\lambda \mathbf{x} = A\mathbf{x}.$$

In particular

$$\lambda x_i = \sum_j a_{ij} x_j.$$

Let $|x_i|$ be the largest of all the $|x_j|$. Then explain why

$$|\lambda|\,|x_i| \leq ||A||\,n\,|x_i|.$$

(9) ↑The **spectral radius** of A, denoted by $\rho(A)$ is defined as

$$\rho(A) = \max\{|\lambda| : \lambda \in \sigma(A)\},$$

In words, it is the largest of the absolute values of the eigenvalues of A. Use the Jordan form of a matrix and block multiplication to show that if $|\lambda| > \rho(A)$, then for all k sufficiently large, $\left|\left|\frac{A^k}{\lambda^k}\right|\right|^{1/k} \leq 1$. Thus for any $\varepsilon > 0$, it follows that for k sufficiently large,

$$\left|\left|A^k\right|\right|^{1/k} \leq \rho(A) + \varepsilon.$$

Hint: Recall that there exists an invertible matrix P such that $A = P^{-1}JP$ where J is Jordan form. Thus J is block diagonal having blocks of the form $\lambda_k I + N_k$ where N_k is upper triangular having ones or zeros down the super diagonal and zeros elsewhere. Since N_k and $\lambda_k I$ commute, you can apply the binomial theorem to the product

$$\left(\frac{\lambda_k I + N_k}{\lambda}\right)^m.$$

Using some elementary calculus involving the ratio test or the root test, verify that the entries of these blocks raised to the mth power converge to 0 as $m \to \infty$. Thus $\left|\left|\frac{J^m}{\lambda^m}\right|\right| \to 0$ as $m \to 0$. Explain why

$$\left|\left|\frac{A^m}{\lambda^m}\right|\right| = \left|\left|P^{-1}\frac{J^m}{\lambda^m}P\right|\right| \leq C_n\,||P^{-1}||\,||P||\left|\left|\frac{J^m}{\lambda^m}\right|\right| \leq 1.$$

whenever m is large enough. Here C_n is a constant which depends on n.

(10) ↑For A a complex $n \times n$ matrix, a famous formula of Gelfand says that

$$\rho(A) = \lim_{p \to \infty} ||A^p||^{1/p}$$

where $||B||$ is the norm of the matrix. Here we will define this norm as[1]

$$||B|| = \max\{|a_{ij}|, i, j = 1, \ldots, n\}.$$

Hint: If $\lambda \in \sigma(A)$, then $\lambda^p \in \sigma(A^p)$ by Problem 6. Hence by Problem 8, $|\lambda|^p = |\lambda^p| \leq n ||A^p||$. Finally use Problem 9 to explain why for each $\varepsilon > 0$, there exists a P_ε such that if $p \geq P_\varepsilon$,

$$n^{1/p} ||A^p||^{1/p} \geq \rho(A) \geq ||A^p||^{1/p} - \varepsilon.$$

Using some elementary calculus and the above inequality, show that

$$\lim_{p \to \infty} ||A^p||^{1/p}.$$

exists and equals $\rho(A)$.[2]

(11) Explain how to find the Jordan form of a complex $n \times n$ matrix whenever the eigenvalues can be found exactly. **Hint:** The proof of its existence actually gives a way to construct it. It might help to consider how the chains were constructed in Proposition 8.1. If you have a basis for W $\{w_1, \ldots, w_m\}$, and if $N^p x \neq 0$ for some $x \in W$, explain why $N^p w_k \neq 0$ for some w_k. Thus you could consider $w_k, Nw_k, N^2 w_k, \ldots$ for each w_k and one of these must have maximal length.

(12) Let A be an $n \times n$ matrix and let J be its Jordan canonical form. Recall J is a block diagonal matrix having blocks $J_k(\lambda)$ down the diagonal. Each of these blocks is of the form

$$J_k(\lambda) = \begin{pmatrix} \lambda & 1 & & 0 \\ & \lambda & \ddots & \\ & & \ddots & 1 \\ 0 & & & \lambda \end{pmatrix}.$$

Now for $\varepsilon > 0$ given, let the diagonal matrix D_ε be given by

$$D_\varepsilon = \begin{pmatrix} 1 & & & 0 \\ & \varepsilon & & \\ & & \ddots & \\ 0 & & & \varepsilon^{k-1} \end{pmatrix}.$$

---

[1] It turns out that it makes absolutely no difference in this formula how you define the norm provided it satisfies certain algebraic properties. This is just a convenient norm which was used in the above problems.

[2] This amazing theorem is normally done using the theory of Laurent series from complex analysis. However, the Jordan form allows this much more elementary treatment in the case of finite dimensions.

Show that $D_\varepsilon^{-1} J_k(\lambda) D_\varepsilon$ has the same form as $J_k(\lambda)$ but instead of ones down the super diagonal, there is ε down the super diagonal. That is, $J_k(\lambda)$ is replaced with

$$\begin{pmatrix} \lambda & \varepsilon & & 0 \\ & \lambda & \ddots & \\ & & \ddots & \varepsilon \\ 0 & & & \lambda \end{pmatrix}.$$

Now show that for A an $n \times n$ matrix, it is similar to one which is just like the Jordan canonical form except instead of the blocks having 1 down the super diagonal, it has ε.

(13) Proposition 8.1 gave the existence of a canonical form of a certain sort. Show that this matrix described there is unique. Then give a sense in which the Jordan canonical form is uniquely determined. **Hint:** Suppose you have two such matrices which are of the sort described in the proposition J and J'. Then since these are similar matrices, $R\left(J^k\right) = R\left(J'^k\right)$ for each $k = 1, 2, \ldots$. However, this condition would be violated if there is a discrepancy in the size of the blocks when viewed from the upper left toward lower right. Explain why. As to the Jordan form, the blocks corresponding to each eigenvalue must be of the same size, because this size equals the algebraic multiplicity of the eigenvalues due to the fact that the two Jordan forms are similar. Next pick an eigenvalue and subtract $\lambda_k I$ from both of the two Jordan forms and explain why the rank of the resulting matrices is determined by the rank of the blocks corresponding to λ_k. Now use the Proposition about nilpotent matrices.

(14) Let A be an $n \times n$ matrix whose entries are from a field F. It was shown above that the Jordan canonical form of A exists provided the minimal polynomial $p_F(\lambda)$ splits. Recall the definition of the minimal polynomial $p_F(\lambda)$ as the monic polynomial having coefficients in the field of scalars F with the property that $p_F(A) = 0$ and $p_F(\lambda)$ has smallest degree out of all such polynomials. Using Problem 21 on Page 123, show that there exists a possibly larger field G such that $p_F(\lambda)$ splits in this larger field. Of course A is a matrix whose entries are in G, and one can consider the minimal polynomial $p_G(\lambda)$. Since $p_F(A) = 0$, explain why $p_G(\lambda)$ must divide $p_F(\lambda)$. Now explain why there exists a field G containing F such that $p_G(\lambda)$ splits. Thus, in this larger field, A has a Jordan canonical form. Next show that if a matrix has a Jordan canonical form, then the minimal polynomial must divide the characteristic polynomial. Finally, note that the characteristic polynomial of A described above does not change when the fields are enlarged. What does this say about the generality of the Cayley-Hamilton theorem? In fact, show that the minimal polynomial of L **always** divides the characteristic polynomial of L, for L any linear transformation from a vector space to itself.

(15) To find the Jordan canonical form, you must be able to exactly determine

the eigenvalues and corresponding eigenvectors. However, this is a very hard problem and typically cannot be done. Various ways of approximating the eigenvalues and eigenvectors have been used. Currently the best method is the QR algorithm, discussed briefly later. An earlier method was based on the LU factorization. Consider the following algorithm. $A = LU, UL = L_1U_1 = A_1$. Continuing this way, you obtain $A_n = L_nU_n$ and $U_nL_n = A_{n+1}$. Show that $A_{n+1} = U_n^{-1}A_nU_n$. Iterating this, show that the sequence of matrices $\{A_n\}$ are each similar to the original matrix A. Now it can be shown that if B is close to C in the sense that corresponding entries are close, then the eigenvalues of B will be close to the eigenvalues of C when they are counted according to multiplicity. This follows from Rouché's theorem in complex analysis. See also the set of problems beginning with Problem 27 below on Page 288. Then if the matrices A_n are close to a given matrix A which has a form suitable to easily finding its eigenvalues, it follows that we could find them and thereby have approximate values for the eigenvalues of A. This method, based on an LU factorization, is inferior to the QR algorithm because of stability considerations, among other issues such as a given matrix not having an LU factorization. Nevertheless, it does work, at least on simple examples.

(16) Show that a given $n \times n$ matrix is nondefective (Jordan form is a diagonal matrix) if and only if the minimal polynomial splits and has no repeated roots. **Hint:** If the Jordan form is a diagonal matrix

$$\begin{pmatrix} \lambda_1 I_1 & & \\ & \ddots & \\ & & \lambda_r I_r \end{pmatrix}$$

argue that the minimal polynomial is of the form $(\lambda - \lambda_1) \cdots (\lambda - \lambda_r)$. Next, if the minimal polynomial is of this form, recall that $F^n = \ker(A - \lambda_1 I) \oplus \cdots \oplus \ker(A - \lambda_r I)$.

(17) Suppose you have two polynomials, $p(\lambda)$ and $q(\lambda)$. Give an algorithm which will compute the greatest common divisor of these polynomials and use your algorithm on the two polynomials

$$\lambda^3 + 4\lambda^2 + 5\lambda + 2,$$
$$\lambda^3 + 3\lambda^2 - 4.$$

Hint: Use the Euclidean algorithm for polynomials to write

$$q(\lambda) = p(\lambda) m(\lambda) + r(\lambda)$$

where $r(\lambda) = 0$ or else it has degree less than the degree of $p(\lambda)$. Argue that the greatest common divisor of $q(\lambda)$ and $p(\lambda)$ is the same as the greatest common divisor of $r(\lambda)$ and $q(\lambda)$.

(18) For the field of scalars equal to $\mathbb{C}$, show that an $n \times n$ matrix is non defective if and only if, for $p(\lambda)$ its minimal polynomial, $p(\lambda)$ and its derivative $p'(\lambda)$ are relatively prime. Would this work for an arbitrary field?

(19) Let A be an $n \times n$ matrix. Describe a fairly simple method based on row operations for computing the minimal polynomial of A. Recall, that this is a monic polynomial $p(\lambda)$ such that $p(A) = 0$ and it has smallest degree of all such monic polynomials. **Hint:** Consider $I, A^2, \cdots$. Regard each as a vector in $\mathbb{F}^{n^2}$ and consider taking the row reduced echelon form or something like this. You might also use the Cayley-Hamilton theorem to note that you can stop the above sequence at A^n. Explain how to determine whether a given matrix is defective through a routine sequence of steps.

Chapter 9

Some applications

Chapter summary

This chapter is on applications of the above theory. To begin with, the existence of the Jordan canonical form is used to consider the limits of powers of stochastic matrices, those matrices which have nonnegative entries and the sum of every column equals one. This is used to consider Markov processes including the case of random walks and gambler's ruin problems in which there are absorbing states. After this, a general proof of the Cayley-Hamilton theorem is presented which is not limited to the field of scalars being either $\mathbb{R}$ or $\mathbb{C}$. This is further generalized in the exercises to include all situations. The Cayley-Hamilton theorem is then used to give a completely general treatment of first order systems of linear ordinary differential equations having constant coefficients which is based entirely on algebraic techniques. This includes consideration of the existence of fundamental matrices and existence and uniqueness of solutions. Finally a simple algebraic technique is presented which is useful for computing solutions using a table of Laplace transforms.

9.1 Limits of powers of stochastic matrices

The existence of the Jordan form is the basis for the proof of limit theorems for certain kinds of matrices called stochastic matrices.

Definition 9.1. An $n \times n$ matrix $A = (a_{ij})$, is a **transition matrix** if $a_{ij} \geq 0$ for all i, j and

$$\sum_i a_{ij} = 1.$$

It may also be called a Markov matrix or a stochastic matrix. Such a matrix is called **regular** if some power of A has all entries strictly positive. A vector $\mathbf{v} \in \mathbb{R}^n$ is a **steady state** if $A\mathbf{v} = \mathbf{v}$.

Lemma 9.1. *The property of being a transition matrix is preserved by taking products.*

Proof. Suppose the sum over a column equals 1 for A and B. Then letting the entries be denoted by (a_{ij}) and (b_{ij}) respectively, the sum of the kth column of AB is given by

$$\sum_i \left(\sum_j a_{ij} b_{jk} \right) = \sum_j b_{jk} \left(\sum_i a_{ij} \right) = 1.$$

It is obvious that when the product is taken, if each $a_{ij}, b_{ij} \geq 0$, then the same will be true of sums of products of these numbers.

Theorem 9.1. *Let A be a $p \times p$ transition matrix and suppose the distinct eigenvalues of A are $\{1, \lambda_2, \ldots, \lambda_m\}$ where each $|\lambda_j| < 1$. Then $\lim_{n\to\infty} A^n = A_\infty$ exists in the sense that $\lim_{n\to\infty} a_{ij}^n = a_{ij}^\infty$, the ij^{th} entry of A_∞. Also, if $\lambda = 1$ has algebraic multiplicity r, then the Jordan block corresponding to $\lambda = 1$ is just the $r \times r$ identity.*

Proof. By the existence of the Jordan form for A, it follows that that there exists an invertible matrix P such that

$$P^{-1}AP = \begin{pmatrix} I+N & & & \\ & J_{r_2}(\lambda_2) & & \\ & & \ddots & \\ & & & J_{r_m}(\lambda_m) \end{pmatrix} = J$$

where I is $r \times r$ for r the multiplicity of the eigenvalue 1, and N is a nilpotent matrix for which $N^r = 0$. We will show that because the sum over every column is 1, $N = 0$.

First of all,

$$J_{r_i}(\lambda_i) = \lambda_i I + N_i$$

where N_i satisfies $N_i^{r_i} = 0$ for some $r_i > 0$. It is clear that $N_i(\lambda_i I) = (\lambda_i I) N$ because a multiple of the identity commutes with any matrix, and so, by the binomial theorem,

$$(J_{r_i}(\lambda_i))^n = \sum_{k=0}^{n} \binom{n}{k} N_i^k \lambda_i^{n-k} = \sum_{k=0}^{r_i} \binom{n}{k} N_i^k \lambda_i^{n-k}$$

which converges to 0 due to the assumption that $|\lambda_i| < 1$. There are finitely many terms and a typical one is a matrix whose entries are no larger than an expression of the form

$$|\lambda_i|^{n-k} C_k n(n-1)\cdots(n-k+1) \leq C_k |\lambda_i|^{n-k} n^k.$$

This expression converges to 0 because, by the root test, the series $\sum_{n=1}^{\infty} |\lambda_i|^{n-k} n^k$ converges. Thus for each $i = 2, \ldots, p$,

$$\lim_{n\to\infty} (J_{r_i}(\lambda_i))^n = 0.$$

By Lemma 9.1, A^n is also a transition matrix. Therefore, since the entries are nonnegative and each column sums to 1, the entries of A^n must be bounded independent of n.

It follows easily from

$$\overbrace{P^{-1}APP^{-1}APP^{-1}AP\cdots P^{-1}AP}^{n \text{ times}} = P^{-1}A^nP$$

that

$$P^{-1}A^nP = J^n. \tag{9.1}$$

Hence J^n must also have bounded entries as $n \to \infty$. However, this requirement is incompatible with an assumption that $N \neq 0$.

If $N \neq 0$, then $N^s \neq 0$ but $N^{s+1} = 0$ for some $1 \leq s \leq r$. Then

$$(I+N)^n = I + \sum_{k=1}^{s} \binom{n}{k} N^k.$$

One of the entries of N^s is nonzero by the definition of s. Let this entry be n_{ij}^s. Then this implies that one of the entries of $(I+N)^n$ is of the form $\binom{n}{s} n_{ij}^s$. This entry dominates the ij^{th} entries of $\binom{n}{k} N^k$ for all $k < s$ because

$$\lim_{n\to\infty} \binom{n}{s} / \binom{n}{k} = \infty.$$

Therefore, the entries of $(I+N)^n$ cannot all be bounded independent of n. From block multiplication,

$$P^{-1}A^nP = \begin{pmatrix} (I+N)^n & & & \\ & (J_{r_2}(\lambda_2))^n & & \\ & & \ddots & \\ & & & (J_{r_m}(\lambda_m))^n \end{pmatrix},$$

and this is a contradiction because entries are bounded on the left and unbounded on the right.

Since $N = 0$, the above equation implies $\lim_{n\to\infty} A^n$ exists and equals

$$P \begin{pmatrix} I & & & \\ & 0 & & \\ & & \ddots & \\ & & & 0 \end{pmatrix} P^{-1}.$$

This proves the theorem. □

What is a convenient condition which will imply a transition matrix has the properties of the above theorem? It turns out that if $a_{ij} > 0$, not just ≥ 0, then the eigenvalue condition of the above theorem is valid.

Lemma 9.2. *Suppose $A = (a_{ij})$ is a transition matrix. Then $\lambda = 1$ is an eigenvalue. Also if λ is an eigenvalue, then $|\lambda| \leq 1$. If $a_{ij} > 0$ for all i, j, then if μ is an eigenvalue of A, either $|\mu| < 1$ or $\mu = 1$. In addition to this, if $A^t\mathbf{v} = \mathbf{v}$ for a nonzero vector $\mathbf{v} \in \mathbb{R}^n$, then $v_jv_i \geq 0$ for all $i, j,$, so the components of $\mathbf{v}$ have the same sign.*

Proof. Since each column sums to 1, we have $I - A \sim$

$$\begin{pmatrix} 0 & 0 & \cdots & 0 \\ -a_{21} & 1 - a_{21} & \cdots & -a_{2n} \\ \vdots & \ddots & \ddots & \vdots \\ -a_{n1} & -a_{n2} & \cdots & 1 - a_{nn} \end{pmatrix}$$

on adding the other rows to the first row. Therefore, the determinant of $I - A = 0$.

Suppose then that μ is an eigenvalue. Are the only two cases $|\mu| < 1$ and $\mu = 1$? As noted earlier, μ is also an eigenvalue of A^t. Let $\mathbf{v}$ be an eigenvector for A^t and let $|v_i|$ be the largest of the $|v_j|$. Then

$$\mu v_i = \sum_j a_{ij}^t v_j.$$

Now multiply both sides by $\overline{\mu v_i}$ to obtain

$$|\mu|^2 |v_i|^2 = \sum_j a_{ij}^t v_j \overline{v_i \mu} = \sum_j a_{ij}^t \Re \left(v_j \overline{v_i \mu} \right) \leq \sum_j a_{ij}^t |\mu| |v_i|^2 = |\mu| |v_i|^2$$

Therefore, $|\mu| \leq 1$.

Now assume each $a_{ij} > 0$. If $|\mu| = 1$, then equality must hold in the above, and so $v_j \overline{v_i \mu}$ must be real and nonnegative for each j. In particular, this holds for $j = 1$ which shows $\overline{\mu}$ and hence μ are real and nonnegative. Thus, in this case, $\mu = 1$. The only other case is where $|\mu| < 1$. □

The next lemma says that a transition matrix conserves the sum of the entries of the vectors.

Lemma 9.3. *Let A be any transition matrix and let $\mathbf{v}$ be a vector having all its components non negative with $\sum_i v_i = c$. Then if $\mathbf{w} = A\mathbf{v}$, it follows that $w_i \geq 0$ for all i and $\sum_i w_i = c$.*

Proof. From the definition of $\mathbf{w}$,

$$w_i \equiv \sum_j a_{ij} v_j \geq 0.$$

Also

$$\sum_i w_i = \sum_i \sum_j a_{ij} v_j = \sum_j \sum_i a_{ij} v_j = \sum_j v_j = c. \square$$

The following theorem about limits is now easy to obtain.

Theorem 9.2. *Suppose A is a transition matrix in which $a_{ij} > 0$ for all i, j and suppose $\mathbf{w}$ is a vector. Then for each i,*

$$\lim_{k \to \infty} \left(A^k \mathbf{w} \right)_i = v_i$$

where $A\mathbf{v} = \mathbf{v}$. In words, $A^k \mathbf{w}$ always converges to a steady state. In addition to this, if the vector, $\mathbf{w}$ satisfies $w_i \geq 0$ for all i and $\sum_i w_i = c$, Then the vector, $\mathbf{v}$ will also satisfy the conditions, $v_i \geq 0$, $\sum_i v_i = c$.

Proof. By Lemma 9.2, since each $a_{ij} > 0$, the eigenvalues are either 1 or have absolute value less than 1. Therefore, the claimed limit exists by Theorem 9.1. The assertion that the components are nonnegative and sum to c follows from Lemma 9.3. That $A\mathbf{v} = \mathbf{v}$ follows from

$$\mathbf{v} = \lim_{n\to\infty} A^n \mathbf{w} = \lim_{n\to\infty} A^{n+1}\mathbf{w} = A \lim_{n\to\infty} A^n \mathbf{w} = A\mathbf{v}.$$

This proves the theorem. □

This theorem and the following corollary are special cases of the Perron - Frobenius theorem applied to transition matrices. To see different proofs from those presented here, see the book by Nobel and Daniel [13]. In the next section, we will present an important example of a transition matrix involving a Markov chain with absorbing states in which these theorems cannot be applied.

Corollary 9.1. *Suppose A is a regular transition matrix and suppose $\mathbf{w}$ is a vector. Then for each i,*

$$\lim_{n\to\infty} (A^n \mathbf{w})_i = v_i$$

where $A\mathbf{v} = \mathbf{v}$. In words, $A^n\mathbf{w}$ always converges to a steady state. In addition to this, if the vector, $\mathbf{w}$ satisfies $w_i \geq 0$ for all i and $\sum_i w_i = c$, Then the vector, $\mathbf{v}$ will also satisfy the conditions, $v_i \geq 0$, $\sum_i v_i = c$.

Proof. Let the entries of A^k be all positive. Now suppose that $a_{ij} \geq 0$ for all i, j and $A = (a_{ij})$ is a transition matrix. Then if $B = (b_{ij})$ is a transition matrix with $b_{ij} > 0$ for all ij, it follows that BA is a transition matrix which has strictly positive entries. The ij^{th} entry of BA is

$$\sum_k b_{ik} a_{kj} > 0,$$

Thus, from Lemma 9.2, A^k has an eigenvalue equal to 1 for all k sufficiently large, and all the other eigenvalues have absolute value strictly less than 1. The same must be true of A, for if λ is an eigenvalue of A with $|\lambda| = 1$, then λ^k is an eigenvalue for A^k and so, for all k large enough, $\lambda^k = 1$ which is absurd unless $\lambda = 1$. By Theorem 9.1, $\lim_{n\to\infty} A^n\mathbf{w}$ exists. The rest follows as in Theorem 9.2. This proves the corollary. □

9.2 Markov chains

We now give an application of Corollary 9.1. The mathematical model is a simple one, but it is the starting point for more sophisticated probability models (stochastic processes).

We are given a family of objects $\mathcal{F}$, each of which is in one of n possible states, State $1, \ldots,$ State n. The objects can move from State i to State j. We record the number of objects in each state at time intervals of (say) one year. We find that the probability p_{ij} of an object moving from State j to State i is the same year after year. We now have a Markov chain with $n \times n$ transition matrix $A = (p_{ij})$.

Example 9.1. $\mathcal{F}$ is the set of fields under cultivation in an area that grows only peas (State 1), beans (State 2) and potatoes (State 3). If the transition matrix is

$$A = \begin{pmatrix} 0.2 & 0.5 & 0.1 \\ 0.3 & 0.4 & 0.2 \\ 0.5 & 0.1 & 0.7 \end{pmatrix},$$

then the fraction of fields switching from potatoes to peas each year is 0.1; the fraction of fields that stick with potatoes each year is 0.7.

You can easily make up further examples. We concentrate here on a phenomenon which occurs unless the transition matrix is a rather unusual one: there is a *steady state*, that is, in the long run, the proportion of objects in State j approaches a limiting value p_j.

Definition 9.2. A **probability vector** is a vector.

$$\mathbf{p} = \begin{pmatrix} p_1 \\ \vdots \\ p_2 \end{pmatrix}$$

with $p_j \geq 0$ $(j = 1, \ldots, n)$ and

$$p_1 + p_2 + \cdots + p_n = 1.$$

Thus a **transition matrix** is an $n \times n$ matrix $A = \begin{pmatrix} \mathbf{p}_1 & \cdots & \mathbf{p}_n \end{pmatrix}$ whose columns $\mathbf{p}_j$ are probability vectors.

The conditions in the definition of a transition matrix arise in an obvious way. Once j is fixed, the sum over $i = 1, \ldots, n$ of the fractions p_{ij} of objects in $\mathcal{F}$ switching from State j to State i must be 1.

We can use a probability vector $\mathbf{q}_h$ to represent year h, with the ith entry showing the fraction of objects in state i that year. We have the simple relation

$$\mathbf{q}_{h+1} = A\mathbf{q}_h.$$

To see this, let $\mathbf{q}_h = \begin{pmatrix} a_1 \\ \vdots \\ a_n \end{pmatrix}$, $\mathbf{q}_{h+1} = \begin{pmatrix} b_1 \\ \vdots \\ b_n \end{pmatrix}$. Then

$$\begin{aligned} b_j &= \text{fraction of objects in State } j \text{ in year } h+1 \\ &= \sum_{k=1}^{n} (\text{fraction of objects in State } k \text{ in year } h) \\ &\qquad \times (\text{probability of moving from State } k \text{ to State } j) \\ &= \sum_{k=1}^{n} a_k p_{jk}, \text{ or } \mathbf{q}_{h+1} = A\mathbf{q}_h. \end{aligned}$$

We call $\mathbf{q}_0$ the **initial probability vector**, measured in year 0. Now

$$\mathbf{q}_1 = A\mathbf{q}_0, \mathbf{q}_2 = A\mathbf{q}_1 = A^2\mathbf{q}_0, \mathbf{q}_3 = A\mathbf{q}_2 = A^3\mathbf{q}_0$$

and so on. Evidently $\mathbf{q}_h = A^h\mathbf{q}_0$.

We have the following corollary of Theorem 9.1 and Corollary 9.1.

Corollary 9.2. *Suppose that the eigenvalues λ_j of the transition matrix A are either 1 or satisfy $|\lambda_j| < 1$. Then the vector $\mathbf{q}_h$ representing year h has a limit $\mathbf{p}$ as $h \to \infty$. Moreover,*

(i) $\mathbf{p}$ *is a probability vector.*
(ii) $A\mathbf{p} = \mathbf{p}$.

The above condition on the eigenvalues will hold if some power of the transition matrix has all positive entries.

Proof. In the case where the transition matrix A is regular, the corollary follows from Corollary 9.1. Assume the first condition about the eigenvalues, which may hold even if the condition on the power of the matrix does not. By Theorem 9.1,

$$\lim_{h\to\infty} A^h = C$$

exists. So

$$\lim_{h\to\infty} \mathbf{q}_n = \lim_{h\to\infty} A^h\mathbf{q}_0 = C\mathbf{q}_0$$

exists. Let $C\mathbf{q}_0 = \mathbf{p} = \begin{pmatrix} p_1 & \cdots & p_n \end{pmatrix}^t$.

Since p_i is the limit of entry i of $\mathbf{q}_h$,

$$p_i \geq 0.$$

Also, $p_1 + \cdots + p_n$ is the limit of the sequence $1, 1, 1, \ldots$ (the sum of the entries of $\mathbf{q}_h$ is 1). So $p_1 + \cdots + p_n = 1$. Thus $\mathbf{p}$ is a probability vector.

Why is $\mathbf{p}$ an eigenvector of A with eigenvalue 1? We have

$$A\mathbf{q}_h = \mathbf{q}_{h+1}.$$

Take the limit on both sides to obtain

$$A\mathbf{p} = \mathbf{p}. \ \square$$

The vector $\mathbf{p}$ in the Corollary is a **steady state vector**.

Recall it was shown in Lemma 9.2 that the eigenvalues of any transition matrix are in absolute value no larger than 1 and as discussed above, regular transition matrices, as well as many which are not, have the property that the only eigenvalue having magnitude 1 is $\lambda = 1$. However, it is easy to find transition matrices having the eigenvalue -1, so that the Corollary fails to apply.

Example 9.2. Let A be a transition matrix with column $i = \mathbf{e}_j$ and column $j = \mathbf{e}_i$, where $i \neq j$. Let $\mathbf{q} = a\mathbf{e}_i + b\mathbf{e}_j$. Then

$$A\mathbf{q} = a\mathbf{e}_j + b\mathbf{e}_i.$$

This shows that

(i) -1 is an eigenvalue (take $a = 1, b = -1$),
(ii) there is a probability vector $\mathbf{q}_0 = 1/3\mathbf{e}_i + 2/3\mathbf{e}_j$ (for instance) such that $A^h\mathbf{q}_i$ oscillates between two values for odd and even h. (There is no steady state vector.)

The following example is much more typical.

Example 9.3. Find the steady state vector in Example 9.1.

Notice that the value of $\mathbf{q}_0$ has not been provided! The matrix of this example is a regular transition matrix, so all the above theory applies, and one only needs to find the steady state which is an eigenvector for $\lambda = 1$ because from the above, all other eigenvalues have absolute value less than 1. This eigenvector, scaled so that the sum of its components is 1, will be the limit. Thus, to find the limit vector $\mathbf{p}$, we compute the eigenspace for $\lambda = 1$:

$$\ker(I - A) = \ker \begin{pmatrix} 0.8 & -0.5 & -0.1 \\ -0.3 & 0.6 & -0.2 \\ -0.5 & -0.1 & 0.3 \end{pmatrix}$$

After row operations, this equals

$$= \ker \begin{pmatrix} 1 & 0 & -16/33 \\ 0 & 1 & -19/33 \\ 0 & 0 & 0 \end{pmatrix}$$

The corresponding set of equations reads $x_1 = 16/33x_3, x_2 = 19/33x_3$. The eigenspace is the line $\text{span}\{(16, 19, 33)\}$. Since $\mathbf{p}$ is a probability vector on this line,

$$\mathbf{p} = (16/68, 19/68, 33/68).$$

Notice that the argument shows that when the transition matrix has eigenvalues which are either 1 or have absolute value less than 1, then the limit must exist and the limit will be an eigenvector for $\lambda = 1$. If the eigenspace for $\lambda = 1$ is one dimensional as in this case, then the steady state vector *does not depend on* $\mathbf{q}_0$. It is the unique vector in the one-dimensional space $\ker(I - A)$ whose coordinates add up to 1.

Example 9.4. Let $A = \begin{pmatrix} 0.3 & 0.9 & 0.2 & 0.1 \\ 0.2 & 0.1 & 0.6 & 0.3 \\ 0.4 & 0 & 0.1 & 0.5 \\ 0.1 & 0 & 0.1 & 0.1 \end{pmatrix}$. It is a regular transition matrix because A^2 has all positive entries. Maple gives the approximate eigenvalues $1, -0.18+0.39i, -0.18-0.39i, -0.04$, conforming to the above theory. Thus $\lim_{n\to\infty} A^n\mathbf{q}_0$ exists and is an eigenvector for $\lambda = 1$. The eigenspace corresponding to 1 is

$$\text{span}\{(684, 437, 369, 117)\},$$

which is one dimensional, and so the limiting vector is (dividing by the sum of the coordinates)

$$\mathbf{p} = \frac{1}{1607}(684, 437, 369, 117).$$

It might be interesting to note that this theory of transition matrices is the basis for the way which google orders web sites. A transition matrix is determined according to number of times a link to a web site is followed from one web site to another. This defines a transition matrix which is very large. Then a limiting vector is obtained which ranks the web sites. This is discussed in Moler, Cleve, The world's largest matrix computation. MATLAB News and notes, The Mathworks, Natick, MA, October 2002.

9.3 Absorbing states

There is a different kind of Markov chain containing so called **absorbing states** which result in transition matrices which are not regular. However, Theorem 9.1 may still apply. One such example is the Gambler's ruin problem. There is a total amount of money denoted by b. The Gambler starts with an amount $j > 0$ and gambles till he either loses everything or gains everything. He does this by playing a game in which he wins with probability p and loses with probability q. When he wins, the amount of money he has increases by 1 and when he loses, the amount of money he has, decreases by 1. Thus the states are the integers from 0 to b. Let p_{ij} denote the probability that the gambler has i at the end of a game given that he had j at the beginning. Let p_{ij}^n denote the probability that the gambler has i after n games given that he had j initially. Thus

$$p_{ij}^{n+1} = \sum_k p_{ik} p_{kj}^n,$$

and so p_{ij}^n is the ij^{th} entry of P^n where P is the transition matrix. The above description indicates that this transition probability matrix is of the form $P =$

$$\begin{pmatrix} 1 & q & 0 & \cdots & 0 \\ 0 & 0 & \ddots & & 0 \\ 0 & p & \ddots & q & \vdots \\ \vdots & & \ddots & 0 & 0 \\ 0 & \cdots & 0 & p & 1 \end{pmatrix} \tag{9.2}$$

The absorbing states are 0 and b. In the first, the gambler has lost everything and hence has nothing else to gamble, so the process stops. In the second, he has won everything and there is nothing else to gain, so again the process stops.

To consider the eigenvalues of this matrix, we give the following lemma.

Lemma 9.4. *Let $p, q > 0$ and $p + q = 1$. Then the eigenvalues of*

$$\begin{pmatrix} 0 & q & 0 & \cdots & 0 \\ p & 0 & q & \cdots & 0 \\ 0 & p & 0 & \ddots & \vdots \\ \vdots & 0 & \ddots & \ddots & q \\ 0 & \vdots & 0 & p & 0 \end{pmatrix}$$

have absolute value less than 1.

Proof. By Gerschgorin's theorem, (See Page 221) if λ is an eigenvalue, then $|\lambda| \leq 1$. Now suppose $\mathbf{v}$ is an eigenvector for λ. Then

$$A\mathbf{v} = \begin{pmatrix} qv_2 \\ pv_1 + qv_3 \\ \vdots \\ pv_{n-2} + qv_n \\ pv_{n-1} \end{pmatrix} = \lambda \begin{pmatrix} v_1 \\ v_2 \\ \vdots \\ v_{n-1} \\ v_n \end{pmatrix}.$$

Suppose $|\lambda| = 1$. Then the top row shows that $q|v_2| = |v_1|$, so $|v_1| < |v_2|$. Suppose $|v_1| < |v_2| < \cdots < |v_k|$ for some $k < n$. Then

$$|\lambda v_k| = |v_k| \leq p|v_{k-1}| + q|v_{k+1}| < p|v_k| + q|v_{k+1}|,$$

and so subtracting $p|v_k|$ from both sides,

$$q|v_k| < q|v_{k+1}|$$

showing $\{|v_k|\}_{k=1}^n$ is an increasing sequence. Now a contradiction results on the last line which requires $|v_{n-1}| > |v_n|$. Therefore, $|\lambda| < 1$ for any eigenvalue of the above matrix and this proves the lemma.

Now consider the eigenvalues of (9.2). For P given there,

$$P - \lambda I = \begin{pmatrix} 1-\lambda & q & 0 & \cdots & 0 \\ 0 & -\lambda & \ddots & & 0 \\ 0 & p & \ddots & q & \vdots \\ \vdots & & \ddots & -\lambda & 0 \\ 0 & \cdots & 0 & p & 1-\lambda \end{pmatrix}$$

and so, expanding the determinant of the matrix along the first column and then along the last column yields

$$(1-\lambda)^2 \det \begin{pmatrix} -\lambda & q & & \\ p & \ddots & \ddots & \\ & \ddots & -\lambda & q \\ & & p & -\lambda \end{pmatrix}.$$

The roots of the polynomial after $(1-\lambda)^2$ have absolute value less than 1 because they are just the eigenvalues of a matrix of the sort in Lemma 9.4. It follows that the conditions of Theorem 9.1 apply and therefore, $\lim_{n\to\infty} P^n$ exists.

Of course, the above transition matrix, models many other kinds of problems. It is called a Markov process with two absorbing states.

Of special interest is $\lim_{n\to\infty} p_{0j}^n$, the probability that the gambler loses everything given that he starts with an amount j. The determination of this limit is left for the exercises.

9.4 Exercises

(1) Suppose

$$A = \begin{pmatrix} 1 & 0 & .7 \\ 0 & 1 & 0 \\ 0 & 0 & .3 \end{pmatrix}.$$

Find $\lim_{n\to\infty} A^n \mathbf{e}_1$. Now find $\lim_{n\to\infty} A^n \mathbf{e}_2$, and $\lim_{n\to\infty} A^n \mathbf{e}_3$. If you did it right, you see that this matrix does not produce a single steady state vector.

(2) Here are some transition matrices. Determine whether the given transition matrix is regular. Next find whether the transition matrix has a unique steady state probability vector. If it has one, find it.

(a) $\begin{pmatrix} .5 & .7 \\ .5 & .3 \end{pmatrix}$

(b) $\begin{pmatrix} .1 & .2 & .4 \\ 0 & .5 & .6 \\ .9 & .3 & 0 \end{pmatrix}$

(c) $\begin{pmatrix} 0 & 0 & 0 \\ .1 & .5 & .2 \\ .9 & .5 & .8 \end{pmatrix}$

(d) $\begin{pmatrix} .1 & 0 & .1 \\ .1 & .5 & .2 \\ .8 & .5 & .7 \end{pmatrix}$

(3) Suppose A is a transition matrix which has eigenvalues $\{1, \lambda_{r+1}, \ldots, \lambda_n\}$ where each $|\lambda_j| < 1$ and 1 has algebraic multiplicity r. Let V denote the set of vectors $\mathbf{v}$ which satisfy

$$\mathbf{v} = \lim_{n\to\infty} A^n \mathbf{w}$$

for some $\mathbf{w}$. Show that V is a subspace of $\mathbb{C}^n$ and find its dimension. **Hint:** Consider the proof of Theorem 9.1.

(4) In the city of Ichabod, it has been determined that if it does not rain, then with probability .95 it will not rain the next day. On the other hand, if it

does rain, then with probability .7 it will rain the following day. What is the probability of rain on a typical day?

(5) In a certain country, the following table describes the transition between being married (M) and being single (S) from one year to the next.

	S	M
S	.7	.1
M	.3	.9

The top row is the state of the initial marital situation of the person and the side row is the marital situation of the person the following year. Thus the probability of going from married to single is .1. Determine the probability of a person being married.

(6) In the city of Nabal, there are three political persuasions, republicans (R), democrats (D), and neither one (N). The following table shows the transition probabilities between the political parties, the top row being the initial political party and the side row being the political affiliation the following year.

	R	D	N
R	.9	.1	.1
D	.1	.7	.1
N	0	.2	.8

Find the probabilities that a person will be identified with the various political persuasions. Which party will end up being most important?

(7) The University of Poohbah offers three degree programs, scouting education (SE), dance appreciation (DA), and engineering (E). It has been determined that the probabilities of transferring from one program to another are as in the following table.

	SE	DA	E
SE	.8	.1	.3
DA	.1	.7	.5
E	.1	.2	.2

where the number indicates the probability of transferring from the top program to the program on the left. Thus the probability of going from DA to E is .2. Find the probability that a student is enrolled in the various programs. Which program should be eliminated if there is a budget problem?

(8) From the transition matrix for the gambler's ruin problem, it follows that

$$\begin{aligned} p_{0j}^n &= qp_{0(j-1)}^{n-1} + pp_{0(j+1)}^{n-1} \text{for } j \in [1, b-1], \\ p_{00}^n &= 1, \text{ and } p_{0b}^n = 0. \end{aligned}$$

Assume here that $p \neq q$. The reason for the top equation is that the ij^{th} entry of P^n is

$$p_{0j}^n = \sum_k p_{0k}^{n-1} p_{kj} = qp_{0(j-1)}^{n-1} + pp_{0(j+1)}^{n-1}$$

Now it was shown above that $\lim_{n\to\infty} p_{0j}^n$ exists. Denote by P_j this limit. Then the above becomes much simpler if written as

$$\begin{aligned} P_j &= qP_{j-1} + pP_{j+1} \text{ for } j \in [1, b-1], & (9.3) \\ P_0 &= 1 \text{ and } P_b = 0. & (9.4) \end{aligned}$$

It is only required to find a solution to the above difference equation with boundary conditions. To do this, look for a solution in the form $P_j = r^j$and use the difference equation with boundary conditions to find the correct values of r. Show first that the difference equation is satisfied by $r =$

$$\frac{1}{2p}\left(1 + \sqrt{1-4pq}\right), \frac{1}{2p}\left(1 - \sqrt{1-4pq}\right)$$

Next show that $\sqrt{1-4pq} = \sqrt{1-4p+4p^2} = 1-2p$. Show that the above expressions simplify to 1 and $\frac{q}{p}$. Therefore, for any choice of $C_i, i = 1, 2,$

$$C_1 + C_2\left(\frac{q}{p}\right)^j$$

will solve the difference equation. Now choose C_1, C_2 to satisfy the boundary conditions. Show that this requires that the solution to the difference equation with boundary conditions is

$$P_j = \frac{q^b - p^{b-j}q^j}{q^b - p^b} = q^j\left(\frac{q^{b-j} - p^{b-j}}{q^b - p^b}\right).$$

(9) To find the solution in the case of a fair game, it is reasonable to take the $\lim_{p\to 1/2}$ of the above solution. Show that this yields

$$P_j = \frac{b-j}{b}.$$

Find a solution directly in the case where $p = q = 1/2$ in (9.3) and (9.4) by verifying directly that $P_j = 1$ and $P_j = j$ are two solutions to the difference equation and proceeding as before.

(10) Using similar techniques to the above, find the probability that in the gambler's ruin problem, he eventually wins all the money.

(11) In the gambler's ruin problem, show that with probability 1, one of the two absorbing states is eventually attained.

9.5 The Cayley-Hamilton theorem

As another application of the Jordan canonical form, we give another proof of the Cayley-Hamilton theorem. This is a better proof because it can be generalized (See Problem 14 on Page 232) to include the case of an arbitrary field of scalars, in contrast to the proof of Theorem 7.3 which was based on taking a limit as $|\lambda| \to \infty$, which option might not be available for some fields, such as a field of residue classes.

However, we are mainly interested in the case where the field of scalars is the real or complex numbers, so the earlier version is entirely adequate for our purposes.

Theorem 9.3. *Let A be an $n \times n$ matrix with entries in a field F. Suppose the minimal polynomial of A (see Problem 21 on Page 151) factors completely into linear factors.*[1] *Then if $q(\lambda)$ is the characteristic polynomial,*

$$\det(\lambda I - A) = q(\lambda),$$

it follows that

$$q(A) = 0.$$

Proof. Since the minimal polynomial factors, it follows that A has a Jordan canonical form. Thus there exists a matrix P such that

$$P^{-1}AP = J$$

where J is of the form

$$J = \begin{pmatrix} J_1 & & \\ & \ddots & \\ & & J_p \end{pmatrix}.$$

Here J_i is an $r_i \times r_i$ matrix of the form

$$J_i = \lambda_i I_{r_i} + N_i, \ N_i^{r_i} = 0.$$

with $\lambda_1, \ldots, \lambda_p$ the distinct eigenvalues of A. Therefore, the characteristic polynomial of A is of the form

$$\prod_{i=1}^{p} (\lambda - \lambda_i)^{r_i} = \lambda^n + a_{n-1}\lambda^{n-1} + \ldots + a_1\lambda + a_0.$$

It follows from block multiplication that

$$q(J) = \begin{pmatrix} q(J_1) & & \\ & \ddots & \\ & & q(J_p) \end{pmatrix}.$$

It follows that

$$q(J_j) = \prod_{i=1}^{p} \left(J_j - \lambda_i I_{r_j}\right)^{r_i} = \prod_{i \neq j} \left(J_j - \lambda_i I_{r_j}\right)^{r_i} N_i^{r_i} = 0.$$

It follows that $q(J) = 0$ and so

$$q(A) = q\left(PJP^{-1}\right) = Pq(J)P^{-1} = 0.$$

This proves the Cayley-Hamilton theorem. □

[1] In case $F = \mathbb{C}$, this is automatic because of the fundamental theorem of algebra. In general, when F is an arbitrary field, one can enlarge the field sufficiently to obtain a proof of the Cayley-Hamilton theorem in complete generality.

9.6 Systems of linear differential equations

We give a further application. For a set of functions $y_1(x), \ldots, y_n(x)$ in $S_{\mathbb{C}}[a,b]$ we write

$$\mathbf{y}' = (y_1', \ldots, y_n')^t \tag{9.5}$$

for the vector of derivatives. Consider a system of linear differential equations of the form

$$\begin{aligned} y_1' &= a_{11}y_1 + \cdots + a_m y_n \\ &\vdots \\ y_n' &= a_{n1}y_1 + \cdots + a_{1n}y_n, \end{aligned}$$

where $A = [a_{ij}]$ is a matrix over $\mathbb{C}$. We may write this

$$\mathbf{y}' = A\mathbf{y}.$$

A way to find a solution to this system of equations is to look for one which is of the form

$$\mathbf{y}(t) = \mathbf{v}e^{\lambda t}.$$

Then this will be a solution to the equation if and only if

$$\lambda e^{\lambda t}\mathbf{v} = A\mathbf{v}e^{\lambda t}.$$

Cancelling the $e^{\lambda t}$ on both sides, this shows that

$$\lambda \mathbf{v} = A\mathbf{v},$$

so that $(\lambda, \mathbf{v})$ is an eigenvalue with corresponding eigenvector.

Example 9.5. Solve the linear system $\mathbf{x}' = A\mathbf{x}$ using Maple, in the case

$$A = \begin{pmatrix} 1 & 3 & 3 \\ -3 & -5 & -3 \\ 3 & 3 & 1 \end{pmatrix}$$

Using Maple, we find the eigenvalues and eigenvectors are of the form

$$\begin{pmatrix} -1 \\ 1 \\ -1 \end{pmatrix} \leftrightarrow 1, \begin{pmatrix} 1 \\ 0 \\ -1 \end{pmatrix}, \begin{pmatrix} 0 \\ 1 \\ -1 \end{pmatrix} \leftrightarrow -2$$

Therefore, we can obtain three solutions to the first order system,

$$\begin{pmatrix} -1 \\ 1 \\ -1 \end{pmatrix} e^t, \begin{pmatrix} 1 \\ 0 \\ -1 \end{pmatrix} e^{-2t}, \text{ and } \begin{pmatrix} 0 \\ 1 \\ -1 \end{pmatrix} e^{-2t}.$$

It follows easily that any linear combination of these vector valued functions will be a solution to the system of equations $\mathbf{x}' = A\mathbf{x}$.

The main question is whether such linear combinations yield all possible solutions. In this case the answer is yes. This happens because the three eigenvectors are linearly independent, which will imply that if $\mathbf{x}(t)$ is any solution to the system of equations, then there exist constants c_i such that

$$\mathbf{x}(0) = c_1 \begin{pmatrix} -1 \\ 1 \\ -1 \end{pmatrix} + c_2 \begin{pmatrix} 1 \\ 0 \\ -1 \end{pmatrix} + c_3 \begin{pmatrix} 0 \\ 1 \\ -1 \end{pmatrix}$$

It will follow that both $\mathbf{x}(t)$ and

$$c_1 \begin{pmatrix} -1 \\ 1 \\ -1 \end{pmatrix} e^t + c_2 \begin{pmatrix} 1 \\ 0 \\ -1 \end{pmatrix} e^{-2t} + c_3 \begin{pmatrix} 0 \\ 1 \\ -1 \end{pmatrix} e^{-2t} \tag{9.6}$$

are solutions to the initial value problem

$$\mathbf{y}' = A\mathbf{y}, \ \mathbf{y}(0) = \mathbf{x}(0).$$

It will be shown below that there is a unique solution to this system, and so (9.6) must equal $\mathbf{x}(t)$.

Note that this general solution can also be written in the form

$$\begin{pmatrix} -e^t & e^{-2t} & 0 \\ e^t & 0 & e^{-2t} \\ -e^t & -e^{-2t} & -e^{-2t} \end{pmatrix} \mathbf{c} \tag{9.7}$$

where $\mathbf{c}$ is a constant vector in $\mathbb{R}^3$.

9.7 The fundamental matrix

In the above, it was easy to come up with this general solution because in this example, the matrix A had a basis of eigenvectors. However, such a basis of eigenvectors does not always exist. Furthermore, even in the case where there exists such a basis, it may be impossible to find the eigenvalues and eigenvectors exactly. Therefore, it is important to have another way to obtain solutions of this problem which will be completely general. The matrix in (9.7) is sometimes called a fundamental matrix. We will be a little more specific about this term, however.

Definition 9.3. If $\Phi(t)$ is a matrix with differentiable entries,

$$\Phi(t) = \begin{pmatrix} \mathbf{q}_1(t) & \mathbf{q}_2(t) & \cdots & \mathbf{q}_n(t) \end{pmatrix}$$

then $\Phi'(t)$ is defined as

$$\Phi'(t) = \begin{pmatrix} \mathbf{q}_1'(t) & \mathbf{q}_2'(t) & \cdots & \mathbf{q}_n'(t) \end{pmatrix}.$$

Let A be an $n \times n$ matrix. Then $\Phi(t)$ is called a **fundamental matrix** for A if

$$\Phi'(t) = A\Phi(t), \ \Phi(0) = I, \tag{9.8}$$

and $\Phi(t)^{-1}$ exists for all $t \in \mathbb{R}$.

The fundamental matrix plays the same role in solving linear systems of differential equations of the form

$$\mathbf{y}'(t) = A\mathbf{y}(t) + \mathbf{f}(t), \ \mathbf{y}(0) = \mathbf{y}_0$$

as the inverse matrix does in solving an algebraic system of the form

$$A\mathbf{y} = \mathbf{b}.$$

Once you know the fundamental matrix, you can routinely solve every such system.

We will show that **every** complex $n \times n$ matrix A has a unique fundamental matrix. When this has been done, it is completely routine to give the variation of constants formula for the solution to an arbitrary initial value problem

$$\mathbf{y}' = A\mathbf{y} + \mathbf{f}(t), \ \mathbf{y}(0) = \mathbf{y}_0.$$

To begin with, consider the special case for a complex valued function y satisfying

$$y' = ay + f(t), \ y(0) = y_0. \tag{9.9}$$

By Proposition 5.3, there is a unique solution to the initial value problem

$$y'(t) = (a + ib)\, y(t), \ y(0) = 1,$$

and this solution is $e^{(a+ib)t}$ with the definition

$$e^{(a+ib)t} = e^{at}(\cos(bt) + i \sin(bt)).$$

Proposition 9.1. Let $f(t), a, y_0$ be complex valued with $f(t)$ continuous. Then there exists a unique solution to (9.9) and it is given by a simple formula.

Proof. Using Proposition 5.3 and the rules of differentiation,

$$\frac{d}{dt}\left(e^{-at}y\right) = e^{-at} f(t).$$

Then after an integration,

$$e^{-at}y(t) - y_0 = \int_0^t e^{-as} f(s)\, ds,$$

and so

$$y(t) = e^{at}y_0 + \int_0^t e^{a(t-s)} f(s)\, ds. \ \square$$

Now here is the main result, due to Putzer [15]. This result is outstanding because it gives a good approximation to the fundamental matrix, even if the eigenvalues are not known exactly. This is in stark contrast to the usual situation presented in ordinary differential equations which relies on finding the exact eigenvalues. It also has no dependence on the existence of the Jordan form, depending only on the much easier Cayley-Hamilton theorem for complex vector spaces.

Theorem 9.4. *Let A be a complex $n \times n$ matrix whose eigenvalues are $\{\lambda_1, \cdots, \lambda_n\}$ repeated according to multiplicity. Define*

$$P_k(A) = \prod_{m=1}^{k} (A - \lambda_m I), \ P_0(A) = I,$$

and let the scalar valued functions $r_k(t)$ be the solutions to the following initial value problem

$$\begin{pmatrix} r_0'(t) \\ r_1'(t) \\ r_2'(t) \\ \vdots \\ r_n'(t) \end{pmatrix} = \begin{pmatrix} 0 \\ \lambda_1 r_1(t) + r_0(t) \\ \lambda_2 r_2(t) + r_1(t) \\ \vdots \\ \lambda_n r_n(t) + r_{n-1}(t) \end{pmatrix}, \quad \begin{pmatrix} r_0(0) \\ r_1(0) \\ r_2(0) \\ \vdots \\ r_n(0) \end{pmatrix} = \begin{pmatrix} 0 \\ 1 \\ 0 \\ \vdots \\ 0 \end{pmatrix}$$

Now define

$$\Phi(t) = \sum_{k=0}^{n-1} r_{k+1}(t) P_k(A).$$

Then

$$\Phi'(t) = A\Phi(t), \ \Phi(0) = I. \tag{9.10}$$

Note that the equations in the system are of the type in Proposition 9.1.

Proof. This is an easy computation.

$$\Phi(0) = \sum_{k=0}^{n-1} r_{k+1}(0) P_k(A) = r_1(0) P_0(A) = I.$$

Next consider the differential equation.

$$\Phi'(t) = \sum_{k=0}^{n-1} r_{k+1}'(t) P_k(A) = \sum_{k=0}^{n-1} (\lambda_{k+1} r_{k+1}(t) + r_k(t)) P_k(A).$$

Also

$$A\Phi(t) = \sum_{k=0}^{n-1} r_{k+1}(t) A P_k(A).$$

By the Cayley-Hamilton theorem, $P_n(A) = 0$. Then the above equals

$$\begin{aligned} &= \sum_{k=0}^{n-1} r_{k+1}(t)(A - \lambda_{k+1} I) P_k(A) + \sum_{k=0}^{n-1} r_{k+1}(t) \lambda_{k+1} P_k(A) \\ &= \sum_{k=1}^{n} r_k(t) P_k(A) + \sum_{k=0}^{n-1} r_{k+1}(t) \lambda_{k+1} P_k(A) \\ &= \sum_{k=0}^{n-1} r_k(t) P_k(A) + \sum_{k=0}^{n-1} r_{k+1}(t) \lambda_{k+1} P_k(A) \end{aligned}$$

because $r_0(t) = 0$. This is the same thing as $\Phi'(t)$. This proves the theorem. □

This theorem establishes an important theoretical conclusion, that every $n \times n$ matrix has a fundamental matrix of the sort described above. However, in finding the fundamental matrix, we usually use more ad hoc methods. Suppose

$$\Phi(t) = \begin{pmatrix} \mathbf{x}_1(t) & \cdots & \mathbf{x}_n(t) \end{pmatrix}$$

and A can be diagonalized. This often happens. (Recall Lemma 8.2 which implies that if the eigenvalues are distinct, then the matrix can be diagonalized.) When it does, the computation of the fundamental matrix is particularly simple. This is because, for some invertible P,

$$\Phi'(t) = \overbrace{P^{-1}DP}^{A}\Phi(t), \; P\Phi(0) = P,$$

where D is a diagonal matrix. Therefore, letting $\Psi(t) = P\Phi(t)$,

$$\Psi'(t) = D\Psi(t), \; \Psi(0) = P, \tag{9.11}$$

$$D = \begin{pmatrix} \lambda_1 & & \\ & \ddots & \\ & & \lambda_n \end{pmatrix}.$$

It is particularly easy to solve this initial value problem. In fact, a solution to (9.11) is

$$\Psi(t) = \begin{pmatrix} e^{\lambda_1 t} & & \\ & \ddots & \\ & & e^{\lambda_n t} \end{pmatrix} P,$$

because

$$\Psi'(t) = \begin{pmatrix} \lambda_1 e^{\lambda_1 t} & & \\ & \ddots & \\ & & \lambda_n e^{\lambda_n t} \end{pmatrix} P =$$

$$\begin{pmatrix} \lambda_1 & & \\ & \ddots & \\ & & \lambda_n \end{pmatrix} \begin{pmatrix} e^{\lambda_1 t} & & \\ & \ddots & \\ & & e^{\lambda_n t} \end{pmatrix} P = D\Psi(t),$$

and

$$\Psi(0) = P.$$

Then the fundamental matrix is $\Phi(t)$ where

$$\Phi(t) = P^{-1}\Psi(t).$$

We illustrate with the following example.

Example 9.6. Find the fundamental matrix for

$$A = \begin{pmatrix} -3 & -2 & -8 \\ 3 & 2 & 6 \\ 1 & 1 & 3 \end{pmatrix}$$

This matrix can be diagonalized. To do so, you find the eigenvalues and then let the columns of P^{-1} consist of eigenvectors corresponding to these eigenvalues. As explained earlier, PAP^{-1} is then a diagonal matrix. Maple and other computer algebra systems are adept at diagonalizing matrices when this can be done. To do so, you ask for the Jordan form of the matrix. If the matrix is diagonalizable, the Jordan form will be a diagonal matrix. In this case

$$\begin{pmatrix} -3 & -2 & -8 \\ 3 & 2 & 6 \\ 1 & 1 & 3 \end{pmatrix} = \overbrace{\begin{pmatrix} 1 & 2 & -2 \\ -1 & 0 & 1 \\ 0 & -1 & 1 \end{pmatrix}}^{P^{-1}} \begin{pmatrix} -1 & 0 & 0 \\ 0 & 1 & 0 \\ 0 & 0 & 2 \end{pmatrix} \overbrace{\begin{pmatrix} 1 & 0 & 2 \\ 1 & 1 & 1 \\ 1 & 1 & 2 \end{pmatrix}}^{P}$$

Thus $\Psi(t) =$

$$\begin{pmatrix} e^{-t} & 0 & 0 \\ 0 & e^{t} & 0 \\ 0 & 0 & e^{2t} \end{pmatrix} \begin{pmatrix} 1 & 0 & 2 \\ 1 & 1 & 1 \\ 1 & 1 & 2 \end{pmatrix} = \begin{pmatrix} e^{-t} & 0 & 2e^{-t} \\ e^{t} & e^{t} & e^{t} \\ e^{2t} & e^{2t} & 2e^{2t} \end{pmatrix},$$

and so $\Phi(t) =$

$$\begin{pmatrix} 1 & 2 & -2 \\ -1 & 0 & 1 \\ 0 & -1 & 1 \end{pmatrix} \begin{pmatrix} e^{-t} & 0 & 2e^{-t} \\ e^{t} & e^{t} & e^{t} \\ e^{2t} & e^{2t} & 2e^{2t} \end{pmatrix}$$

$$= \begin{pmatrix} e^{-t} + 2e^{t} - 2e^{2t} & 2e^{t} - 2e^{2t} & 2e^{-t} + 2e^{t} - 4e^{2t} \\ -e^{-t} + e^{2t} & e^{2t} & -2e^{-t} + 2e^{2t} \\ -e^{t} + e^{2t} & -e^{t} + e^{2t} & -e^{t} + 2e^{2t} \end{pmatrix}.$$

Checking this, $\Phi'(t) =$

$$\begin{pmatrix} -e^{-t} + 2e^{t} - 4e^{2t} & 2e^{t} - 4e^{2t} & -2e^{-t} + 2e^{t} - 8e^{2t} \\ e^{-t} + 2e^{2t} & 2e^{2t} & 2e^{-t} + 4e^{2t} \\ -e^{t} + 2e^{2t} & -e^{t} + 2e^{2t} & -e^{t} + 4e^{2t} \end{pmatrix}$$

and $A\Phi(t) =$

$$\begin{pmatrix} -3 & -2 & -8 \\ 3 & 2 & 6 \\ 1 & 1 & 3 \end{pmatrix} \begin{pmatrix} e^{-t} + 2e^{t} - 2e^{2t} & 2e^{t} - 2e^{2t} & 2e^{-t} + 2e^{t} - 4e^{2t} \\ -e^{-t} + e^{2t} & e^{2t} & -2e^{-t} + 2e^{2t} \\ -e^{t} + e^{2t} & -e^{t} + e^{2t} & -e^{t} + 2e^{2t} \end{pmatrix}$$

$$= \begin{pmatrix} -e^{-t} + 2e^{t} - 4e^{2t} & 2e^{t} - 4e^{2t} & -2e^{-t} + 2e^{t} - 8e^{2t} \\ e^{-t} + 2e^{2t} & 2e^{2t} & 2e^{-t} + 4e^{2t} \\ -e^{t} + 2e^{2t} & -e^{t} + 2e^{2t} & -e^{t} + 4e^{2t} \end{pmatrix}$$

which is the same thing. The multiplications are tedious but routine. They were accomplished here through the use of a computer algebra system.

We need to justify the use of the definite article in referring to **the** fundamental matrix. To do this, we need a simple result which depends on calculus. It is called **Gronwall's inequality**.

Lemma 9.5. *Let $u(t)$ be a continuous nonnegative function which satisfies*

$$u(t) \leq a + \int_0^t ku(s)\,ds, \; k \geq 0.$$

for $t \geq 0$. Then $u(t) \leq ae^{kt}$ for all $t \geq 0$.

Proof. Let $w(s) = \int_0^s u(\tau)\, d\tau$. Then $w(s) \geq 0,\ w(0) = 0$. By the fundamental theorem of calculus, $w'(s)$ exists and equals $u(s)$. Hence

$$w'(s) - kw(s) \leq a$$

Multiply both sides by e^{-ks}. Then using the chain rule and product rule,

$$\frac{d}{ds}\left(e^{-ks} w(t)\right) \leq ae^{-ks}$$

Applying the integral $\int_0^t ds$ to both sides,

$$0 \leq e^{-kt} w(t) \leq a \int_0^t e^{-ks} ds = a \frac{1 - e^{-tk}}{k}.$$

It follows that $w(t) \leq a\left(\frac{1}{k} e^{kt} - \frac{1}{k}\right)$, and so $a + kw(t) \leq ae^{-kt}$. Therefore, from the assumptions,

$$0 \leq u(t) \leq a + kw(t) \leq ae^{-kt}. \ \square$$

Definition 9.4. Let A be a complex $m \times n$ matrix. Let

$$||A|| = \max\left\{|A_{ij}|, i \in \{1, \cdots, m\}, j \in \{1, \cdots, n\}\right\}.$$

The integral of a matrix whose entries are functions of t is defined as the matrix in which each entry is integrated. Thus

$$\left(\int_0^t \Phi(s)\, ds\right)_{ij} = \int_0^t \Phi(s)_{ij}\, ds.$$

Lemma 9.6. *Let A be an $m \times n$ matrix and let B be a $n \times p$ matrix. Then*

$$||AB|| \leq n\,||A||\,||B||.$$

Also,

$$\left|\left|\int_0^t \Phi(s)\, ds\right|\right| \leq \int_0^t ||\Phi(s)||\, ds.$$

Proof. This follows from the definition of matrix multiplication.

$$\left|(AB)_{ij}\right| \leq \left|\sum_k A_{ik} B_{kj}\right| \leq \sum_k |A_{ik}|\,|B_{kj}| \leq n\,||A||\,||B||.$$

Since ij is arbitrary, the inequality follows. Consider the claim about the integral.

$$\left|\left(\int_0^t \Phi(s)\, ds\right)_{ij}\right| = \left|\int_0^t \Phi(s)_{ij}\, ds\right| \leq \int_0^t \left|\Phi(s)_{ij}\right| ds \leq \int_0^t ||\Phi(s)||\, ds.$$

Hence the result follows. $\square$

Lemma 9.7. *Let Φ be a matrix which satisfies*

$$\Phi'(t) = A\Phi(t),\ \Phi(0) = 0. \tag{9.12}$$

Then $\Phi(t) = 0$. Also, the solution to (9.10) is unique.

Proof. Integrating both sides, it follows that

$$\Phi(t) = \int_0^t A\Phi(s)\,ds.$$

Therefore,

$$||\Phi(t)|| = \left|\left|\int_0^t A\Phi(s)\,ds\right|\right| \leq \int_0^t n\,||A||\,||\Phi(s)||\,ds.$$

It follows from Lemma 9.5 with $a = 0$ that $\Phi(t) = 0$ for all $t \geq 0$. This proves the first part. To prove the second part, if Φ_i is a solution to (9.10) for $i = 1, 2$, then $\Psi(t) = \Phi_1(t) - \Phi_2(t)$ solves (9.12). Therefore, from what was just shown, $\Psi(t) = 0$ for all $t \geq 0$. □

Corollary 9.3. *The fundamental matrix* $\Phi(t)$ *of (9.10) has an inverse for all* t. *This inverse is* $\Phi(-t)$. *Furthermore,* $\Phi(t)A = A\Phi(t)$ *for all* $t \geq 0$. *In addition,*

$$\Phi(t-s) = \Phi(t)\Phi(-s). \tag{9.13}$$

Proof. First consider the claim that A and $\Phi(t)$ commute. Let $\Psi(t) = \Phi(t)A - A\Phi(t)$. Then

$$\begin{aligned}\Psi'(t) &= \Phi'(t)A - A\Phi'(t) = A\Phi(t)A - A^2\Phi(t) \\ &= A(\Phi(t)A - A\Phi(t)) = A\Psi(t).\end{aligned}$$

Also $\Psi(0) = A - A = 0$. Therefore, by Lemma 9.7, $\Psi(t) = 0$ for all $t \geq 0$.

Now consider the claim about the inverse of $\Phi(t)$. For $t \geq 0$,

$$\begin{aligned}(\Phi(t)\Phi(-t))' &= \Phi'(t)\Phi(-t) - \Phi(t)\Phi'(-t) \\ &= A\Phi(t)\Phi(-t) - \Phi(t)A\Phi(-t) \\ &= \Phi(t)A\Phi(-t) - \Phi(t)A\Phi(-t) = 0.\end{aligned}$$

Also $\Phi(0)\Phi(0) = I$, and each entry of $\Phi(t)\Phi(-t)$ is a constant. Therefore, for all $t \geq 0$,

$$\Phi(t)\Phi(-t) = I.$$

Recall that this shows $\Phi(t)^{-1}$ exists and equals $\Phi(-t)$ for all $t \in \mathbb{R}$.

To verify the last claim, fix s and let $\Psi(t) = \Phi(t-s) - \Phi(t)\Phi(-s)$. Then $\Psi(0) = 0$. Moreover,

$$\begin{aligned}\Psi'(t) &= \Phi'(t-s) - \Phi'(t)\Phi(-s) \\ &= A(\Phi(t-s) - \Phi(t)\Phi(-s)) = A\Psi(t).\end{aligned}$$

By Lemma 9.7, it follows that $\Psi(t) = 0$ for all $t \geq 0$. □

9.8 Existence and uniqueness

With Corollary 9.3 and Theorem 9.4, it is now easy to solve the general first order system of equations.

$$\mathbf{x}'(t) = A\mathbf{x}(t) + \mathbf{f}(t), \ \mathbf{x}(0) = \mathbf{x}_0 \in \mathbb{R}^n. \tag{9.14}$$

Theorem 9.5. *Let $\mathbf{f}(t)$ be a continuous function defined on $\mathbb{R}$. Then there exists a unique solution to the first order system (9.14), given by the formula*

$$\mathbf{x}(t) = \Phi(t)\mathbf{x}_0 + \int_0^t \Phi(t-s)\mathbf{f}(s)\,ds \tag{9.15}$$

where $\Phi(t)$ is the fundamental matrix of Theorem 9.4.

Proof. From Corollary 9.3, the right side of the above formula for $\mathbf{x}(t)$ is of the form

$$\Phi(t)\mathbf{x}_0 + \Phi(t)\int_0^t \Phi(-s)\mathbf{f}(s)\,ds$$

Using the fundamental theorem of calculus and the product rule, the derivative of this is

$$\begin{aligned} & A\Phi(t)\mathbf{x}_0 + A\Phi(t)\int_0^t \Phi(-s)\mathbf{f}(s)\,ds + \Phi(t)\Phi(-t)\mathbf{f}(t) \\ = \; & A\left(\Phi(t)\mathbf{x}_0 + \Phi(t)\int_0^t \Phi(-s)\mathbf{f}(s)\,ds\right) + \mathbf{f}(t) = A\mathbf{x}(t) + \mathbf{f}(t). \end{aligned}$$

Therefore, the right side of (9.15) satisfies the differential equation of (9.14). Evaluating at $t = 0$ yields $\mathbf{x}(0) = \mathbf{x}_0$. Therefore, there exists a solution to (9.14).

In fact, this is the only solution. If $\mathbf{y}(t)$ is a solution to (9.14), then

$$\mathbf{y}'(t) - A\mathbf{y}(t) = \mathbf{f}(t).$$

Multiplying by $\Phi(-t)$, it follows from the chain rule that

$$\begin{aligned} \frac{d}{dt}(\Phi(-t)\mathbf{y}(t)) &= -\Phi'(-t)\mathbf{y}(t) + \Phi(-t)\mathbf{y}'(t) \\ &= -A\Phi(-t)\mathbf{y}(t) + \Phi(-t)\mathbf{y}'(t) \\ &= \Phi(-t)(\mathbf{y}'(t) - A\mathbf{y}(t)). \end{aligned}$$

Therefore,

$$\frac{d}{dt}(\Phi(-t)\mathbf{y}(t)) = \Phi(-t)\mathbf{f}(t).$$

Now apply $\int_0^t$ to both sides to obtain

$$\Phi(-t)\mathbf{y}(t) - \mathbf{x}_0 = \int_0^t \Phi(-s)\mathbf{f}(s)\,ds$$

which yields (9.15). $\square$

This formula is called the **variation of constants formula**. Since every n^{th} order linear ordinary differential equation for a scalar valued function, can be considered as a first order system like the above, it follows that the above section has included the mathematical theory of most of an ordinary differential equations course as a special case, and has also resolved the difficult questions about existence and uniqueness of solutions, using methods which are mainly algebraic in nature. In addition, the above methods are not dependent on finding the exact values of the eigenvalues in order to obtain meaningful solutions.

9.9 The method of Laplace transforms

There is a popular algebraic technique which may be the fastest way to find closed form solutions to the initial value problem. This method of Laplace transforms succeeds so well because of the algebraic technique of partial fractions and the fact that the Laplace transform is a linear mapping.

We will ignore all analytical questions and emphasize only the algebraic procedures. The analytical questions are not trivial and for this reason are never discussed in undergraduate courses on differential equations anyway.

Definition 9.5. Let f be a function defined on $[0, \infty)$ which has **exponential growth**, meaning that

$$|f(t)| \leq Ce^{\lambda t}$$

for some real λ. Then the **Laplace transform** of f, denoted by $\mathfrak{L}(f)$ is defined as

$$\mathfrak{L}f(s) = \int_0^\infty e^{-ts} f(t)\, dt$$

for all s sufficiently large. It is customary to write this transform as $F(s)$ and the function as $f(t)$. In case $\mathbf{f}(t)$ is a vector valued function, its Laplace transform is the vector valued function obtained by replacing each entry by its Laplace transform.

We leave as an exercise the verification that $\mathfrak{L}$ is a linear mapping.

The usefulness of this method in solving differential equations, comes from the following observation.

$$\begin{aligned}\int_0^\infty x'(t)\, e^{-ts} dt &= x(t)\, e^{-st}|_0^\infty + \int_0^\infty se^{-st} x(t)\, dt \\ &= -x(0) + s\mathfrak{L}x(s).\end{aligned}$$

Doing this for each component, it follows that after taking the Laplace transform of

$$\mathbf{x}' = A\mathbf{x} + \mathbf{f}(t), \ \mathbf{x}(0) = \mathbf{x}_0,$$

we obtain the following for all s large enough.

$$s\mathbf{X}(s) - \mathbf{x}_0 = A\mathbf{X}(s) + \mathbf{F}(s)$$

where $\mathbf{X}(s) = \mathfrak{L}(\mathbf{x})(s)$. Thus for all s large enough,

$$(sI - A)\mathbf{X}(s) = \mathbf{x}_0 + \mathbf{F}(s),$$

and so

$$\mathbf{X}(s) = (sI - A)^{-1}(\mathbf{x}_0 + \mathbf{F}(s)).$$

Now you examine the component functions of $\mathbf{X}(s)$ and look in the table to find the functions of t which result in these component functions. This is then the solution to the initial value problem.

Here is a simple table.

Table of Laplace Transforms

$f(t) = \mathfrak{L}^{-1}\{F(s)\}$	$F(s) = \mathfrak{L}\{f(t)\}$
1.) 1	$1/s$
2.) e^{at}	$1/(s-a)$
3.) t^n, n positive integer	$\frac{n!}{s^{n+1}}$
4.) $t^p,\ p > -1$	$\frac{\Gamma(p+1)}{s^{p+1}}$
5.) $\sin at$	$\frac{a}{s^2+a^2}$
6.) $\cos at$	$\frac{s}{s^2+a^2}$
7.) e^{iat}	$\frac{s+ia}{s^2+a^2}$
8.) $\sinh at$	$\frac{a}{s^2-a^2}$
9.) $\cosh at$	$\frac{s}{s^2-a^2}$
10.) $e^{at}\sin bt$	$\frac{b}{(s-a)^2+b^2}$
11.) $e^{at}\cos bt$	$\frac{s-a}{(s-a)^2+b^2}$
12.) $t^n e^{at}$, n positive integer	$\frac{n!}{(s-a)^{n+1}}$
13.) $u_c(t)$	$\frac{e^{-cs}}{s}$
14.) $u_c(t) f(t-c)$	$e^{-cs}F(s)$
15.) $e^{ct}f(t)$	$F(s-c)$
16.) $f(ct)$	$\frac{1}{c}F\left(\frac{s}{c}\right)$
17.) $f * g = \int_0^t f(t-u)\, g(u)\, du$	$F(s)\, G(s)$
18.) $\delta(t-c)$	e^{-cs}
19.) $f'(t)$	$sF(s) - f(0)$
20.) $(-t)^n f(t)$	$\frac{d^n F}{ds^n}(s)$

In this table, $\Gamma(p+1)$ denotes the gamma function

$$\Gamma(p+1) = \int_0^\infty e^{-t} t^p dt$$

The function $u_c(t)$ denotes the step function which equals 1 for $t > c$ and 0 for $t < c$. The expression in formula 18.) is explained in Exercise 12 below. It models an impulse and is sometimes called the Dirac delta function.

Here is an example.

Example 9.7. Solve the initial value problem

$$\mathbf{x}' = \begin{pmatrix} 1 & 1 \\ 1 & 2 \end{pmatrix} \mathbf{x} + \begin{pmatrix} 1 \\ t \end{pmatrix}, \ \mathbf{x}(0) = \begin{pmatrix} 1 \\ 0 \end{pmatrix}.$$

You do $\mathfrak{L}$ to both sides. Thus, from the above discussion,

$$s\mathbf{X}(s) - \mathbf{X}(0) = s\mathbf{X}(s) - \begin{pmatrix} 1 \\ 0 \end{pmatrix} = \begin{pmatrix} 1 & 1 \\ 1 & 2 \end{pmatrix} \mathbf{X}(s) + \begin{pmatrix} 1/s \\ 1/s^2 \end{pmatrix}$$

Thus, solving for $\mathbf{X}(s)$,

$$\left(s \begin{pmatrix} 1 & 0 \\ 0 & 1 \end{pmatrix} - \begin{pmatrix} 1 & 1 \\ 1 & 2 \end{pmatrix} \right) \mathbf{X}(s) = \begin{pmatrix} 1 \\ 0 \end{pmatrix} + \begin{pmatrix} 1/s \\ 1/s^2 \end{pmatrix}$$

Of course, for large s the matrix on the left has an inverse and so you can solve for $\mathbf{X}(s)$ for all s large enough and obtain

$$\mathbf{X}(s) = \left(s \begin{pmatrix} 1 & 0 \\ 0 & 1 \end{pmatrix} - \begin{pmatrix} 1 & 1 \\ 1 & 2 \end{pmatrix} \right)^{-1} \left(\begin{pmatrix} 1 \\ 0 \end{pmatrix} + \begin{pmatrix} 1/s \\ 1/s^2 \end{pmatrix} \right).$$

Therefore, $\mathbf{X}(s)$ equals

$$\begin{pmatrix} \frac{s^3 - s^2 - 2s + 1}{(s^2 - 3s + 1)s^2} \\ \frac{s^2 + 2s - 1}{(s^2 - 3s + 1)s^2} \end{pmatrix}.$$

Now it is just a matter of using the table to find the functions whose Laplace transforms equal these functions. However, the entries in the above vector are not listed in the table. Therefore, we take the partial fractions expansion of these entries to find things which are in the table. Then the above vector equals

$$\begin{pmatrix} \frac{1}{s^2} + \frac{1}{s} + \frac{1}{\sqrt{5}} \frac{1}{\left(s - \left(\frac{3}{2} + \frac{\sqrt{5}}{2}\right)\right)} - \frac{1}{\sqrt{5}} \frac{1}{\left(s - \left(\frac{3}{2} - \frac{\sqrt{5}}{2}\right)\right)} \\ -\frac{1}{s^2} - \frac{1}{s} + \frac{\sqrt{5}+5}{10} \frac{1}{s - \left(\frac{3+\sqrt{5}}{2}\right)} + \frac{5-\sqrt{5}}{10} \frac{1}{s - \left(\frac{3}{2} - \frac{\sqrt{5}}{2}\right)} \end{pmatrix}.$$

Each of these terms is in the table. Hence the solution to the initial value problem is

$$\mathbf{x}(t) = \begin{pmatrix} t + 1 + \frac{1}{\sqrt{5}} e^{\left(\frac{3+\sqrt{5}}{2}\right)t} - \frac{1}{\sqrt{5}} e^{\left(\frac{3-\sqrt{5}}{2}\right)t} \\ -t - 1 + \frac{\sqrt{5}+5}{10} e^{\left(\frac{3+\sqrt{5}}{2}\right)t} + \frac{5-\sqrt{5}}{10} e^{\left(\frac{3-\sqrt{5}}{2}\right)t} \end{pmatrix}.$$

Note that the eigenvalues of the matrix in the initial value problem are

$$\frac{3+\sqrt{5}}{2}, \frac{3-\sqrt{5}}{2}.$$

It is not a coincidence that these are occurring as in the exponents for the formula for the solution.

Example 9.8. Use the method of Laplace transforms to solve the following initial value problem.

$$\mathbf{x}' = \begin{pmatrix} 3 & 3 & 2 \\ -1 & 0 & -1 \\ 0 & -1 & 1 \end{pmatrix} \mathbf{x} + \begin{pmatrix} 1 \\ t \\ \sin(2t) \end{pmatrix}, \ \mathbf{x}(0) = \begin{pmatrix} 0 \\ 1 \\ 0 \end{pmatrix}$$

The matrix cannot be diagonalized.

From the above technique, take the Laplace transform of both sides. This yields

$$s\mathbf{X}(s) - \begin{pmatrix} 0 \\ 1 \\ 0 \end{pmatrix} = \begin{pmatrix} 3 & 3 & 2 \\ -1 & 0 & -1 \\ 0 & -1 & 1 \end{pmatrix} \mathbf{X}(s) + \begin{pmatrix} 1/s \\ 1/s^2 \\ 2/\left(s^2+4\right) \end{pmatrix}$$

Then, solving for $\mathbf{X}(s)$ and then finding the partial fraction expansion of each entry yields $\mathbf{X}(s) =$

$$\begin{pmatrix} \frac{2}{s-2} + \frac{27}{5(s-1)^2} - \frac{189}{25(s-1)} + \frac{21}{4s} + \frac{5}{2s^2} + \frac{1}{100}\frac{-34+31s}{s^2+4} \\ \frac{27}{5(s-1)} - \frac{2}{s-2} - \frac{9}{4s} - \frac{3}{2s^2} - \frac{1}{20}\frac{3s-2}{s^2+4} \\ \frac{2}{s-2} - \frac{27}{5(s-1)^2} - \frac{15}{4s} + \frac{54}{25(s-1)} - \frac{3}{2s^2} - \frac{1}{100}\frac{26+41s}{s^2+4} \end{pmatrix}.$$

It follows from going backwards in the table that $\mathbf{x}(t) =$

$$\begin{pmatrix} 2e^{2t} + \frac{27}{5}te^t - \frac{189}{25}e^t + \frac{21}{4} + \frac{5}{2}t + \frac{1}{100}\left(-17\sin(2t) + 31\cos(2t)\right) \\ \frac{27}{5}e^t - 2e^{2t} - \frac{9}{4} - \frac{3}{2}t - \frac{1}{20}\left(3\cos(2t) - \sin(2t)\right) \\ 2e^{2t} - \frac{27}{5}te^t - \frac{15}{4} + \frac{54}{25}e^t - \frac{3}{2}t - \frac{1}{100}\left(13\sin(2t) + 41\cos(2t)\right) \end{pmatrix}.$$

We used a computer algebra system to accomplish these manipulations, but they can all be done by hand with more effort.

We have now included all important techniques for solving linear ordinary differential equations with constant coefficients.

9.10 Exercises

(1) Abel's formula was proved for a complex $n \times n$ matrix which can be diagonalized in Problem 28 on Page 220. Extend this result to an arbitrary complex $n \times n$ matrix using Problem 3 on Page 229. **Hint:** Let A_k be matrices which

can be diagonalized such that the entries of A_k all converge to the corresponding entries of A. Let $\Phi_k(t)$ correspond to $\Phi(t)$ in Problem 28 on Page 220 with A replaced with A_k. Now it follows that

$$||\Phi(t) - \Phi_k(t)|| \leq \int_0^t ||(A - A_k)\Phi(s)||\, ds.$$

Apply Gronwall's inequality to obtain that

$$\lim_{k\to\infty} \Phi_k(t) = \Phi(t),$$

and use the formula of Problem 28 on Page 220.

(2) In the theoretical treatment above, the fundamental matrix $\Phi(t)$ satisfies

$$\Phi'(t) = A\Phi(t),\ \Phi(0) = I. \tag{9.16}$$

With such a fundamental matrix, the solution to

$$\mathbf{x}' = A\mathbf{x} + \mathbf{f},\ \mathbf{x}(0) = \mathbf{x}_0,$$

was given by

$$\mathbf{x}(t) = \Phi(t)\mathbf{x}_0 + \int_0^t \Phi(t-s)\mathbf{f}(s)\, ds.$$

However, it may be less trouble to simply insist that $\Phi(0)^{-1}$ exists. When this weaker requirement is made, we will refer to $\Phi(t)$ as **a** fundamental matrix. Then by Abel's formula in Problem 1, $\Phi(t)^{-1}$ exists for all t. Find a formula for the solution to the initial value Problem 9.16 in terms of a fundamental matrix.

(3) Suppose $\mathbf{x}_1(t), \cdots, \mathbf{x}_n(t)$ each are solutions of the differential equation

$$\mathbf{y}' = A\mathbf{y}.$$

Then let

$$\Phi(t) = \begin{pmatrix} \mathbf{x}_1(t) & \cdots & \mathbf{x}_n(t) \end{pmatrix}$$

and suppose that at some t_0, $\det(\Phi(t_0)) \neq 0$. Using Problem 1, show that $\det(\Phi(t)) \neq 0$ for all t and that $\Phi(t)$ is a fundamental matrix.

(4) If $\mathbf{v}$ is an eigenvector of A corresponding to an eigenvalue λ, show that $\mathbf{v}e^{\lambda t}$ is a solution to $\mathbf{x}' = A\mathbf{x}$. We say that $\mathbf{w}$ is a generalized eigenvector for the eigenvalue λ if for some positive integer m,

$$(\lambda I - A)^m \mathbf{w} = \mathbf{0}.$$

Thus a generalized eigenvector is on the top of a chain based on an eigenvector as in the proof of the existence of the Jordan canonical form. Suppose $(\lambda I - A)\mathbf{w} = \mathbf{v}$ where $\mathbf{v}$ is an eigenvector of A, so that $(\lambda I - A)^2 \mathbf{w} = \mathbf{0}$. Show that

$$\mathbf{w}e^{\lambda t} + t\mathbf{v}e^{\lambda t}$$

is a solution of the equation $\mathbf{x}' = A\mathbf{x}$. Now suppose that $\mathbf{v}, \mathbf{w}_1, \mathbf{w}_2$ is a chain satisfying $(\lambda I - A)\mathbf{w}_2 = \mathbf{w}_1, (\lambda I - A)\mathbf{w}_1 = \mathbf{v}$. Show that

$$\frac{t^2}{2}\mathbf{v}e^{\lambda t} + t\mathbf{w}_1 e^{\lambda t} + \mathbf{w}_2 e^{\lambda t}$$

is a solution of $\mathbf{x}' = A\mathbf{x}$. Generalize to obtain a formula for a chain of arbitrary length. Describe how to use this to find a fundamental matrix. **Hint:** Use Problem 1 and show that every chain is linearly independent. Why do there exist enough solutions of the various kinds just described to yield a fundamental matrix?

(5) Let A be an $n \times n$ complex matrix and suppose also that it can be diagonalized,

$$A = PDP^{-1}$$

where

$$D = \begin{pmatrix} \lambda_1 & & \\ & \ddots & \\ & & \lambda_n \end{pmatrix}.$$

Show that the fundamental matrix is

$$P\begin{pmatrix} e^{\lambda_1 t} & & \\ & \ddots & \\ & & e^{\lambda_n t} \end{pmatrix}P^{-1}.$$

(6) Here is a matrix whose eigenvalues are $1, -1, 0$.

$$A = \begin{pmatrix} 1 & 1 & 2 \\ -4 & -6 & -10 \\ 2 & 3 & 5 \end{pmatrix}$$

(a) Find a fundamental matrix for this matrix.
(b) Using this fundamental matrix, give the solution to the initial value problem

$$\mathbf{x}' = A\mathbf{x} + \mathbf{f}(t), \ \mathbf{x}(0) = (1, 0, 0)^t,$$

where

$$\mathbf{f}(t) = \begin{pmatrix} 1 \\ t \\ t^2 \end{pmatrix}.$$

(7) Solve the above problem using the method of Laplace transforms.

(8) Here is a matrix whose eigenvalues are $1 + i, 1 - i$ and 1.

$$\begin{pmatrix} 1 & -1 & 0 \\ 1 & 1 & 0 \\ 0 & 0 & 1 \end{pmatrix}$$

(a) Find a fundamental matrix for this matrix.
(b) Using this fundamental matrix, give the solution to the initial value problem

$$\mathbf{x}' = A\mathbf{x} + \mathbf{f}(t), \ \mathbf{x}(0) = (0,0,1)^t$$

where

$$\mathbf{f}(t) = \begin{pmatrix} 1 \\ 1 \\ t \end{pmatrix}.$$

(9) Find the solution to the above problem using the method of Laplace transforms.

(10) Recall that the power series for e^x given by

$$1 + x + \frac{x^2}{2!} + \ldots = \sum_{k=0}^{\infty} \frac{x^k}{k!}.$$

What if you formally placed an $n \times n$ matrix in place of x? The partial sums would all make sense and would yield a matrix which is an $n \times n$ matrix. Show that the partial sums obtained by this formal substitution are a sequence of $n \times n$ matrices having the property that for $S(p)_{ij}$ denoting the ij^{th} entry of the pth partial sum in the above series, $\left\{S(p)_{ij}\right\}_{p=1}^{\infty}$ is a Cauchy sequence and therefore, converges. This series

$$\sum_{k=0}^{\infty} \frac{A^k}{k!}$$

is denoted by e^A, not surprisingly. Go ahead and assume[2] the series

$$\sum_{k=0}^{\infty} \frac{t^k A^k}{k!}$$

can be differentiated term by term. Show that if this is so, then this series yields the fundamental matrix $\Phi(t)$. **Hint:** For $||A||$ given as above, show that it suffices to show that for every $\varepsilon > 0$ there exists p such that for $k, l > p$,

$$||S(k) - S(l)|| < \varepsilon.$$

(11) Consider the following initial value problem.

$$y''' - y'' - y' + y = t, \ y(0) = 1, y'(0) = 0, y''(0) = 0.$$

This can be considered as a first order system in the following way. Let

$$\begin{aligned} y''(t) &= x_3(t) \\ y'(t) &= x_2(t) \\ y(t) &= x_1(t) \end{aligned}$$

[2] This is not too hard to show but involves more analysis than we wish to include in this book.

Then show the above scalar system can be described as follows

$$\begin{pmatrix} x_1'(t) \\ x_2'(t) \\ x_3'(t) \end{pmatrix} = \begin{pmatrix} x_2(t) \\ x_3(t) \\ x_3(t) + x_2(t) - x_1(t) + t \end{pmatrix},$$

$$\begin{pmatrix} x_1(0) \\ x_2(0) \\ x_3(0) \end{pmatrix} = \begin{pmatrix} 1 \\ 0 \\ 0 \end{pmatrix}.$$

Now finish writing this in the form

$$\mathbf{x}' = A\mathbf{x} + \mathbf{f}, \ \mathbf{x}(0) = \mathbf{x}_0.$$

All such linear equations for a scalar function can be considered similarly as a first order system. However, it does not go the other way. You cannot consider an arbitrary first order system as a higher order equation for a scalar function. Explain why.

(12) Verify that the Laplace transform $\mathfrak{L}$ is a linear transformation in the sense that $\mathfrak{L}$ acting on a linear combination equals the linear combination of the transformed functions. Also verify the entries of the table. To get 5.) and 6.) you ought to just do 7.) and then take real and imaginary parts to save a lot of trouble. A similar trick works well with 10.) and 11.) and in fact, if you have done these two, you can simply let $a = 0$ to get many of these entries all at once. In the table, $\delta(t-c)$ denotes the Dirac measure having unit mass at c. The way to think of this is as something which does the following: $\int_0^\infty \delta(t-c) f(t) \, dt = f(c)$. It is an idealization of a large positive function which has integral equal to 1 which is only nonzero on a small interval. An example of such a thing would be the force exerted by a hammer on a nail during the short time it is in contact with the head of the nail.

(13) Use the method of Laplace transforms and Problem 5 to obtain the variation of constants formula in the case where A can be diagonalized.

(14) Show that if $\Phi(t)$ is the fundamental matrix satisfying

$$\Phi'(t) = A\Phi(t), \ \Phi(0) = I,$$

then the Laplace transform $\Phi(s)$ of $\Phi(t)$ equals

$$(sI - A)^{-1}.$$

Compare the second line of the table of Laplace transforms.

(15) Using the above problem, find the fundamental matrix for the differential equation

$$\mathbf{x}' = A\mathbf{x}$$

where

$$A = \begin{pmatrix} 3 & 4 & 4 \\ 2 & 3 & 4 \\ -3 & -4 & -5 \end{pmatrix}.$$

This is a **defective** matrix which has eigenvalues $1, 1, -1$. The term defective, when applied to a matrix, means that the matrix is not similar to a diagonal matrix.

Chapter 10

Unitary, orthogonal, Hermitian and symmetric matrices

Chapter summary

This chapter is on the inner product and its relation to linear transformations. It begins with a description of unitary matrices and describes them algebraically and geometrically. Special kinds of unitary matrices are given. The QR factorization is also discussed. Schur's theorem is presented which states that every matrix is unitarily similar to an upper triangular matrix. Then the spectral theory of Hermitian matrices is considered, including the major theorem that every Hermitian matrix is unitarily similar to a diagonal matrix. Real quadratic forms are discussed next. These are based on the spectral theory of Hermitian matrices. Other applications are considered in the exercises.

10.1 Unitary and orthogonal matrices

Given a complex matrix $A = (a_{ij})$, we write

$$\overline{A} = (\bar{a}_{ij})$$

for the matrix obtained by complex conjugation of the entries. We recall Definition 6.3: the **adjoint** of A is

$$A^* = \left(\overline{A}\right)^t.$$

We noted in Chapter 6 that

$$(A^*)^* = A,\ (A+B)^* = A^* + B^*,\ (AB)^* = B^*A^*$$

(for matrices permitting the sum or product to be formed).

Definition 10.1. The $n \times n$ matrix A is said to be **unitary** if

$$AA^* = I. \tag{10.1}$$

A real unitary matrix is said to be an **orthogonal** matrix.

Thus an orthogonal matrix is a real $n \times n$ matrix with $AA^t = I$.

The following is a simple but important observation.

Observation 10.1. A matrix A is unitary if and only if $A^*A = I$. A matrix is orthogonal if and only if $A^tA = I$.

This follows right away from the result presented earlier that for a square matrix, right inverses, left inverses, and inverses are all the same thing. See Theorem 3.4 on Page 77.

Example 10.1. The matrix

$$\begin{pmatrix} \frac{3+i}{\sqrt{11}} & \frac{1+i}{\sqrt{22}} \\ \frac{1}{\sqrt{11}} & \frac{-4-2i}{\sqrt{22}} \end{pmatrix} = (\mathbf{a}_1 \ \mathbf{a}_2)$$

is unitary. For with a little calculation, we obtain $|\mathbf{a}_1| = |\mathbf{a}_2| = 1, \langle \mathbf{a}_1, \mathbf{a}_2 \rangle = 0$. Hence the i, j entry of $A^t\overline{A}$ is

$$\mathbf{a}_j^t \overline{\mathbf{a}_j} = \langle \mathbf{a}_i, \mathbf{a}_j \rangle = \begin{cases} 1 \text{ if } i = j \\ 0 \text{ if } i \neq j. \end{cases}$$

This shows that $A^t\overline{A} = I$. Conjugating,

$$A^*A = I$$

and A^* is the inverse of A.

The same argument shows that a unitary matrix is a square matrix A satisfying

$$\langle \mathbf{a}_i, \mathbf{a}_j \rangle = \begin{cases} 1 \text{ if } i = j \\ 0 \text{ if } i \neq j. \end{cases}$$

That is, a unitary matrix is a square matrix over $\mathbb{C}$ *whose columns are an orthonormal set in* $\mathbb{C}^n$. In the real case, an orthogonal matrix is a square matrix over $\mathbb{R}$ whose columns form an orthonormal set in $\mathbb{R}^n$.

Example 10.2. The matrix

$$\begin{pmatrix} \frac{1}{\sqrt{3}} & \frac{1}{\sqrt{6}} & -\frac{1}{\sqrt{2}} \\ -\frac{1}{\sqrt{3}} & \frac{2}{\sqrt{6}} & 0 \\ \frac{1}{\sqrt{3}} & \frac{1}{\sqrt{6}} & \frac{1}{\sqrt{2}} \end{pmatrix} = \begin{pmatrix} \mathbf{a}_1 & \mathbf{a}_2 & \mathbf{a}_3 \end{pmatrix}$$

is orthogonal. It is simply a matter of checking that the columns have unit length and are orthogonal in pairs.

A linear mapping $T : \mathbb{R}^n \to \mathbb{R}^n$ is said to be **orthogonal** if the matrix of T is orthogonal. A linear mapping $T : \mathbb{C}^n \to \mathbb{C}^n$ is **unitary** if the matrix of T is unitary.

Example 10.3. A rotation $T : \mathbb{R}^2 \to \mathbb{R}^2$ is orthogonal.

For the matrix of T,

$$\begin{pmatrix} \cos\theta & -\sin\theta \\ \sin\theta & \cos\theta \end{pmatrix}$$

is easily seen to be orthogonal.

Example 10.4. A **reflection** $T : \mathbb{R}^2 \to \mathbb{R}^2$ is a linear mapping of the form

$$T(x_1\mathbf{w}_1 + x_2\mathbf{w}_2) = x_1\mathbf{w}_1 - x_2\mathbf{w}_2.$$

Here $\mathbf{w}_1, \mathbf{w}_2$ is an orthonormal basis of $\mathbb{R}^2$.

The matrix of T in the basis $\mathbf{w}_1, \mathbf{w}_2$ is certainly orthogonal: it is

$$\begin{pmatrix} 1 & 0 \\ 0 & -1 \end{pmatrix}.$$

This suggests that we should look at orthogonal (or unitary) mappings in a 'coordinate-free' way.

Lemma 10.1. *Let $T : \mathbb{C}^n \to \mathbb{C}^n$ be linear. The following statements are equivalent.*

(i) T is unitary.
(ii) $\langle T\mathbf{u}, T\mathbf{v}\rangle = \langle \mathbf{u}, \mathbf{v}\rangle$ always holds.
(iii) There is an orthonormal basis $\mathbf{v}_1, \ldots, \mathbf{v}_n$ of $\mathbb{C}^n$ such that $T\mathbf{v}_1, \ldots, T\mathbf{v}_n$ is an orthonormal basis of $\mathbb{C}^n$.

Proof. It suffices to prove that $(i) \Rightarrow (ii) \Rightarrow (i)$ and $(ii) \Rightarrow (iii) \Rightarrow (ii)$.

Suppose that (i) holds. Let A be the matrix of T. As we saw earlier, $A^t\overline{A} = I$. Hence

$$\langle T\mathbf{u}, T\mathbf{v}\rangle = (A\mathbf{u})^t\overline{A\mathbf{v}} = \mathbf{u}^t A^t \overline{A}\bar{\mathbf{v}} = \mathbf{u}^t\bar{\mathbf{v}} = \langle \mathbf{u}, \mathbf{v}\rangle.$$

Suppose that (ii) holds. In particular,

$$\langle T\mathbf{e}_i, T\mathbf{e}_j\rangle = \langle \mathbf{e}_i, \mathbf{e}_j\rangle.$$

So the columns of the matrix of T are an orthonormal set and (i) holds.

Suppose that (ii) holds. Pick any orthonormal basis $\mathbf{v}_1, \ldots, \mathbf{v}_n$ of $\mathbb{C}^n$. Then

$$\langle T\mathbf{v}_i, T\mathbf{v}_j\rangle = \langle \mathbf{v}_i, \mathbf{v}_j\rangle = \begin{cases} 1 \text{ if } i = j \\ 0 \text{ if } i \neq j. \end{cases}$$

This proves (iii).

Suppose that (iii) holds. For any $\mathbf{u} = a_1\mathbf{v}_1 + \cdots + a_n\mathbf{v}_n$, $\mathbf{v} = b_1\mathbf{v}_1 + \cdots + b_n\mathbf{v}_n$ in $\mathbb{C}^n$,

$$\begin{aligned} \langle T\mathbf{u}, T\mathbf{v}\rangle &= \left\langle \sum_{i=1}^n a_i T\mathbf{v}_i, \sum_{j=1}^n b_j T\mathbf{v}_j \right\rangle \\ &= \sum_{j=1}^n a_j\bar{b}_j = \langle \mathbf{u}, \mathbf{v}\rangle, \end{aligned}$$

which establishes (ii) □

Lemma 10.1 remains true if we replace $\mathbb{C}$ by $\mathbb{R}$ and 'unitary' by 'orthogonal'. We can now say the reflection in Example 10.4 is orthogonal, because (iii) is applicable: $T\mathbf{w}_1, T\mathbf{w}_2$ is the orthonormal set $\mathbf{w}_1, -\mathbf{w}_2$.

Example 10.5. Fix an orthonormal basis $\mathbf{w}_1, \mathbf{w}_2, \mathbf{w}_3$ for $\mathbb{R}^3$. **A rotation of** $\mathbb{R}^3$ about the axis $\mathbf{w}_1$ is a mapping T defined by

$$T(x_1\mathbf{w}_1 + x_2\mathbf{w}_2 + x_3\mathbf{w}_3) = (x_1\cos\theta - x_2\sin\theta)\mathbf{w}_1 + (x_1\sin\theta + x_2\cos\theta)\mathbf{w}_2 + x_3\mathbf{w}_3.$$

(If, for example $\mathbf{w}_1, \mathbf{w}_2, \mathbf{w}_3$ point east, north and vertically, this is a rotation through θ about the vertical axis, as one can easily see.) The rotation is an orthogonal mapping, again using (iii) of Lemma 10.1;

$$T\mathbf{w}_1 = (\cos\theta)\mathbf{w}_1 + (\sin\theta)\mathbf{w}_2, T\mathbf{w}_2 = (-\sin\theta)\mathbf{w}_1 + (\cos\theta)\mathbf{w}_2, T\mathbf{w}_3 = \mathbf{w}_3.$$

Likewise a **reflection** in the plane of $\mathbf{w}_1, \mathbf{w}_2$, defined by

$$S(x_1\mathbf{w}_1 + x_2\mathbf{w}_2 + x_3\mathbf{w}_3) = x_1\mathbf{w}_1 + x_2\mathbf{w}_2 - x_3\mathbf{w}_3$$

is an orthogonal mapping.

Note that $|T\mathbf{u}| = |\mathbf{u}|$ for a unitary mapping $T : \mathbb{C}^n \to \mathbb{C}^n$ or an orthogonal mapping $T : \mathbb{R}^n \to \mathbb{R}^n$. This follows from (ii) of Lemma 10.1 with $\mathbf{u} = \mathbf{v}$.

Lemma 10.1 suggests how to define a unitary mapping in an infinite dimensional inner product space; we can use (ii) for the definition. We will not pursue this idea here.

10.2 *QR* factorization

The following factorization is useful in solving large scale systems of equations, in particular in the method of least squares.

Lemma 10.2. *Let A be a real $m \times n$ matrix of rank n. We can write A as*

$$A = QR, \tag{10.2}$$

where Q is an $m \times n$ matrix whose columns form an orthonormal basis for Col(A), *and R is an $n \times n$ upper triangular matrix with positive entries on the diagonal.*

Proof. Let $A = \begin{pmatrix} \mathbf{a}_1 & \cdots & \mathbf{a}_n \end{pmatrix}$. Apply the Gram-Schmidt process to $\mathbf{a}_1, \dots, \mathbf{a}_n$ to produce an orthonormal basis $\mathbf{u}_1, \dots, \mathbf{u}_n$ of Col(A). We recall that

$$\mathbf{a}_j \in \text{span}\{\mathbf{u}_1, \dots, \mathbf{u}_j\},$$

but $\mathbf{a}_j \notin \text{span}\{\mathbf{a}_1, \dots, \mathbf{a}_{j-1}\}$. It follows that

$$\mathbf{a}_j = \langle \mathbf{a}_j, \mathbf{u}_1 \rangle \mathbf{u}_1 + \cdots + \langle \mathbf{a}_j, \mathbf{u}_j \rangle \mathbf{u}_j, \langle \mathbf{a}_j, \mathbf{u}_j \rangle \neq 0.$$

We may assume that $\langle \mathbf{a}_j, \mathbf{u}_j \rangle > 0$ (if necessary, replace $\mathbf{u}_j$ by $-\mathbf{u}_j$). Now (10.2) holds with

$$Q = \begin{pmatrix} \mathbf{u}_1 & \cdots & \mathbf{u}_n \end{pmatrix},$$

$$R = \begin{pmatrix} \langle \mathbf{a}_1, \mathbf{u}_1 \rangle & \langle \mathbf{a}_2, \mathbf{u}_1 \rangle & \cdots & \langle \mathbf{a}_n, \mathbf{u}_1 \rangle \\ & \langle \mathbf{a}_2, \mathbf{u}_2 \rangle & & \langle \mathbf{a}_n, \mathbf{u}_2 \rangle \\ & & \ddots & \vdots \\ & & & \langle \mathbf{a}_n, \mathbf{u}_n \rangle \end{pmatrix}$$

because column j of the product QR is

$$\langle \mathbf{a}_j, \mathbf{u}_1 \rangle \mathbf{u}_1 + \cdots + \langle \mathbf{a}_j, \mathbf{u}_j \rangle \mathbf{u}_j = \mathbf{a}_j. \quad \square$$

This is also called the thin QR factorization by Golub and Van Loan.

Example 10.6. Find a QR factorization of

$$A = \begin{pmatrix} 2 & 0 & 0 \\ 2 & 1 & 0 \\ 2 & 1 & 3 \\ 2 & 1 & 3 \end{pmatrix} = (\mathbf{a}_1 \mathbf{a}_2 \mathbf{a}_3).$$

We let, rescaling for convenience,

$$\mathbf{v}_1 = (1/2)\,\mathbf{a}_1 = (1,1,1,1)^t,$$

$$\mathbf{v}_2' = \mathbf{a}_2 - \frac{\langle \mathbf{a}_2, \mathbf{v}_1 \rangle}{\langle \mathbf{v}_1, \mathbf{v}_1 \rangle}\mathbf{v}_1 = (0,1,1,1)^t - 3/4(1,1,1,1)^t,$$

$$\mathbf{v}_2 = 4\mathbf{v}_2' = (-3,1,1,1)^t,$$

$$\begin{aligned} \mathbf{v}_3' &= \mathbf{a}_3 - \frac{\langle \mathbf{a}_3, \mathbf{v}_1 \rangle}{\langle \mathbf{v}_1, \mathbf{v}_1 \rangle}\mathbf{v}_1 - \frac{\langle \mathbf{a}_3, \mathbf{v}_2 \rangle}{\langle \mathbf{v}_2, \mathbf{v}_2 \rangle}\mathbf{v}_2 \\ &= 3((0,0,1,1)^t - 2/4(1,1,1,1)^t - 2/12(-3,1,1,1)^t) \\ &= 1/2((0,0,6,6)^t - (3,3,3,3)^t - (-3,1,1,1)^t) \\ &= 1/2(0,-4,2,2)^t \end{aligned}$$

$$\mathbf{v}_3 = (0,-2,1,1)^t.$$

Normalizing, we take

$$Q = \begin{pmatrix} 1/2 & \frac{-3}{\sqrt{12}} & 0 \\ 1/2 & \frac{1}{\sqrt{12}} & \frac{-2}{\sqrt{6}} \\ 1/2 & \frac{1}{\sqrt{12}} & \frac{1}{\sqrt{6}} \\ 1/2 & \frac{1}{\sqrt{12}} & \frac{1}{\sqrt{6}} \end{pmatrix}.$$

Since Q^tQ is the 3×3 identity $I, QR = A$ implies

$$\begin{aligned} R &= Q^tA \\ &= \begin{pmatrix} 1/2 & 1/2 & 1/2 & 1/2 \\ \frac{-3}{\sqrt{12}} & \frac{1}{\sqrt{12}} & \frac{1}{\sqrt{12}} & \frac{1}{\sqrt{12}} \\ 0 & \frac{-2}{\sqrt{6}} & \frac{1}{\sqrt{6}} & \frac{1}{\sqrt{6}} \end{pmatrix} \begin{pmatrix} 2 & 0 & 0 \\ 2 & 1 & 0 \\ 2 & 1 & 3 \\ 2 & 1 & 3 \end{pmatrix} \\ &= \begin{pmatrix} 4 & 3/2 & 3 \\ 0 & \frac{3}{\sqrt{12}} & \frac{6}{\sqrt{12}} \\ 0 & 0 & \frac{6}{\sqrt{6}} \end{pmatrix}. \end{aligned}$$

Example 10.7. For larger examples, Maple produces the QR factorization under 'Solvers and forms'. Note that this is a way of doing the Gram-Schmidt process via Maple.

Example 10.8. Use Maple to redo Example 6.7 of Chapter 6, i.e. find an orthonormal basis of Col(A),

$$A = \begin{pmatrix} -1 & 0 & 1 \\ 1 & 2 & 3 \\ 0 & 1 & 2 \\ 1 & 3 & 0 \end{pmatrix}.$$

Within 'QR decomposition', select 'Unitary factor'. We obtain

$$\begin{pmatrix} (-1/3)\sqrt{3} & (5/51)\sqrt{51} & \frac{1}{102}\sqrt{102} \\ (1/3)\sqrt{3} & (1/51)\sqrt{51} & \frac{7}{102}\sqrt{102} \\ 0 & (1/17)\sqrt{51} & \frac{2}{51}\sqrt{102} \\ (1/3)\sqrt{3} & (4/51)\sqrt{51} & -\frac{1}{17}\sqrt{102} \end{pmatrix}.$$

The columns of this matrix are the vectors $\mathbf{u}_1, \mathbf{u}_2, \mathbf{u}_3$ obtained previously (Chapter 6), but written a little differently.

Example 10.9. Use Maple to find an orthonormal basis of Col(A), where

$$A = \begin{pmatrix} 3 & 2 & 1 \\ 6 & 3 & 1 \\ 1 & 1 & -1 \\ 5 & -1 & 2 \\ 4 & 2 & 6 \end{pmatrix}.$$

Solution. One enters the matrix in Maple. Then you right click on "solvers and forms" and then select and click on (Q,R). This yields for Q the following matrix.

$$\begin{pmatrix} \frac{1}{29}\sqrt{87} & \frac{30}{25\,201}\sqrt{87}\sqrt{869} & -\frac{433}{17\,142\,763}\sqrt{869}\sqrt{19\,727} \\ \frac{2}{29}\sqrt{87} & \frac{31}{25\,201}\sqrt{87}\sqrt{869} & -\frac{1693}{17\,142\,763}\sqrt{869}\sqrt{19\,727} \\ \frac{1}{87}\sqrt{87} & \frac{59}{75\,603}\sqrt{87}\sqrt{869} & -\frac{1317}{17\,142\,763}\sqrt{869}\sqrt{19\,727} \\ \frac{5}{87}\sqrt{87} & -\frac{227}{75\,603}\sqrt{87}\sqrt{869} & -\frac{250}{17\,142\,763}\sqrt{869}\sqrt{19\,727} \\ \frac{4}{87}\sqrt{87} & \frac{62}{75\,603}\sqrt{87}\sqrt{869} & \frac{3506}{17\,142\,763}\sqrt{869}\sqrt{19\,727} \end{pmatrix}$$

and the columns are an orthonormal basis for the column space of the above matrix. This is more horrible than might naively be expected. Of course, one would not typically try to obtain the exact orthonormal basis. To obtain numbers in terms of decimals, you write some number in the matrix as a decimal. This tells Maple that you want the answer in terms of decimals. This yields

$$\begin{pmatrix} 0.321\,63 & 0.327\,32 & -0.104\,58 \\ 0.643\,27 & 0.338\,23 & -0.408\,90 \\ 0.107\,21 & 0.214\,58 & -0.318\,09 \\ 0.536\,06 & -0.825\,57 & -6.038\,1 \times 10^{-2} \\ 0.428\,85 & 0.225\,49 & 0.846\,78 \end{pmatrix}$$

which is not quite as ugly.

The usefulness of the QR factorization (which is relatively cheap in computer time) emerges when we want to solve a linear system

$$A\mathbf{x} = \mathbf{b},$$

where A is $m \times n$ of rank n. We write $A = QR$ as above, and solve the equivalent system

$$R\mathbf{x} = Q^t\mathbf{b}. \tag{10.3}$$

From (10.3), given the triangular shape of R, we can easily read off $x_n, x_{n-1}, \ldots, x_1$ in succession. In particular, least squares computations which have the shape

$$A^t A\hat{\mathbf{x}} = A^t\mathbf{b}.$$

are generally done in this way if the $m \times n$ matrix A has rank n.

It is not necessary in order to obtain a QR factorization for A to have rank n. Nor is it necessary to assume A is a real matrix, if you replace Q orthogonal with Q unitary. We illustrate the technique for A a square, possibly complex, matrix. The generalization to rectangular matrices is slightly different than the above. The idea here is to always have Q be a unitary matrix and R an upper triangular matrix in the sense that $r_{ij} = 0$ whenever $j < i$. The proof in the following proposition yields a QR factorization even if A is not square.

Proposition 10.1. Let A be any $n \times n$ complex matrix. Then there exists a unitary matrix Q, and an upper triangular matrix R which has all non-negative entries on the main diagonal such that $A = QR$.

Proof. Letting $A = \begin{pmatrix} \mathbf{a}_1 & \cdots & \mathbf{a}_n \end{pmatrix}$ be the matrix with the $\mathbf{a}_j$ the columns, each a vector in $\mathbb{C}^n$, let Q_1 be a unitary matrix which maps $\mathbf{a}_1$ to $|\mathbf{a}_1|\,\mathbf{e}_1$ in the case that $\mathbf{a}_1 \neq \mathbf{0}$. If $\mathbf{a}_1 = \mathbf{0}$, let $Q_1 = I$. Why does such a unitary matrix exist? From the Gram-Schmidt process, let

$$\{\mathbf{a}_1 / |\mathbf{a}_1|, \mathbf{u}_2, \cdots, \mathbf{u}_n\}$$

be an orthonormal basis, and let $Q_1\left(\frac{\mathbf{a}_1}{|\mathbf{a}_1|}\right) = \mathbf{e}_1, Q_1(\mathbf{u}_2) = \mathbf{e}_2$ etc. Extend Q_1 linearly. Then Q_1 preserves lengths because it maps an orthonormal basis to an orthonormal basis and is therefore unitary by Lemma 10.1. Then

$$\begin{aligned} Q_1 A &= \begin{pmatrix} Q_1\mathbf{a}_1 & Q_1\mathbf{a}_2 & \cdots & Q_1\mathbf{a}_n \end{pmatrix} \\ &= \begin{pmatrix} |\mathbf{a}_1|\,\mathbf{e}_1 & Q_1\mathbf{a}_2 & \cdots & Q_1\mathbf{a}_n \end{pmatrix}, \end{aligned}$$

which is a matrix of the form

$$\begin{pmatrix} |\mathbf{a}_1| & \mathbf{b} \\ \mathbf{0} & A_1 \end{pmatrix}.$$

Now do the same thing for A_1 obtaining an $n-1 \times n-1$ unitary matrix Q_2' which, when multiplied on the left of A_1, yields something of the form

$$\begin{pmatrix} a & \mathbf{b}_1 \\ \mathbf{0} & A_2 \end{pmatrix}.$$

Then multiplying A on the left by the product

$$\begin{pmatrix} 1 & \mathbf{0} \\ \mathbf{0} & Q_2' \end{pmatrix} Q_1 \equiv Q_2 Q_1,$$

yields a matrix which is upper triangular with respect to the first two columns. Continuing this way,

$$Q_n Q_{n-1} \cdots Q_1 A = R,$$

where R is upper triangular having all positive entries on the main diagonal. Then the desired unitary matrix is

$$Q = (Q_n Q_{n-1} \cdots Q_1)^* . \square$$

In the following exercises, you might want to save time by using Maple or some other computer algebra system. To use Maple, you first open Maple, then click on the matrix tab on the left of the screen. Then select the size of the matrix in the two lines below this tab. Next click on insert matrix. This produces a template for a matrix. Fill in the symbols with the desired numbers. Then right click on the matrix you just entered, select solvers and forms, and finally click on QR. Maple will do all the tedious computations for you.

10.3 Exercises

(1) If Q is an orthogonal matrix, show that the only possible values of $\det(Q)$ are ± 1. When this value is 1, the orthogonal matrix is called proper. Show that rotations in $\mathbb{R}^2$ are proper and that reflections about an axis are not proper.

(2) ↑If Q is a proper 3×3 orthogonal matrix, show that Q must have an eigenvalue equal to 1. **Hint:** Show that the absolute value of all eigenvalues must equal 1. Then explain why the determinant of a complex matrix equals the product of the eigenvalues.

(3) Using the proof of Proposition 10.1 generalize the result to obtain one which is valid for A an $m \times n$ matrix, which may have $m \neq n$, and in which there is no assumption made about rank.

(4) If Q is an orthogonal matrix, show that the columns are an orthonormal set. That is, show that for

$$Q = \begin{pmatrix} \mathbf{q}_1 & \cdots & \mathbf{q}_n \end{pmatrix},$$

it follows that $\mathbf{q}_i \cdot \mathbf{q}_j = \delta_{ij}$. Also show that any orthonormal set of vectors is linearly independent.

(5) For real matrices, a QR factorization is often accomplished through the use of Householder reflections. Using Problem 34 and the problem right after this one on Page 135, describe how to use Householder reflections to obtain a QR factorization for an arbitrary $m \times n$ real matrix. **Hint:** Follow the procedure of the proof of Proposition 10.1, except obtain Q_1 as the Householder reflection which maps $\mathbf{a}_1$ to $|\mathbf{a}_1| \mathbf{e}_1$.

(6) Find a QR factorization for the matrix

$$\begin{pmatrix} 1 & 2 & 1 \\ 3 & -2 & 1 \\ 1 & 0 & 2 \end{pmatrix}.$$

(7) Find a QR factorization for the matrix

$$\begin{pmatrix} 1 & 2 & 1 & 0 \\ 3 & 0 & 1 & 1 \\ 1 & 0 & 2 & 1 \end{pmatrix}.$$

(8) If you had a QR factorization, $A = QR$, for Q a unitary matrix, describe how you could use it to solve the equation $A\mathbf{x} = \mathbf{b}$ for A an arbitrary $m \times n$ matrix.

(9) Show that you can't expect uniqueness for QR factorizations in the case where A is not invertible. Consider

$$\begin{pmatrix} 0 & 0 & 0 \\ 0 & 0 & 1 \\ 0 & 0 & 1 \end{pmatrix}$$

and verify this equals

$$\begin{pmatrix} 0 & 1 & 0 \\ \frac{1}{2}\sqrt{2} & 0 & \frac{1}{2}\sqrt{2} \\ \frac{1}{2}\sqrt{2} & 0 & -\frac{1}{2}\sqrt{2} \end{pmatrix} \begin{pmatrix} 0 & 0 & \sqrt{2} \\ 0 & 0 & 0 \\ 0 & 0 & 0 \end{pmatrix}$$

and also

$$\begin{pmatrix} 1 & 0 & 0 \\ 0 & 1 & 0 \\ 0 & 0 & 1 \end{pmatrix} \begin{pmatrix} 0 & 0 & 0 \\ 0 & 0 & 1 \\ 0 & 0 & 1 \end{pmatrix}.$$

Show that if A is an invertible real or complex $n \times n$ matrix, then there is only one QR factorization of A. **Hint:** Show that if Q is unitary and R is upper triangular having all positive entries on the main diagonal and if $Q = R$, then both matrices equal I. Show that you can reduce to this case.

(10) *Suppose that the entries of $Q_n R_n$ converge to the corresponding entries of an orthogonal matrix Q where Q_n is orthogonal and R_n is upper triangular having all positive entries on the main diagonal. Show that then the entries of Q_n converge to the corresponding entries of Q and the entries of R_n converge to the corresponding entries of the identity. **Hint:** Letting a_{ij}^n denote the ij^{th} entry of A^n, show first that $r_{11}^k \mathbf{q}_1^k \to \mathbf{q}_1$ where the ith column of Q_k is $\mathbf{q}_i^k$ and the ith column of Q is $\mathbf{q}_i$. Then argue that this implies $r_{11}^k \to 1$ and consequently, that $\mathbf{q}_1^k \to \mathbf{q}_1$. Next consider the second column of the product $Q_k R_k$. Explain why

$$r_{12}^k \mathbf{q}_1^k + r_{22}^k \mathbf{q}_2^k \to \mathbf{q}_2$$

and take the inner product of both sides with $\mathbf{q}_1^k$ to conclude $r_{12}^k \to 0$ and that $r_{22}^k \to 1$.

(11) Let A be an $n \times n$ complex matrix. Let $A_0 = A$. Suppose that A_{k-1} has been found. To find A_k let

$$A_{k-1} = Q_k R_k, \ A_k = R_k Q_k,$$

where $Q_k R_k$ is a QR factorization of A_{k-1}. Show that every pair of matrices in this sequence are similar. Thus they all have the same eigenvalues. This is the QR algorithm for finding the eigenvalues of a matrix. Although the sequence $\{A_k\}$ may fail to converge, it is nevertheless often the case that for large k, A_k is of the form

$$\begin{pmatrix} B_k & & * \\ & \ddots & \\ e & & B_r \end{pmatrix}$$

where the B_i are blocks which run down the diagonal of the matrix, and all of the entries below this block diagonal are very small. Then letting T_B denote the matrix obtained by setting all of these sub-diagonal entries equal to zero, one can argue, using methods of complex analysis, that the eigenvalues of A_k are close to the eigenvalues of T_B and that each of these eigenvalues is an eigenvalue of a block B_i. (For an approach to the approximation of the eigenvalues not dependent on complex analysis, see Problem 29 on 288) To see more on this algorithm, consult Golub and Van Loan [3]. For an explanation of why the algorithm works see Wilkinson [18].

(12) ↑Show that the two matrices $A = \begin{pmatrix} 0 & -1 \\ 4 & 0 \end{pmatrix}$ and $B = \begin{pmatrix} 0 & -2 \\ 2 & 0 \end{pmatrix}$ are similar; that is, there exists a matrix S such that $A = S^{-1}BS$. Now show that there is no orthogonal matrix Q such that $Q^t BQ = A$. Show that the QR algorithm does converge for the matrix B although it fails to do so for A. **Hint:** You should use Maple or some other computer algebra system to find the sequence of matrices in the QR algorithm. The claim will be obvious as soon as you do this. To verify that there can be no orthogonal matrix such that $Q^t BQ = A$, you might use the fact that orthogonal matrices preserve distances.

10.4 Schur's theorem

Letting A be a complex $n \times n$ matrix, it was shown in Section 8.3 that there exists an invertible matrix P such that $P^{-1}AP = J$ where J is an upper triangular matrix which has a very special structure. Schur's theorem is similar, but instead of a general invertible matrix P one can use a unitary matrix U.

Theorem 10.1. *Let A be a complex $n \times n$ matrix. Then there exists a unitary matrix U such that*

$$U^* AU = T, \tag{10.4}$$

where T is an upper triangular matrix having the eigenvalues of A on the main diagonal, listed with multiplicity[1].

Proof. The theorem is clearly true if A is a 1×1 matrix. Just let $U = 1$, the 1×1 matrix which has entry 1. Suppose it is true for $(n-1) \times (n-1)$ matrices and let A be an $n \times n$ matrix. Then let $\mathbf{v}_1$ be a unit eigenvector for A. Then there exists λ_1 such that

$$A\mathbf{v}_1 = \lambda_1 \mathbf{v}_1, \ |\mathbf{v}_1| = 1.$$

Extend $\{\mathbf{v}_1\}$ to a basis and then use the Gram - Schmidt process Lemma 6.1 to obtain $\{\mathbf{v}_1, \cdots, \mathbf{v}_n\}$, an orthonormal basis of $\mathbb{C}^n$. Let U_0 be a matrix whose i^{th} column is $\mathbf{v}_i$. Then from the definition of a unitary matrix Definition 10.1, it follows that U_0 is unitary. Consider $U_0^* A U_0$.

$$U_0^* A U_0 = \begin{pmatrix} \mathbf{v}_1^* \\ \vdots \\ \mathbf{v}_n^* \end{pmatrix} \begin{pmatrix} A\mathbf{v}_1 & \cdots & A\mathbf{v}_n \end{pmatrix} = \begin{pmatrix} \mathbf{v}_1^* \\ \vdots \\ \mathbf{v}_n^* \end{pmatrix} \begin{pmatrix} \lambda_1 \mathbf{v}_1 & \cdots & A\mathbf{v}_n \end{pmatrix}$$

Thus $U_0^* A U_0$ is of the form

$$\begin{pmatrix} \lambda_1 & \mathbf{a} \\ \mathbf{0} & A_1 \end{pmatrix}$$

where A_1 is an $n-1 \times n-1$ matrix. Now by induction, there exists an $(n-1) \times (n-1)$ unitary matrix $\widetilde{U}_1$ such that

$$\widetilde{U}_1^* A_1 \widetilde{U}_1 = T_{n-1},$$

an upper triangular matrix. Consider

$$U_1 \equiv \begin{pmatrix} 1 & \mathbf{0} \\ \mathbf{0} & \widetilde{U}_1 \end{pmatrix}.$$

An application of block multiplication shows that U_1 is a unitary matrix and also that

$$U_1^* U_0^* A U_0 U_1 = \begin{pmatrix} 1 & \mathbf{0} \\ \mathbf{0} & \widetilde{U}_1^* \end{pmatrix} \begin{pmatrix} \lambda_1 & * \\ \mathbf{0} & A_1 \end{pmatrix} \begin{pmatrix} 1 & \mathbf{0} \\ \mathbf{0} & \widetilde{U}_1 \end{pmatrix} = \begin{pmatrix} \lambda_1 & * \\ \mathbf{0} & T_{n-1} \end{pmatrix} = T$$

where T is upper triangular. Then let $U = U_0 U_1$. Since $(U_0 U_1)^* = U_1^* U_0^*$, it follows that A is similar to T and that $U_0 U_1$ is unitary. Hence A and T have the same characteristic polynomials, and since the eigenvalues of T are the diagonal entries listed with multiplicity, this proves the theorem. □

The same argument yields the following corollary in the case where A has real entries. The only difference is the use of the real inner product instead of the complex inner product.

Corollary 10.1. *Let A be a real $n \times n$ matrix which has only real eigenvalues. Then there exists a real orthogonal matrix Q such that*

$$Q^t A Q = T$$

[1]'Listed with multiplicity' means that the diagonal entries are repeated according to their multiplicity as roots of the characteristic equation.

where T is an upper triangular matrix having the eigenvalues of A on the main diagonal, listed with multiplicity.

Proof. This follows by observing that if all eigenvalues are real, then corresponding to each real eigenvalue, there exists a real eigenvector. Thus the argument of the above theorem applies with the real inner product. □

10.5 Hermitian and symmetric matrices

A complex $n \times n$ matrix A with $A^* = A$ is said to be **Hermitian**. A real $n \times n$ matrix A with $A^t = A$ is said to be **symmetric**. In either case, note that

$$\langle A\mathbf{u}, \mathbf{v}\rangle = (A\mathbf{u})^t \bar{\mathbf{v}} = \mathbf{u}^t A^t \bar{\mathbf{v}} = \mathbf{u}^t \overline{A} \bar{\mathbf{v}} = \langle \mathbf{u}, A\mathbf{v}\rangle.$$

Thus, as a numerical example, the matrix

$$\begin{pmatrix} 1 & 1-i \\ 1+i & 2 \end{pmatrix}$$

is Hermitian, while

$$\begin{pmatrix} 1 & -1 & -2 \\ -1 & 2 & 4 \\ -2 & 4 & 3 \end{pmatrix}$$

is symmetric. Hermitian matrices are named in honor of the French mathematician Charles Hermite (1822–1901).

With Schur's theorem, the theorem on diagonalization of a Hermitian matrix follows.

Theorem 10.2. *Let A be Hermitian. Then the eigenvalues of A are all real, and there exists a unitary matrix U such that*

$$U^* A U = D,$$

a diagonal matrix whose diagonal entries are the eigenvalues of A listed with multiplicity. In case A is symmetric, U may be taken to be an orthogonal matrix. The columns of U form an orthonormal basis of eigenvectors of A.

Proof. By Schur's theorem and the assumption that A is Hermitian, there exists a triangular matrix T, whose diagonal entries are the eigenvalues of A listed with multiplicity, and a unitary matrix U such that

$$T = U^* A U = U^* A^* U = (U^* A U)^* = T^*.$$

It follows from this that T is a diagonal matrix and has all real entries down the main diagonal. Hence the eigenvalues of A are real. If A is symmetric (real and Hermitian) it follows from Corollary 10.1 that U may be taken to be orthogonal.

That the columns of U form an orthonormal basis of eigenvectors of A, follows right away from the definition of matrix multiplication which implies that if $\mathbf{u}_i$ is a column of U, then $A\mathbf{u}_i =$ column i of $(UD) = \lambda_i \mathbf{u}_i$. $\square$

Example 10.10. The eigenvectors for the Hermitian matrix

$$\begin{pmatrix} 1 & 2i \\ -2i & 1 \end{pmatrix}$$

are $\begin{pmatrix} i \\ 1 \end{pmatrix}$ corresponding to the eigenvalue 3 and $\begin{pmatrix} -i \\ 1 \end{pmatrix}$ corresponding to the eigenvalue -1. Thus

$$A = P \begin{pmatrix} 3 & 0 \\ 0 & -1 \end{pmatrix} P^*$$

where the columns of the unitary matrix

$$P = \begin{pmatrix} \frac{i}{\sqrt{2}} & \frac{-i}{\sqrt{2}} \\ \frac{i}{\sqrt{2}} & \frac{i}{\sqrt{2}} \end{pmatrix}$$

are obtained from these eigenvectors by normalizing.

Example 10.11. Let

$$A = \begin{pmatrix} 3 & 2 & 2 \\ 2 & 3 & 2 \\ 2 & 2 & 3 \end{pmatrix}.$$

Find an orthonormal basis of $\mathbb{R}^3$ consisting of eigenvectors of A. Write A in the form

$$A = PDP^t,$$

where P is an orthogonal matrix and D is a diagonal matrix.

One can use Maple to find the eigenvalues, but in fact the factorization of $P_A(\lambda)$ is fairly obvious:

$$\det(\lambda I - A) = \det \begin{pmatrix} \lambda - 3 & -2 & -2 \\ -2 & \lambda - 3 & -2 \\ -2 & -2 & \lambda - 3 \end{pmatrix} = \det \begin{pmatrix} \lambda - 1 & 1 - \lambda & 0 \\ -2 & \lambda - 3 & -2 \\ 0 & \lambda - 1 & \lambda - 1 \end{pmatrix}$$

(by subtracting row 2 from each of the other rows). Taking out the factor $\lambda - 1$ twice,

$$\det(\lambda I - A) = (\lambda - 1)^2 \det \begin{pmatrix} 1 & -1 & 0 \\ -2 & \lambda - 3 & -2 \\ 0 & -1 & 1 \end{pmatrix} = (\lambda - 1)^2(\lambda - 7)$$

(expanding by row 1).

The eigenspace for $\lambda = 1$ is

$$\ker\begin{pmatrix} -2 & -2 & -2 \\ -2 & -2 & -2 \\ -2 & -2 & -2 \end{pmatrix} = \ker\begin{pmatrix} 1 & 1 & 1 \\ 0 & 0 & 0 \\ 0 & 0 & 0 \end{pmatrix}$$

The corresponding system of equations is $x_1 + x_2 + x_3 = 0$, with general solution

$$\begin{pmatrix} -x_2 - x_3 \\ x_2 \\ x_3 \end{pmatrix} = x_2\begin{pmatrix} -1 \\ 1 \\ 0 \end{pmatrix} + x_3\begin{pmatrix} -1 \\ 0 \\ 1 \end{pmatrix}.$$

A basis of the eigenspace is $\mathbf{w}_1 = (-1,1,0), \mathbf{w}_2(-1,0,1)$. Since we want an orthonormal basis, we use the Gram-Schmidt process: $\mathbf{v}_1 = \mathbf{w}_1$,

$$\mathbf{v}_2 = \mathbf{w}_2 - \frac{\langle \mathbf{w}_2, \mathbf{v}_1 \rangle}{\langle \mathbf{w}_1, \mathbf{v}_1 \rangle}\mathbf{v}_1 = (-1,0,1) - 1/2(-1,1,0) = (-1/2,-1/2,1).$$

This is quickly converted into an orthonormal basis

$$\mathbf{u}_1 = \left(-\frac{1}{\sqrt{2}}, \frac{1}{\sqrt{2}}, 0\right), \mathbf{u}_2 = \left(-\frac{1}{\sqrt{6}}, \frac{-1}{\sqrt{6}}, \frac{2}{\sqrt{6}}\right).$$

The eigenspace for $\lambda = 7$ is obtained from doing row operations and equals

$$\ker\begin{pmatrix} 4 & -2 & -2 \\ -2 & 4 & -2 \\ -2 & -2 & 4 \end{pmatrix} = \ker\begin{pmatrix} 1 & 1 & -2 \\ 0 & 6 & -6 \\ 0 & -6 & 6 \end{pmatrix} = \ker\begin{pmatrix} 1 & 0 & -1 \\ 0 & -1 & 1 \\ 0 & 0 & 0 \end{pmatrix}.$$

The corresponding system of equations reads $x_1 = x_3, x_2 = x_3$. A basis of the eigenspace is $\mathbf{y}_3 = \left(\frac{1}{\sqrt{3}}, \frac{1}{\sqrt{3}}, \frac{1}{\sqrt{3}}\right)$. It follows that $A = PDP^t$ with

$$P = \begin{pmatrix} -\frac{1}{\sqrt{2}} & -\frac{1}{\sqrt{6}} & \frac{1}{\sqrt{3}} \\ \frac{1}{\sqrt{2}} & -\frac{1}{\sqrt{6}} & \frac{1}{\sqrt{3}} \\ 0 & \frac{2}{\sqrt{6}} & \frac{1}{\sqrt{3}} \end{pmatrix} \text{ orthogonal, } D = \begin{pmatrix} 1 & & \\ & 1 & \\ & & 7 \end{pmatrix}.$$

10.6 Real quadratic forms

A **real quadratic form** is a function $Q : \mathbb{R}^n \to \mathbb{R}$ of the form

$$Q(x_1, \ldots, x_n) = Q(\mathbf{x}) = \sum_{i=1}^{n}\sum_{j=1}^{n} a_{ij}x_ix_j \qquad (a_{ij} = a_{ji} \in \mathbb{R}).$$

The **matrix of** Q is the symmetric matrix

$$A = (a_{ij}).$$

For example, $Q(\mathbf{x}) = x_1^2 + 3x_1x_2 + 2x_2^2$ has $a_{11} = 1, a_{22} = 2, a_{12} = a_{21} = 3/2$. The matrix of Q is

$$\begin{pmatrix} 1 & 3/2 \\ 3/2 & 2 \end{pmatrix}.$$

Quadratic forms occur in a variety of problems. For instance, let $f(x_1, \dots, x_n)$ be a smooth real function on $\mathbb{R}^n$ with a stationary point at $\mathbf{c}$, that is,

$$\frac{\partial f}{\partial x_j} = 0 \text{ at } \mathbf{x} = \mathbf{c} \qquad (j = 1, \dots, n).$$

A satisfactory approximation to $f(\mathbf{c} + \mathbf{y})$ for small $\mathbf{y}$ is

$$f(\mathbf{c}) + Q(\mathbf{y}),$$

where Q is the quadratic form with matrix

$$a_{ij} = \frac{1}{2} \frac{\partial^2 f}{\partial x_i \partial x_j} \text{ at } \mathbf{x} = \mathbf{c}.$$

Thus we can predict whether $\mathbf{c}$ is a local maximum, a local minimum or a *saddle point* (i.e. neither a local maximum nor a local minimum) by studying the quadratic form Q. See Edwards (1994) for further discussion.

As another instance, we mention that a good model for a planetary surface centered at $\mathbf{0}$ is an *ellipsoid*, which with a correctly chosen set of axes has equation

$$\frac{x_1^2}{a^2} + \frac{x_2^2}{b^2} + \frac{x_3^2}{c^2} = 1.$$

If the axes are chosen differently the equation would be

$$Q(\mathbf{y}) = 1$$

where Q is a *positive definite* quadratic form $Q = Q(y_1, y_2, y_3)$.

Definition 10.2. Let $Q(x_1, \dots, x_n)$ be a real quadratic form. We say that Q is **positive definite** if $Q(\mathbf{x}) > 0$ whenever $\mathbf{x} \neq \mathbf{0}$; Q is **negative definite** if $Q(\mathbf{x}) < 0$ whenever $\mathbf{x} \neq \mathbf{0}$. If Q is neither positive definite nor negative definite, then Q is **indefinite**.

A simple example of an indefinite quadratic form is $Q(\mathbf{x}) = (ax_1 + bx_2)(cx_1 + dx_2)$ where $(a, b) \neq (0, 0)$ and $(c, d) \notin \text{span}\{(a, b)\}$. You should be able to visualize the regions where $Q > 0$ and $Q < 0$.

For one class of quadratic forms, the *diagonal* forms, it is easy to decide between positive definiteness, negative definiteness and indefiniteness. A **diagonal form** has the shape

$$Q(\mathbf{x}) = b_1 x_1^2 + \cdots + b_n x_n^2.$$

The matrix of Q is of course diagonal. Clearly Q is positive definite if $b_1 > 0, \dots, b_n > 0$, negative definite if $b_1 < 0, \dots, b_n < 0$, and indefinite in all remaining cases.

Conveniently, a change of basis can be used to bring *any* Q into diagonal form. In fact, we can find an orthonormal basis β of $\mathbb{R}^n$, depending on Q, such that

$$Q(P\mathbf{x}) = \sum_{i=1}^{n} b_i x_i^2,$$

where the columns of P are the vectors of β.

Lemma 10.3. *Let $\mathbf{v}_1, \ldots, \mathbf{v}_n$ be a basis of $\mathbb{R}^n$, and let*

$$\mathbf{y} = P\mathbf{x}$$

give the relation between coordinates in the bases $\beta : \mathbf{e}_1, \ldots, \mathbf{e}_n([\mathbf{v}]\beta = \mathbf{y})$ and $\delta : \mathbf{v}_1, \ldots, \mathbf{v}_n([\mathbf{v}]\delta = \mathbf{x})$.

The quadratic form $Q(\mathbf{y}) = \mathbf{y}^t A\mathbf{y}$ with matrix A can be written as

$$Q(P\mathbf{x}) = \mathbf{x}^t(P^t AP)\mathbf{x}.$$

That is, Q has the matrix $P^t AP$ in the coordinates $x_1, \ldots, x_n$.

Proof. This is almost trivial. We first note that

$$\mathbf{y}^t A\mathbf{y} = \sum_{j=1}^{n} y_j \sum_{k=1}^{n} a_{jk} y_k = Q(\mathbf{y})$$

(again interpreting a 1×1 matrix as a real number). Hence

$$Q(P\mathbf{x}) = (P\mathbf{x})^t A(P\mathbf{x}) = \mathbf{x}^t(P^t AP)\mathbf{x}. \ \square$$

Corollary 10.2. *Let $Q(\mathbf{y})$ be a quadratic form in $(y_1, \ldots, y_n)$ with matrix A. There is an orthonormal basis $\mathbf{v}_1, \ldots, \mathbf{v}_n$ of $\mathbb{R}^n$ such that*

$$Q(P\mathbf{x}) = \sum_{i=1}^{n} \lambda_i x_i^2, \ P = \begin{pmatrix} \mathbf{v}_1 & \cdots & \mathbf{v}_n \end{pmatrix}.$$

The λ_i are the eigenvalues of A, and $A\mathbf{v}_i = \lambda_i \mathbf{v}_i$.

Proof. From Theorem 10.2,

$$A = PDP^t$$

where D is diagonal,

$$D = \begin{pmatrix} \lambda_1 & & \\ & \ddots & \\ & & \lambda_n \end{pmatrix},$$

$P = \begin{pmatrix} \mathbf{v}_1 & \cdots & \mathbf{v}_n \end{pmatrix}$ is orthogonal, and $A\mathbf{v}_i = \lambda_i \mathbf{v}_i$. Since

$$Q(P\mathbf{x}) = \mathbf{x}^t P^t AP\mathbf{x} = \mathbf{x}^t D\mathbf{x} = \sum_{i=1}^{n} \lambda_i x_i^2,$$

this gives the desired result. $\square$

Example 10.12. Obtain the representation

$$Q(P\mathbf{x}) = \lambda_1 x_1^2 + \lambda_2 x_2^2$$

of the Corollary when

$$Q(\mathbf{y}) = -5y_1^2 + 4y_1 y_2 - 2y_2^2.$$

The matrix of Q,

$$A = \begin{pmatrix} -5 & 2 \\ 2 & -2 \end{pmatrix},$$

has characteristic polynomial $(\lambda + 5)(\lambda + 2) - 4 = (\lambda + 1)(\lambda + 6)$.

The eigenspace for $\lambda = -1$ is

$$\ker \begin{pmatrix} 4 & -2 \\ -2 & 1 \end{pmatrix} = \ker \begin{pmatrix} 1 & -1/2 \\ 0 & 0 \end{pmatrix}.$$

The corresponding system of equations is $x_1 - \frac{1}{2}x_2 = 0$ and a basis of the eigenspace is $\left(\frac{1}{\sqrt{5}}, \frac{2}{\sqrt{5}}\right)$. Now the eigenspace for $\lambda = -6$ must have basis $\left(\frac{2}{\sqrt{5},}, \frac{-1}{\sqrt{5}}\right)$;

$$A = PDP^t, \text{ or } D = P^tAP,$$

with

$$P = \frac{1}{\sqrt{5}} \begin{pmatrix} 1 & 2 \\ 2 & -1 \end{pmatrix}, D = \begin{pmatrix} -1 & 0 \\ 0 & -6 \end{pmatrix},$$

and

$$Q(P\mathbf{x}) = -x_1^2 - 6x_2^2.$$

10.7 Exercises

(1) Here is a quadratic form.

$$x_1^2 + 4x_1x_2 + x_2^2$$

Find new variables (y_1, y_2) and an orthogonal matrix P such that $\mathbf{x} = P\mathbf{y}$ and the expression in terms of $\mathbf{y}$ is a diagonal quadratic form.

(2) Here is a quadratic form.

$$4x_1^2 + 2x_1x_2 + 4x_2^2$$

Find new variables (y_1, y_2) and an orthogonal matrix P such that $\mathbf{x} = P\mathbf{y}$ and the expression in terms of $\mathbf{y}$ is a diagonal quadratic form.

(3) Here is a quadratic form.

$$-3x_1^2 + 2x_1x_2 - 3x_2^2$$

Find new variables (y_1, y_2) and an orthogonal matrix P such that $\mathbf{x} = P\mathbf{y}$ and the expression in terms of $\mathbf{y}$ is a diagonal quadratic form.

(4) Here is a quadratic form.

$$3x_1^2 - 2x_1x_2 + 3x_2^2 - x_3^2$$

Find new variables (y_1, y_2, y_3) and an orthogonal matrix P such that $\mathbf{x} = P\mathbf{y}$, and the expression in terms of $\mathbf{y}$ is a diagonal quadratic form.

(5) Here is a quadratic form.

$$3x_1^2 - 2x_1x_2 + 3x_2^2 + 2x_3^2$$

Find new variables (y_1, y_2, y_3) and an orthogonal matrix P such that $\mathbf{x} = P\mathbf{y}$, and the expression in terms of $\mathbf{y}$ is a diagonal quadratic form.

(6) Here is a quadratic form.

$$3x_1^2 + 2x_1x_3 + 2x_2^2 + 3x_3^2$$

Find new variables (y_1, y_2, y_3) and an orthogonal matrix P such that $\mathbf{x} = P\mathbf{y}$, and the expression in terms of $\mathbf{y}$ is a diagonal quadratic form.

(7) Show that if A is an Hermitian invertible matrix, then the quadratic form $\mathbf{x} \to \langle A^2\mathbf{x}, \mathbf{x}\rangle$ is positive definite.

(8) If A is an arbitrary matrix, does it follow that $\mathbf{x} \to \langle A^2\mathbf{x}, \mathbf{x}\rangle$ is positive definite? **Hint:** You might consider $\begin{pmatrix} 2 & -2 \\ 0 & 2 \end{pmatrix}$.

(9) An $n \times n$ matrix A is called **normal** if $AA^* = A^*A$. Show that every Hermitian matrix is normal. Show that

$$\begin{pmatrix} 1+i & 1-i \\ 1-i & 1+i \end{pmatrix}$$

is not Hermitian but is normal.

(10) Show that if U is a unitary matrix and A is a normal matrix, then U^*AU is also a normal matrix.

(11) Show that every upper triangular normal matrix must be a diagonal matrix. Explain how this shows with Theorem 10.1 that an $n \times n$ matrix is normal if and only if there exists U unitary such that

$$U^*AU = D$$

where D is a diagonal matrix (maybe D has complex entries). **Hint:** Consider the following. You know

$$TT^* = \begin{pmatrix} a & \mathbf{b}^* \\ \mathbf{0} & T_1 \end{pmatrix} \begin{pmatrix} \bar{a} & \mathbf{0}^* \\ \mathbf{b} & T_1^* \end{pmatrix} = \begin{pmatrix} \bar{a} & \mathbf{0}^* \\ \mathbf{b} & T_1^* \end{pmatrix} \begin{pmatrix} a & \mathbf{b}^* \\ \mathbf{0} & T_1 \end{pmatrix} = T^*T$$

where T_1 is upper triangular. Now block multiply and observe that $\mathbf{b} = \mathbf{0}$ and that T_1 is normal. Then use induction.

(12) Let A be an $n \times n$ complex matrix. Show that A is Hermitian if and only if for some U unitary,

$$U^*AU = D$$

where D is a diagonal matrix having all real entries.

(13) Let F be an $n \times n$ complex matrix. Show F^*F is Hermitian and has all nonnegative eigenvalues.

(14) Suppose A is an $n \times n$ Hermitian matrix which has all nonnegative eigenvalues. Show that there exists an $n \times n$ Hermitian matrix B such that $B^2 = A$, all eigenvalues of B are nonnegative, and B commutes with every matrix which commutes with A. **Hint:** You know that there exists a unitary matrix U such that $U^*AU = D$ where D is a diagonal matrix having all nonnegative entries down the main diagonal. Now consider the problem of taking the square root of D. As to the part about commuting, suppose $CA = AC$. Then $A = UDU^*$ but also, for any C, $C = UC_UU^*$ for a unique C_U. Then show $CA = AC$ if and only if $DC_U = C_UD$. Then show this is the same as having $D^{1/2}C_U = C_UD^{1/2}$.

(15) ↑The above problem deals with the square root of a Hermitian matrix which has nonnegative eigenvalues. Show that there is at most one such Hermitian square root having nonnegative eigenvalues which has the property that it commutes with all matrices which commute with A. **Hint:** Suppose that B, B_1 both satisfy these conditions, explain why for $C = B, B_1$,

$$\langle C\mathbf{x}, \mathbf{x} \rangle \geq 0.$$

Next use this to verify that for any $\mathbf{x}$,

$$\langle B(B - B_1)\mathbf{x}, (B - B_1)\mathbf{x} \rangle \geq 0, \ \langle B_1(B - B_1)\mathbf{x}, (B - B_1)\mathbf{x} \rangle \geq 0. \quad (10.5)$$

Adding these inequalities and using the assumption about commuting, explain why

$$\langle (B^2 - B_1^2)\mathbf{x}, (B - B_1)\mathbf{x} \rangle = 0$$

and why each inequality in (10.5) is actually an equality. Let $\sqrt{B}, \sqrt{B_1}$ denote the square root of B, B_1 whose existence is guaranteed by the above problem. Explain why

$$\sqrt{B}(B - B_1)\mathbf{x} = \mathbf{0} = \sqrt{B_1}(B - B_1)\mathbf{x}.$$

Then conclude that $B(B - B_1)\mathbf{x} = \mathbf{0} = B_1(B - B_1)\mathbf{x}$. Finally,

$$0 = \langle B(B - B_1)\mathbf{x} - B_1(B - B_1)\mathbf{x}, \mathbf{x} \rangle = |(B - B_1)\mathbf{x}|^2.$$

(16) ↑Let F be an $n \times n$ matrix. Explain why there exists a unique square root U for F^*F. Next let $\{U\mathbf{x}_1, \cdots, U\mathbf{x}_r\}$ be an orthonormal basis for $U(\mathbb{C}^n)$ and let $\{U\mathbf{x}_1, \cdots, U\mathbf{x}_r, \mathbf{y}_{r+1}, \cdots, \mathbf{y}_n\}$ be an orthonormal basis for $\mathbb{C}^n$. Verify that $\{F\mathbf{x}_1, \cdots, F\mathbf{x}_r\}$ is also an orthnormal basis. Let $\{F\mathbf{x}_1, \cdots, F\mathbf{x}_r, \mathbf{z}_{r+1}, \cdots, \mathbf{z}_n\}$ be an orthonormal basis for $\mathbb{C}^n$. Now for $\mathbf{x}$ arbitrary, there exist unique scalars c_i, d_i such that

$$\mathbf{x} = \sum_{i=1}^{r} c_i U\mathbf{x}_i + \sum_{i=r+1}^{n} d_i \mathbf{y}_i.$$

Then define $R\mathbf{x}$ by

$$R\mathbf{x} = \sum_{i=1}^{r} c_i F\mathbf{x}_i + \sum_{i=r+1}^{n} d_i \mathbf{z}_i.$$

Verify that R preserves distances, and is therefore unitary. Letting $\mathbf{x}$ be arbitrary, it follows that there exist unique b_i such that

$$U\mathbf{x} = \sum_{i=1}^{r} b_i U\mathbf{x}_i.$$

Verify that $RU\mathbf{x} = \sum_{i=1}^{r} b_i F\mathbf{x}_i = F\left(\sum_{i=1}^{r} b_i \mathbf{x}_i\right)$. Finally verify that

$$F\left(\sum_{i=1}^{r} b_i \mathbf{x}_i\right) = F\mathbf{x}.$$

Thus $F = RU$ where R is unitary and U is Hermitian having nonnegative eigenvalues, and commuting with every matrix which commutes with F^*F. This is called the **right polar decomposition.** In mechanics, when F is a 3×3 real matrix, the matrix U is called the right Cauchy Green strain tensor. It measures the way a three dimensional set is stretched while the R merely preserves distance. Can you extend the result of this theorem to the case where F is an $m \times n$ matrix with $m \geq n$? The answer is yes, and the proof is just like what is given above.

(17) If A is a general $n \times n$ matrix having possibly repeated eigenvalues, show there is a sequence $\{A_k\}$ of $n \times n$ matrices having distinct eigenvalues which has the property that the ij^{th} entry of A_k converges to the ij^{th} entry of A for all ij. **Hint:** Use Schur's theorem.

(18) Prove the Cayley-Hamilton theorem as follows. First suppose A has a basis of eigenvectors $\{\mathbf{v}_k\}_{k=1}^{n}$, $A\mathbf{v}_k = \lambda_k \mathbf{v}_k$. Let $p(\lambda)$ be the characteristic polynomial. Show $p(A)\mathbf{v}_k = p(\lambda_k)\mathbf{v}_k = \mathbf{0}$. Then since $\{\mathbf{v}_k\}$ is a basis, it follows that $p(A)\mathbf{x} = \mathbf{0}$ for all $\mathbf{x}$, and so $p(A) = 0$. Next, in the general case, use Problem 17 to obtain a sequence $\{A_k\}$ of matrices whose entries converge to the entries of A such that A_k has n distinct eigenvalues, and therefore A_k has a basis of eigenvectors. Therefore, from the first part and for $p_k(\lambda)$ the characteristic polynomial for A_k, it follows $p_k(A_k) = 0$. Now explain why and the sense in which

$$\lim_{k \to \infty} p_k(A_k) = p(A).$$

(19) The principal submatrices of an $n \times n$ matrix A are A_k where A_k consists those entries which are in the first k rows and first k columns of A. Suppose A is a real symmetric matrix and that $\mathbf{x} \to \langle A\mathbf{x}, \mathbf{x} \rangle$ is positive definite. This means that if $\mathbf{x} \neq \mathbf{0}$, then $\langle A\mathbf{x}, \mathbf{x} \rangle > 0$. Show that each of the principal submatrices are positive definite. **Hint:** Consider $\begin{pmatrix} \mathbf{x} & \mathbf{0} \end{pmatrix} A \begin{pmatrix} \mathbf{x} \\ \mathbf{0} \end{pmatrix}$ where $\mathbf{x}$ consists of k entries.

(20) $\uparrow$Show that if A is a symmetric positive definite $n \times n$ real matrix, then A has an LU factorization with the property that each entry on the main diagonal in U is positive. This gives an important case of a matrix which always has

an LU factorization. **Hint:** This is pretty clear if A is 1×1. Assume true for $(n-1)\times(n-1)$. Then

$$A = \begin{pmatrix} \hat{A} & \mathbf{a} \\ \mathbf{a}^t & a_{nn} \end{pmatrix}$$

Then as above, $\hat{A}$ is positive definite. Thus it has an LU factorization with all positive entries on the diagonal of U. Notice that, using block multiplication,

$$A = \begin{pmatrix} LU & \mathbf{a} \\ \mathbf{a}^t & a_{nn} \end{pmatrix} = \begin{pmatrix} L & \mathbf{0} \\ \mathbf{0}^t & 1 \end{pmatrix} \begin{pmatrix} U & L^{-1}\mathbf{a} \\ \mathbf{a}^t & a_{nn} \end{pmatrix}$$

Now consider that matrix on the right. Argue that it is of the form $\tilde{L}\tilde{U}$ where $\tilde{U}$ has all positive diagonal entries except possibly for the one in the nth row and n^{th} column. Now explain why $\det(A) > 0$ and argue that in fact all diagonal entries of $\tilde{U}$ are positive.

(21) ↑Let A be a real symmetric $n\times n$ matrix and $A = LU$ where L has all ones down the diagonal and U has all positive entries down the main diagonal. Show that $A = LDH$ where L is lower triangular and H is upper triangular, each having all ones down the diagonal and D a diagonal matrix having all positive entries down the main diagonal. In fact, these are the diagonal entries of U.

(22) ↑Show that if L, L_1 are lower triangular with ones down the main diagonal and H, H_1 are upper triangular with all ones down the main diagonal and D, D_1 are diagonal matrices having all positive diagonal entries, and if $LDH = L_1D_1H_1$, then $L = L_1, H = H_1, D = D_1$. **Hint:** Explain why $D_1^{-1}L_1^{-1}LD = H_1H^{-1}$. Then explain why the right side is upper triangular and the left side is lower triangular. Conclude these are both diagonal matrices. However, there are all ones down the diagonal in the expression on the right. Hence $H = H_1$. Do something similar to conclude that $L = L_1$ and then that $D = D_1$.

(23) ↑Show that if A is a symmetric real matrix such that $\mathbf{x} \to \langle A\mathbf{x}, \mathbf{x}\rangle$ is positive definite, then there exists a lower triangular matrix L having all positive entries down the diagonal such that $A = LL^t$. **Hint:** From the above, $A = LDH$ where L, H are respectively lower and upper triangular having all ones down the diagonal and D is a diagonal matrix having all positive entries. Then argue from the above problem and symmetry of A that $H = L^t$. Now modify L by making it equal to $LD^{1/2}$. This is called the Cholesky factorization.

(24) ↑Here is a positive definite symmetric matrix

$$\begin{pmatrix} 5 & 1 & 2 \\ 1 & 4 & 1 \\ 2 & 1 & 4 \end{pmatrix}$$

Find its Cholesky factorization.

(25) ↑Show that if A is a real symmetric matrix and $\mathbf{x} \to \langle A\mathbf{x}, \mathbf{x}\rangle$ is positive definite, then there exists an upper triangular matrix U, having all positive diagonal entries such that $\langle A\mathbf{x}, \mathbf{x}\rangle = |U\mathbf{x}|^2$.

(26) Letting A be a real $m \times n$ matrix, $m \geq n$ having rank n, consider least squares solutions to $A\mathbf{x} = \mathbf{b}$. Recall this required finding the solutions to $A^t A\mathbf{x} = A^t\mathbf{b}$. How could the Cholesky decomposition be useful in doing this problem?

(27) Let A be an $m \times n$ matrix. Then if you unraveled it, you could consider it as a vector in $\mathbb{C}^{nm}$. The Frobenius inner product on the vector space of $m \times n$ matrices is defined as

$$\langle A, B \rangle \equiv \text{trace}\,(AB^*)$$

Show that this really does satisfy the axioms of an inner product space and that it also amounts to nothing more than considering $m \times n$ matrices as vectors in $\mathbb{C}^{nm}$.

(28) ↑Consider the $n \times n$ unitary matrices. Show that whenever U is such a matrix, it follows that

$$|U|_{\mathbb{C}^{nn}} = \sqrt{n}$$

Next explain why if $\{U_k\}$ is any sequence of unitary matrices, there exists a subsequence $\{U_{k_m}\}_{m=1}^{\infty}$ such that $\lim_{m\to\infty} U_{k_m} = U$ where U is unitary. Here the limit takes place in the sense that the entries of U_{k_m} converge to the corresponding entries of U.

(29) ↑Let A, B be two $n \times n$ matrices. Denote by $\sigma(A)$ the set of eigenvalues of A. Define

$$\text{dist}\,(\sigma(A), \sigma(B)) = \max_{\lambda \in \sigma(A)} \min\{|\lambda - \mu| : \mu \in \sigma(B)\}$$

Explain why $\text{dist}\,(\sigma(A), \sigma(B))$ is small if and only if every eigenvalue of A is close to some eigenvalue of B. Now prove the following theorem using the above problem and Schur's theorem. This theorem says roughly that if A is close to B then the eigenvalues of A are close to those of B in the sense that every eigenvalue of A is close to an eigenvalue of B.

Theorem 10.3. *Suppose* $\lim_{k\to\infty} A_k = A$. *Then*

$$\lim_{k\to\infty} \text{dist}\,(\sigma(A_k), \sigma(A)) = 0$$

Chapter 11

The singular value decomposition

Chapter summary

We discuss the singular value decomposition using the spectral theory of Hermitian matrices. This is related to the problem of best approximation in the Frobenius norm. The relationship of the number of singular values and the rank of a matrix is also discussed. This leads to a discussion of the generalized inverse and its use in obtaining the least squares solution of smallest norm. This generalized inverse is shown to be characterized by the Penrose conditions.

11.1 Properties of the singular value decomposition

The singular value decomposition of an $m \times n$ complex matrix is a useful tool in data compression. In this application, the idea is to approximate a given (very large) matrix by a matrix of small rank. It also has important applications to the problem of least squares and the Moore Penrose inverse.

Before giving the decomposition, we prove a simple lemma.

Lemma 11.1. *Let A be an $m \times n$ real matrix. Then*

$$R(A^*A) = R(A).$$

Also $\ker(A^*A) = \ker(A)$.

Proof. If $A\mathbf{x} = 0$, then evidently $A^*A\mathbf{x} = \mathbf{0}$. On the other hand, if $A^*A\mathbf{x} = \mathbf{0}$, then

$$0 = \mathbf{x}^*A^*A\mathbf{x} = \langle A\mathbf{x}, A\mathbf{x}\rangle,$$

so that $A\mathbf{x} = 0$. Thus A and A^*A have the same kernel, say V. Since A^*A is $n \times n$,

$$R(A^*A) = n - \dim V = R(A).\square$$

Lemma 11.2. *Let A be a complex $m \times n$ matrix. Then A^*A is Hermitian, all its eigenvalues are nonnegative, and if $A \neq 0$, there is at least one nonzero eigenvalue.*

Proof. It is obvious that A^*A is Hermitian because

$$(A^*A)^* = A^*(A^*)^* = A^*A.$$

Suppose $A^*A\mathbf{x} = \lambda\mathbf{x}$ where $\mathbf{x} \neq \mathbf{0}$. Then

$$\lambda|\mathbf{x}|^2 = \langle\lambda\mathbf{x}, \mathbf{x}\rangle = \langle A^*A\mathbf{x}, \mathbf{x}\rangle = \langle A\mathbf{x}, A\mathbf{x}\rangle \geq 0.$$

Hence $\lambda \geq 0$. By Theorem 10.2,

$$A^*A = U^*DU$$

where the entries of the diagonal matrix D are the eigenvalues of A^*A. If all eigenvalues are 0, then $A^*A = 0$ and Lemma 11.1 gives the absurd conclusion that $A = 0$. □

Definition 11.1. Let A be an $m \times n$ complex matrix. The singular values of A are the positive square roots of the positive eigenvalues of A^*A.

The following theorem gives the existence of a singular value decomposition of A.

Theorem 11.1. *Let A be an $m \times n$ complex matrix of rank r. We can write A in the form*

$$A = U\Sigma V^* \tag{11.1}$$

where
(i) U is a unitary $m \times m$ matrix

(ii) V is a unitary $n \times n$ matrix

(iii) $\Sigma = (\sigma_{ij})$ is $m \times n$, and is quasi-diagonal in the sense that $\sigma_{ij} = 0$ for $i \neq j$.

(iv) Writing σ_i instead of σ_{ii}, we have

$$\sigma_1 \geq \sigma_2 \geq \cdots \geq \sigma_r > 0, \sigma_i = 0 \text{ for } i > r.$$

In case A is a real matrix, U, V can be orthogonal matrices.

These are the principal features of the decomposition (11.1), which is called a **singular value decomposition** (SVD) of A. The positive numbers $\sigma_1, \ldots, \sigma_r$ (where $r \leq \min(m,n)$) are called the **singular values** of A. We will see below that this agrees with Definition 11.1. We will note other features later. If $m = 4, n = 3$, then

$$\Sigma = \begin{pmatrix} \sigma_1 & 0 & 0 \\ 0 & \sigma_2 & 0 \\ 0 & 0 & \sigma_3 \\ 0 & 0 & 0 \end{pmatrix},$$

while if $m = 2, n = 3$, then

$$\Sigma = \begin{pmatrix} \sigma_1 & 0 & 0 \\ 0 & \sigma_2 & 0 \end{pmatrix}.$$

Note that once we have an SVD of A, we immediately obtain an SVD of A^*. For (11.1) gives the expression

$$A^* = V\Sigma^t U^*,$$

which has all the desired properties with V, Σ^t, U in place of U, Σ, V.

Proof of Theorem 11.1. By Lemma 11.2 and Theorem 10.2, there exists an orthonormal basis $\{\mathbf{v}_i\}_{i=1}^n$ such that $A^*A\mathbf{v}_i = \sigma_i^2\mathbf{v}_i$ where $\sigma_i^2 > 0$ for $i = 1, \cdots, k$ where $k \geq 1$ and $\sigma_i = 0$ for $i > k$. Thus for $i > k$, $A^*A\mathbf{v}_i = \mathbf{0}$ and so, by Lemma 11.1, $A\mathbf{v}_i = \mathbf{0}$.

For $i = 1, \cdots, k$, define $\mathbf{u}_i \in \mathbb{F}^m$ by

$$\mathbf{u}_i \equiv \sigma_i^{-1} A\mathbf{v}_i.$$

Thus $A\mathbf{v}_i = \sigma_i \mathbf{u}_i$. Now

$$\begin{aligned}
\langle \mathbf{u}_i, \mathbf{u}_j \rangle &= \langle \sigma_i^{-1} A\mathbf{v}_i, \sigma_j^{-1} A\mathbf{v}_j \rangle = \langle \sigma_i^{-1}\mathbf{v}_i, \sigma_j^{-1} A^*A\mathbf{v}_j \rangle \\
&= \langle \sigma_i^{-1}\mathbf{v}_i, \sigma_j^{-1}\sigma_j^2\mathbf{v}_j \rangle = \frac{\sigma_j}{\sigma_i} \langle \mathbf{v}_i, \mathbf{v}_j \rangle = \delta_{ij}.
\end{aligned}$$

Thus $\{\mathbf{u}_i\}_{i=1}^k$ is an orthonormal set of vectors in $\mathbb{C}^m$. Also,

$$AA^*\mathbf{u}_i = AA^*\sigma_i^{-1}A\mathbf{v}_i = \sigma_i^{-1}AA^*A\mathbf{v}_i = \sigma_i^{-1}A\sigma_i^2\mathbf{v}_i = \sigma_i^2\mathbf{u}_i.$$

Now using the Gram-Schmidt process, extend $\{\mathbf{u}_i\}_{i=1}^k$ to an orthonormal basis for all of $\mathbb{C}^m$, $\{\mathbf{u}_i\}_{i=1}^m$ and let the unitary matrices U and V be defined by

$$U = \begin{pmatrix} \mathbf{u}_1 & \cdots & \mathbf{u}_m \end{pmatrix}, \ V = \begin{pmatrix} \mathbf{v}_1 & \cdots & \mathbf{v}_n \end{pmatrix}.$$

Thus U is the matrix which has the $\mathbf{u}_i$ as columns and V is defined as the matrix which has the $\mathbf{v}_i$ as columns. In case A is a real matrix, we may take $\mathbf{v}_i, \mathbf{u}_j$ to be real and have U and V be orthogonal matrices. Then

$$U^*AV = \begin{pmatrix} \mathbf{u}_1^* \\ \vdots \\ \mathbf{u}_k^* \\ \vdots \\ \mathbf{u}_m^* \end{pmatrix} A \begin{pmatrix} \mathbf{v}_1 & \cdots & \mathbf{v}_n \end{pmatrix}$$

$$= \begin{pmatrix} \mathbf{u}_1^* \\ \vdots \\ \mathbf{u}_k^* \\ \vdots \\ \mathbf{u}_m^* \end{pmatrix} \begin{pmatrix} \sigma_1\mathbf{u}_1 & \cdots & \sigma_k\mathbf{u}_k & \mathbf{0} & \cdots & \mathbf{0} \end{pmatrix} = \begin{pmatrix} \sigma & 0 \\ 0 & 0 \end{pmatrix} = \Sigma$$

where σ is the $k \times k$ diagonal matrix of the form

$$\begin{pmatrix} \sigma_1 & & \\ & \ddots & \\ & & \sigma_k \end{pmatrix}.$$

In the above, the zero blocks are the sizes needed in order to make the matrix Σ an $m \times n$ matrix. □

In case A is real, we only use the real inner product in all considerations and so the matrices U, V will be real matrices. The case of real matrices is the one of most interest.

As mentioned above, for A an $m \times n$ matrix, we write

$$\Sigma = \begin{pmatrix} \sigma & 0 \\ 0 & 0 \end{pmatrix}$$

where the zero blocks are the size needed to make the resulting matrix $m \times n$. The matrix σ is uniquely determined as the diagonal matrix which has the singular values of A in descending magnitude from the top left toward the lower right.

Notice that the matrix Σ is uniquely determined by (11.1) and the properties (iii) and (iv). For (11.1) yields

$$A^*A = V\Sigma^t U^* U \Sigma V^* = V\Sigma^t \Sigma V^* = V \begin{pmatrix} \sigma_1^2 & & \\ & \ddots & \\ & & \sigma_n^2 \end{pmatrix} V^*, \tag{11.2}$$

so that $\sigma_1^2, \ldots, \sigma_n^2$ are the eigenvalues of A^*A in non-increasing order, counted with muliplicity. However, if $r < m$, the SVD is not unique; there are many possibilities for $\mathbf{u}_{r+1}, \ldots, \mathbf{u}_m$.

Example 11.1. In (11.1), suppose that $m = n$, A is a real matrix, and that $\sigma_1 > \cdots > \sigma_n > 0$. Show that the SVD is unique apart from the signs of the vectors in V and U.

From (11.2), which in turn follows from (11.1), the columns $\mathbf{v}_i$ of V are eigenvectors of A^*A in order corresponding to the distinct numbers $\sigma_1^2, \ldots, \sigma_n^2$. As the eigenspaces are one-dimensional, the $\mathbf{v}_i$ are unique apart from sign. Moreover,

$$U = A(V^*)^{-1}\Sigma^{-1} = AV \begin{pmatrix} \sigma_1^{-1} & & \\ & \ddots & \\ & & \sigma_n^{-1} \end{pmatrix} = \begin{pmatrix} \frac{A\mathbf{v}_1}{\sigma_1} & \cdots & \frac{A\mathbf{v}_n}{\sigma_n} \end{pmatrix},$$

which is the prescription for U used in the proof of Theorem 11.1.

Example 11.2. Find an SVD of

$$A = \begin{pmatrix} 2 & -1 \\ 2 & 1 \end{pmatrix}.$$

First of all,

$$A^*A = \begin{pmatrix} 8 & 0 \\ 0 & 2 \end{pmatrix}.$$

An orthonormal set of eigenvectors of A^*A is $\mathbf{v}_1 = \mathbf{e}_1, \mathbf{v}_2 = \mathbf{e}_2$, so that $V = I$. Moreover, $\sigma_1^2 = 8, \sigma_2^2 = 2$; so

$$\Sigma = \begin{pmatrix} 2\sqrt{2} & \\ & \sqrt{2} \end{pmatrix}.$$

Now, as prescribed above,

$$\begin{aligned}
\mathbf{u}_1 &= \frac{1}{\sigma_1} A\mathbf{v}_1 = \frac{1}{2\sqrt{2}} \begin{pmatrix} 2 & -1 \\ 2 & 1 \end{pmatrix} \begin{pmatrix} 1 \\ 0 \end{pmatrix} = \begin{pmatrix} \frac{1}{\sqrt{2}} \\ \frac{1}{\sqrt{2}} \end{pmatrix}, \\
\mathbf{u}_2 &= \frac{1}{\sigma_2} A\mathbf{v}_2 = \frac{1}{\sqrt{2}} \begin{pmatrix} 2 & -1 \\ 2 & 1 \end{pmatrix} \begin{pmatrix} 0 \\ 1 \end{pmatrix} = \begin{pmatrix} -\frac{1}{\sqrt{2}} \\ \frac{1}{\sqrt{2}} \end{pmatrix}.
\end{aligned}$$

Example 11.3. Find an SVD of

$$A = \begin{pmatrix} -1/2 & 3/2 & 0 \\ -3/2 & 1/2 & 0 \\ 0 & 0 & 3 \end{pmatrix}.$$

First of all,

$$A^*A = \begin{pmatrix} 5/2 & -3/2 & 0 \\ -3/2 & 5/2 & 0 \\ 0 & 0 & 9 \end{pmatrix}$$

has characteristic polynomial

$$\left((\lambda - 5/2)^2 - 9/4\right)(\lambda - 9) = (\lambda - 9)(\lambda - 4)(\lambda - 1),$$

so that $\sigma_1 = 3, \sigma_2 = 2, \sigma_3 = 1$. An easy computation gives an orthonormal basis $\mathbf{v}_1, \mathbf{v}_2, \mathbf{v}_3$ of eigenvectors of A^*A,

$$A^*A\mathbf{v}_j = \sigma_j^2 \mathbf{v}_j,$$

namely

$$\mathbf{v}_1 = \begin{pmatrix} 0 \\ 0 \\ 1 \end{pmatrix}, \mathbf{v}_2 = \begin{pmatrix} \frac{1}{\sqrt{2}} \\ -\frac{1}{\sqrt{2}} \\ 0 \end{pmatrix}, \mathbf{v}_3 = \begin{pmatrix} \frac{1}{\sqrt{2}} \\ \frac{1}{\sqrt{2}} \\ 0 \end{pmatrix}.$$

We now take

$$\mathbf{u}_1 = \frac{A\mathbf{v}_1}{3} = \begin{pmatrix} 0 \\ 0 \\ 1 \end{pmatrix}, \quad \mathbf{u}_2 = \frac{A\mathbf{v}_2}{2} = \begin{pmatrix} -\frac{1}{\sqrt{2}} \\ -\frac{1}{\sqrt{2}} \\ 0 \end{pmatrix}, \quad \mathbf{u}_3 = A\mathbf{v}_3 = \begin{pmatrix} \frac{1}{\sqrt{2}} \\ -\frac{1}{\sqrt{2}} \\ 0 \end{pmatrix}.$$

So $\Sigma = \begin{pmatrix} 3 & & \\ & 2 & \\ & & 1 \end{pmatrix}, V = \begin{pmatrix} 0 & \frac{1}{\sqrt{2}} & \frac{1}{\sqrt{2}} \\ 0 & -\frac{1}{\sqrt{2}} & \frac{1}{\sqrt{2}} \\ 1 & 0 & 0 \end{pmatrix}, U = \begin{pmatrix} 0 & -\frac{1}{\sqrt{2}} & \frac{1}{\sqrt{2}} \\ 0 & -\frac{1}{\sqrt{2}} & -\frac{1}{\sqrt{2}} \\ 1 & 0 & 0 \end{pmatrix}.$

Example 11.4. Use www.bluebit.gr/matrix-calculator/ to solve Example 11.3.

If we follow the simple procedure at this site and select 9 decimal digits, we get Σ as above.

$$U = \begin{pmatrix} 0 & -a & -a \\ 0 & -a & a \\ 1 & 0 & 0 \end{pmatrix}, V = \begin{pmatrix} 0 & a & -a \\ 0 & -a & -a \\ 1 & 0 & 0 \end{pmatrix}$$

with $a = 0.707106781$. We 'correct' this to $a = \frac{1}{\sqrt{2}}$ by using $|(-a, -a, 0)| = 1$. Note the differences in sign in U, V from the solution that we found above. See the Appendix for a description of how to do this problem using Maple.

The singular value decomposition has as an immediate corollary the following interesting result.

Corollary 11.1. *Let A be an $m \times n$ matrix. Then the rank of A equals the rank of A^* and both equal the number of singular values.*

Proof. Since V and U are unitary, they are each one-to-one and onto, and so it follows that

$$\begin{aligned} \text{rank}\,(A) &= \text{rank}\,(U^* AV) \\ &= \text{rank}\,(\Sigma) \\ &= \text{number of singular values.} \end{aligned}$$

Also since U, V are unitary,

$$\begin{aligned} \text{rank}\,(A^*) &= \text{rank}\,(V^* A^* U) \\ &= \text{rank}\,((U^* AV)^*) \\ &= \text{rank}\,(\Sigma^*) \\ &= \text{number of singular values.} \end{aligned}$$

This proves the corollary. □

A lot of information about subspaces associated with A is 'embedded' in the SVD. This takes its most attractive form in the case where the matrices are all real.

Lemma 11.3. *Let $A = U\Sigma V^t$ be an SVD of an $m \times n$ matrix of rank r. Then*

(i) the first r columns of V are an orthonormal basis for the row space of A,

(ii) the last $n - r$ columns of V are an orthonormal basis for $\ker(A)$,

(iii) the first r columns of U are an orthonormal basis of $\text{Col}\,(A)$,

(iv) the last $m - r$ columns of U are an orthonormal basis of $\ker(A^t)$.

Proof. From $A = U\Sigma V^t$, it follows that

$$U^t A = \Sigma V^t,$$

and so the row space of A equals the row space of $U^t A$ which equals the row space of ΣV^t which equals the column space of the first r columns of V. This proves (i).

(Recall that multiplication on the left by an invertible matrix is the same as doing a sequence of row operations which preserves the row space.)

We noted above that

$$A\mathbf{v}_{r+1} = \cdots = A\mathbf{v}_k = \mathbf{0}. \tag{11.3}$$

Therefore, the space $\ker A$, of dimension $n - r$, contains the orthonormal set $\mathbf{v}_{r+1}, \ldots, \mathbf{v}_n$. Since the rank of A equals r, this implies that $\{\mathbf{v}_{r+1}, \ldots, \mathbf{v}_k\}$ is a basis for $\ker(A)$. This proves (ii).

In $A^t = V\Sigma^t U^t$, the roles of U and V are interchanged. Thus (iii) is a consequence of (i) with A^t, U in place of A, V and similarly (iv) follows from (ii) with A^t, U in place of A, V. □

11.2 Approximation by a matrix of low rank

We now return to the topic raised at the beginning of the chapter. If A is a complex $m \times n$ matrix with rank r, what is the closest matrix to A that has rank l (where $1 \le l < r$)? To measure distance between matrices, we regard the vector space of $m \times n$ complex matrices under addition as coinciding with $\mathbb{C}^{mn}$. Thus

$$\begin{pmatrix} a_{11} & a_{12} & a_{13} \\ a_{21} & a_{22} & a_{23} \end{pmatrix}$$

is identified with

$$(a_{11}, a_{12}, a_{13}, a_{21}, a_{22}, a_{23}),$$

for example. Now the inner product of two $m \times n$ complex matrices $A = (a_{ij})\,, B = (b_{ij})$ is

$$\langle A, B\rangle = \sum_{i=1}^{n}\sum_{j=1}^{n} a_{ij}\overline{b_{ij}},$$

the length or **Frobenius norm** of A is

$$\|A\| = \left(\sum_{i=1}^{m}\sum_{j=1}^{n} |a_{ij}|^2\right)^{1/2}$$

and the **distance** from A to B is $\|A - B\|$. (Other 'distance measures' are also important, see Horn and Johnson (1990). In fact, we used another one in the discussion of first order systems of differential equations.)

We can express $\langle A, B\rangle$ in terms of trace. The **trace** of an $n \times n$ matrix $A = (a_{ij})$ is

$$\operatorname{tr} A = \sum_{i=1}^{n} a_{ii}.$$

Lemma 11.4. *(i) Let A be $m \times n$ and C be $n \times m$. Then* $\operatorname{tr}(AC) = \operatorname{tr}(CA)$.

(ii) Let A and B be $m \times n$. Then

$$\langle A, B\rangle = \operatorname{tr}(AB^*) = \overline{\operatorname{tr}(BA^*)} = \overline{\langle B, A\rangle}.$$

Proof. (i) The ij^{th} entry of AC is

$$\sum_{k=1}^{n} a_{ik}c_{kj}.$$

Hence

$$\mathrm{tr}(AC) = \sum_{i=1}^{m}\sum_{k=1}^{n} a_{ik}c_{ki} = \sum_{k=1}^{n}\sum_{i=1}^{m} c_{ki}a_{ik} = \mathrm{tr}(CA).$$

(ii) $\mathrm{tr}(AB^*) = \sum_{i=1}^{m}\sum_{k=1}^{n} a_{ik}\overline{b_{ik}} = \langle A, B\rangle.$

Interchanging A and B gives

$$\overline{\langle A, B\rangle} = \overline{\mathrm{tr}(AB^*)} = \overline{\sum_{i=1}^{m}\sum_{k=1}^{n} a_{ik}\overline{b_{ik}}} = \sum_{i=1}^{m}\sum_{k=1}^{n} \overline{a_{ik}}b_{ik} = \langle B, A\rangle$$

We are *not* using (i)! □

It turns out that the solution to our problem of minimizing $||A - B||$ (with A given and B of rank $\leq h$) is the following. Let

$$A = U\Sigma V^*,\ \Sigma = \begin{pmatrix} \sigma & 0 \\ 0 & 0 \end{pmatrix}$$

where

$$\sigma = \begin{pmatrix} \sigma_1 & & \\ & \ddots & \\ & & \sigma_r \end{pmatrix}$$

is an SVD of A. Then we take

$$A' = U\begin{pmatrix} \sigma' & 0 \\ 0 & 0 \end{pmatrix}V^*. \tag{11.4}$$

where

$$\sigma' = \begin{pmatrix} \sigma_1 & & \\ & \ddots & \\ & & \sigma_h \end{pmatrix}.$$

In other words, we modify Σ by replacing $\sigma_{h+1}, \ldots, \sigma_r$ by 0.

To prove this, we need the following simple lemma about the Frobenius norm, denoted by $|| \ ||_F$.

Lemma 11.5. *Let A be an $m \times n$ complex matrix with singular matrix*

$$\Sigma = \begin{pmatrix} \sigma & 0 \\ 0 & 0 \end{pmatrix}$$

with σ the diagonal matrix of singular values decreasing from upper left to lower right. Then

$$||\Sigma||_F^2 = ||A||_F^2 . \tag{11.5}$$

Moreover, the following holds for the Frobenius norm. If U, V are unitary and of the right size,

$$||UAV||_F = ||A||_F . \tag{11.6}$$

Proof. From the definition, and letting U, V be unitary and of the right size,
$||UAV||_F^2 \equiv \text{tr}\,(UAVV^*A^*U^*) = \text{tr}\,(UAA^*U^*) = \text{tr}\,(AA^*U^*U) = \text{tr}\,(AA^*) = ||A||_F^2$

Now consider (11.5). From what was just shown,

$$||A||_F^2 = ||U\Sigma V^*||_F^2 = ||\Sigma||_F^2 .$$

This proves the lemma. □

Theorem 11.2. *Let A be a complex $m \times n$ matrix of rank r. Then the matrix of rank h or less which best approximates A in the Frobenius norm is the matrix*

$$A' = U \begin{pmatrix} \sigma' & 0 \\ 0 & 0 \end{pmatrix} V^*$$

where σ' is obtained by replacing σ_i with 0 for all $i \geq h+1$.

Proof. Let B be a matrix which has rank h or less. Then from Lemma 11.5

$$\begin{aligned} ||A - B||_F^2 &= ||U^*(A-B)V||_F^2 = ||U^*AV - U^*BV||_F^2 \\ &= \left\| \begin{pmatrix} \sigma & 0 \\ 0 & 0 \end{pmatrix} - U^*BV \right\|_F^2 \end{aligned}$$

where σ is the diagonal matrix described earlier consisting of the singular values of A arranged in decreasing order from the upper left to lower right. Now U^*BV has rank $\leq h$. Since the singular values of A decrease from the upper left to the lower right, it follows that for B to be as close as possible to A in the Frobenius norm,

$$U^*BV = \begin{pmatrix} \sigma' & 0 \\ 0 & 0 \end{pmatrix},$$

where σ' is obtained from σ by replacing σ_i with 0 for every $i > h$. (Recall that the Frobenius norm involves taking the square root of the sum of the squares of the absolute values of the entries of the matrix.) This implies $B = A'$ above. □

The last part of the above argument is obvious if you look at a simple example. Say

$$\begin{pmatrix} \sigma & 0 \\ 0 & 0 \end{pmatrix} = \begin{pmatrix} 3 & 0 & 0 & 0 \\ 0 & 2 & 0 & 0 \\ 0 & 0 & 0 & 0 \end{pmatrix}.$$

for example. Then what rank 1 matrix would be closest to this one in the Frobenius norm? Obviously

$$\begin{pmatrix} 3 & 0 & 0 & 0 \\ 0 & 0 & 0 & 0 \\ 0 & 0 & 0 & 0 \end{pmatrix}.$$

Example 11.5. Find to two decimal places the closest matrix B of rank 2 to

$$A = \begin{pmatrix} 4 & 6 & 1 & 4 \\ 2 & 1 & 2 & 3 \\ 3 & 1 & 2 & 2 \\ 1 & 4 & 0 & 2 \end{pmatrix}.$$

From bluebit we obtain the SVD

$$A = U\Sigma V^t,$$

where

$$U = \begin{pmatrix} -.780 & .255 & -.298 & -.488 \\ -.345 & -.560 & .730 & -.186 \\ -.341 & -.603 & -.490 & .529 \\ -.396 & .507 & .373 & .669 \end{pmatrix},$$

$$\Sigma = \begin{pmatrix} 10.578 & & & \\ & 3.567 & & \\ & & 1.143 & \\ & & & 0.255 \end{pmatrix} = \begin{pmatrix} \sigma_1 & & & \\ & \sigma_2 & & \\ & & \sigma_3 & \\ & & & \sigma_4 \end{pmatrix},$$

$$V^t = \begin{pmatrix} -.494 & -.657 & -.203 & -.532 \\ -.394 & .671 & -.581 & -.240 \\ -.724 & -.047 & .160 & .670 \\ -.277 & .342 & .772 & .459 \end{pmatrix}.$$

Thus

$$B = U \begin{pmatrix} \sigma_1 & & & \\ & \sigma_2 & & \\ & & 0 & \\ & & & 0 \end{pmatrix} V^t = \begin{pmatrix} 10.578\mathbf{a}_1 & 3.567\mathbf{a}_2 & \mathbf{0} & \mathbf{0} \end{pmatrix} V^t$$

$$= \begin{pmatrix} 3.72 & 6.03 & 1.15 & 4.17 \\ 2.64 & 1.13 & 1.92 & 2.48 \\ 2.63 & 0.93 & 1.98 & 2.44 \\ 1.36 & 3.97 & -0.20 & 1.80 \end{pmatrix} \text{ to two decimal places.}$$

You can also use Maple to find a singular value decomposition of an $m \times n$ matrix. In Maple 15, you enter the matrix and then select the following, "Solvers and Forms", "Singular Value Decomposition" and then "Singular Value Decomposition (USVt)" and press enter. It will give you the orthogonal matrices on either side of a column vector consisting of the singular values. To see more specific directions with pictures, see the appendix.

11.3 The Moore Penrose inverse

Consider the least squares problem

$$A^*A\mathbf{x} = A^*\mathbf{y}.$$

Recall that $A = U\Sigma V^*$, and so $A^* = V\Sigma^t U^*$. Therefore, $\mathbf{x}$ is a least squares solution if and only if

$$V\Sigma^t U^* U\Sigma V^*\mathbf{x} = V\Sigma^t U^*\mathbf{y}$$

Thus

$$V\begin{pmatrix} \sigma^2 & 0 \\ 0 & 0 \end{pmatrix} V^*\mathbf{x} = V\begin{pmatrix} \sigma & 0 \\ 0 & 0 \end{pmatrix} U^*\mathbf{y}$$

which is the same as

$$\begin{pmatrix} \sigma^2 & 0 \\ 0 & 0 \end{pmatrix} V^*\mathbf{x} = \begin{pmatrix} \sigma & 0 \\ 0 & 0 \end{pmatrix} U^*\mathbf{y}. \tag{11.7}$$

It is very easy to spot one solution to the least squares solution,

$$\mathbf{x} = V\begin{pmatrix} \sigma^{-1} & 0 \\ 0 & 0 \end{pmatrix} U^*\mathbf{y}.$$

It turns out that this particular solution to the least squares problem has some very nice properties and is so important that it is given a name $A^+\mathbf{y}$.

Definition 11.2. Let A be an $m \times n$ matrix and let $\begin{pmatrix} \sigma & 0 \\ 0 & 0 \end{pmatrix}$ be the associated singular matrix. Then the **Moore Penrose** inverse of A, denoted by A^+ is defined as

$$A^+ \equiv V\begin{pmatrix} \sigma^{-1} & 0 \\ 0 & 0 \end{pmatrix} U^*.$$

Here

$$U^*AV = \begin{pmatrix} \sigma & 0 \\ 0 & 0 \end{pmatrix}$$

as above.

Thus $A^+\mathbf{y}$ is a solution to the least squares problem to find $\mathbf{x}$ which minimizes $|A\mathbf{x} - \mathbf{y}|$. In fact, one can say more about this. In the following picture $M_{\mathbf{y}}$ denotes the set of least squares solutions $\mathbf{x}$ such that $A^*A\mathbf{x} = A^*\mathbf{y}$.

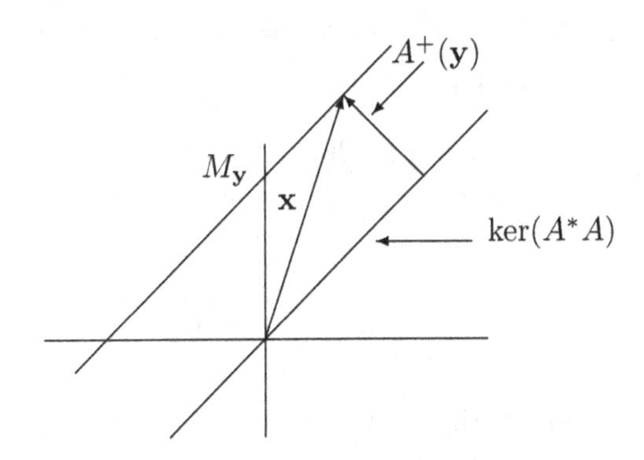

Then $A^+(\mathbf{y})$ is as given in the picture.

Proposition 11.1. $A^+\mathbf{y}$ is the solution to the problem of minimizing $|A\mathbf{x} - \mathbf{y}|$ for all $\mathbf{x}$, which has smallest norm. Thus

$$|AA^+\mathbf{y} - \mathbf{y}| \leq |A\mathbf{x} - \mathbf{y}| \text{ for all } \mathbf{x}$$

and if $\mathbf{x}_1$ satisfies $|A\mathbf{x}_1 - \mathbf{y}| \leq |A\mathbf{x} - \mathbf{y}|$ for all $\mathbf{x}$, then $|A^+\mathbf{y}| \leq |\mathbf{x}_1|$.

Proof. Consider $\mathbf{x}$ satisfying $A^*A\mathbf{x} = A^*\mathbf{y}$. Then as pointed out above,

$$\begin{pmatrix} \sigma^2 & 0 \\ 0 & 0 \end{pmatrix} V^*\mathbf{x} = \begin{pmatrix} \sigma & 0 \\ 0 & 0 \end{pmatrix} U^*\mathbf{y}$$

and it is desired to find the $\mathbf{x}$ satisfying this equation which has smallest norm. This is equivalent to making $|V^*\mathbf{x}|$ as small as possible because V^* is unitary and thus preserves norms.

For $\mathbf{z}$ a vector, denote by $(\mathbf{z})_r$ the vector in $\mathbb{C}^n$ which consists of the first r entries of $\mathbf{z}$. Here σ is $r \times r$. Then if $\mathbf{x}$ is a solution to (11.7),

$$\begin{pmatrix} \sigma^2 (V^*\mathbf{x})_r \\ \mathbf{0} \end{pmatrix} = \begin{pmatrix} \sigma (U^*\mathbf{y})_r \\ \mathbf{0} \end{pmatrix},$$

and so $(V^*\mathbf{x})_r = \sigma^{-1} (U^*\mathbf{y})_r$. Thus the first r entries of $V^*\mathbf{x}$ are determined. In order to make $|V^*\mathbf{x}|$ as small as possible, the remaining $n - r$ entries should equal zero. Therefore,

$$V^*\mathbf{x} = \begin{pmatrix} (V^*\mathbf{x})_r \\ 0 \end{pmatrix} = \begin{pmatrix} \sigma^{-1} (U^*\mathbf{y})_r \\ 0 \end{pmatrix}$$

$$= \begin{pmatrix} \sigma^{-1} & 0 \\ 0 & 0 \end{pmatrix} U^*\mathbf{y},$$

and so

$$\mathbf{x} = V \begin{pmatrix} \sigma^{-1} & 0 \\ 0 & 0 \end{pmatrix} U^*\mathbf{y} \equiv A^+\mathbf{y}.$$

This proves the proposition. □

The four conditions of the following lemma are called the **Penrose conditions.**

Lemma 11.6. *The matrix A^+ satisfies the following conditions.*

$$AA^+A = A,\ A^+AA^+ = A^+,\ A^+A \text{ and } AA^+ \text{ are Hermitian.} \tag{11.8}$$

Proof. This is routine. Recall that

$$A = U \begin{pmatrix} \sigma & 0 \\ 0 & 0 \end{pmatrix} V^*$$

and

$$A^+ = V \begin{pmatrix} \sigma^{-1} & 0 \\ 0 & 0 \end{pmatrix} U^*$$

so you just plug in and verify that it works. □

A much more interesting observation is that A^+ is characterized as being the unique matrix which satisfies (11.8), the Penrose conditions. This is the content of the following Theorem.

Theorem 11.3. *Let A be an $m \times n$ matrix. Then a matrix A_0, is the Moore Penrose inverse of A if and only if A_0 satisfies*

$$AA_0A = A,\ A_0AA_0 = A_0,\ A_0A \text{ and } AA_0 \text{ are Hermitian.} \tag{11.9}$$

Proof. From the above lemma, the Moore Penrose inverse satisfies (11.9). Suppose then that A_0 satisfies (11.9). It is necessary to verify that $A_0 = A^+$. Recall that from the singular value decomposition, there exist unitary matrices U and V such that

$$U^*AV = \Sigma \equiv \begin{pmatrix} \sigma & 0 \\ 0 & 0 \end{pmatrix}, \ A = U\Sigma V^*.$$

Let

$$V^*A_0U = \begin{pmatrix} P & Q \\ R & S \end{pmatrix} \tag{11.10}$$

where P is $k \times k$, the same size as the diagonal matrix σ.

Next use the first equation of (11.9) to write

$$\overbrace{U\Sigma V^*}^{A} \overbrace{V \begin{pmatrix} P & Q \\ R & S \end{pmatrix} U^*}^{A_0} \overbrace{U\Sigma V^*}^{A} = \overbrace{U\Sigma V^*}^{A}.$$

Then multiplying both sides on the left by U^* and on the right by V,

$$\begin{pmatrix} \sigma & 0 \\ 0 & 0 \end{pmatrix} \begin{pmatrix} P & Q \\ R & S \end{pmatrix} \begin{pmatrix} \sigma & 0 \\ 0 & 0 \end{pmatrix} = \begin{pmatrix} \sigma & 0 \\ 0 & 0 \end{pmatrix}.$$

Now this requires

$$\begin{pmatrix} \sigma P \sigma & 0 \\ 0 & 0 \end{pmatrix} = \begin{pmatrix} \sigma & 0 \\ 0 & 0 \end{pmatrix}. \tag{11.11}$$

Therefore, $P = \sigma^{-1}$. From the requirement that AA_0 is Hermitian,

$$\overbrace{U\Sigma V^*}^{A} \overbrace{V \begin{pmatrix} P & Q \\ R & S \end{pmatrix} U^*}^{A_0} = U \begin{pmatrix} \sigma & 0 \\ 0 & 0 \end{pmatrix} \begin{pmatrix} P & Q \\ R & S \end{pmatrix} U^*$$

must be Hermitian. Therefore, it is necessary that

$$\begin{pmatrix} \sigma & 0 \\ 0 & 0 \end{pmatrix} \begin{pmatrix} P & Q \\ R & S \end{pmatrix} = \begin{pmatrix} \sigma P & \sigma Q \\ 0 & 0 \end{pmatrix} = \begin{pmatrix} I & \sigma Q \\ 0 & 0 \end{pmatrix}$$

is Hermitian. Then

$$\begin{pmatrix} I & \sigma Q \\ 0 & 0 \end{pmatrix} = \begin{pmatrix} I & 0 \\ Q^*\sigma & 0 \end{pmatrix}.$$

Thus $Q^*\sigma = 0$, and so, multiplying both sides on the right by σ^{-1}, it follows that $Q^* = 0$ which implies $Q = 0$.

From the requirement that A_0A is Hermitian, it is necessary that

$$\overbrace{V \begin{pmatrix} P & Q \\ R & S \end{pmatrix} U^*}^{A_0} \overbrace{U\Sigma V^*}^{A} = V \begin{pmatrix} P\sigma & 0 \\ R\sigma & 0 \end{pmatrix} V^* = V \begin{pmatrix} I & 0 \\ R\sigma & 0 \end{pmatrix} V^*$$

is Hermitian. Therefore, also

$$\begin{pmatrix} I & 0 \\ R\sigma & 0 \end{pmatrix}$$

is Hermitian. Thus $R = 0$ because this equals

$$\begin{pmatrix} I & 0 \\ R\sigma & 0 \end{pmatrix}^* = \begin{pmatrix} I & \sigma^* R^* \\ 0 & 0 \end{pmatrix}$$

which requires $R\sigma = 0$. Now multiply on the right by σ^{-1} to find that $R = 0$.

Use (11.10) and the second equation of (11.9) to write

$$\overbrace{V\begin{pmatrix} P & Q \\ R & S \end{pmatrix}U^*}^{A_0}\overbrace{U\Sigma V^*}^{A}\overbrace{V\begin{pmatrix} P & Q \\ R & S \end{pmatrix}U^*}^{A_0} = \overbrace{V\begin{pmatrix} P & Q \\ R & S \end{pmatrix}U^*}^{A_0}.$$

which implies

$$\begin{pmatrix} P & Q \\ R & S \end{pmatrix}\begin{pmatrix} \sigma & 0 \\ 0 & 0 \end{pmatrix}\begin{pmatrix} P & Q \\ R & S \end{pmatrix} = \begin{pmatrix} P & Q \\ R & S \end{pmatrix}.$$

This yields from the above in which it was shown that R, Q are both 0,

$$\begin{pmatrix} \sigma^{-1} & 0 \\ 0 & S \end{pmatrix}\begin{pmatrix} \sigma & 0 \\ 0 & 0 \end{pmatrix}\begin{pmatrix} \sigma^{-1} & 0 \\ 0 & S \end{pmatrix} = \begin{pmatrix} \sigma^{-1} & 0 \\ 0 & 0 \end{pmatrix} = \begin{pmatrix} \sigma^{-1} & 0 \\ 0 & S \end{pmatrix}.$$

Therefore, $S = 0$ also, and so

$$V^* A_0 U \equiv \begin{pmatrix} P & Q \\ R & S \end{pmatrix} = \begin{pmatrix} \sigma^{-1} & 0 \\ 0 & 0 \end{pmatrix}$$

which says

$$A_0 = V\begin{pmatrix} \sigma^{-1} & 0 \\ 0 & 0 \end{pmatrix}U^* \equiv A^+. \quad \square$$

The theorem is significant because there is no mention of eigenvalues or eigenvectors in the characterization of the Moore Penrose inverse given in (11.9). It also shows immediately that the Moore Penrose inverse is a generalization of the usual inverse.

In the following exercises, you might want to use some computer algebra system to save yourself trouble.

11.4 Exercises

(1) Find a singular value decomposition for the matrix

$$\begin{pmatrix} 1 & 2 & 3 \\ 0 & -3 & 2 \\ 1 & 2 & 5 \end{pmatrix}.$$

(2) Find the rank 2 matrix which is closest to

$$\begin{pmatrix} 1 & 2 & 3 \\ 4 & -3 & 2 \\ 1 & 6 & 7 \end{pmatrix}.$$

in the Frobenius norm. You should get something like

$$\begin{pmatrix} .955\,078\,958 & 1.964\,583\,04 & 3.037\,036\,98 \\ 4.006\,197\,17 & -2.995\,113\,98 & 1.994\,890\,49 \\ 1.017\,760\,61 & 6.014\,002\,94 & 6.985\,356\,54 \end{pmatrix}.$$

(3) Show that if A is an $n \times n$ matrix which has an inverse then $A^+ = A^{-1}$.

(4) Using the singular value decomposition, show that, for any square matrix A, it follows that A^*A is unitarily similar to AA^*.

(5) Let A be an $m \times n$ matrix. Show

$$||A||_F^2 \equiv (A, A)_F = \sum_j \sigma_j^2$$

where the σ_j are the singular values of A.

(6) Suppose $\{\mathbf{v}_1, \cdots, \mathbf{v}_n\}$ and $\{\mathbf{w}_1, \cdots, \mathbf{w}_n\}$ are two orthonormal bases for $\mathbb{F}^n$ and suppose Q is an $n \times n$ matrix satisfying $Q\mathbf{v}_i = \mathbf{w}_i$. Then show Q is unitary. If $|\mathbf{v}| = 1$, show that there is a unitary transformation which maps $\mathbf{v}$ to $\mathbf{e}_1$.

(7) Let A be a Hermitian matrix, so $A = A^*$ and suppose that all eigenvalues of A are larger than δ^2. Show that

$$\langle A\mathbf{v}, \mathbf{v} \rangle \geq \delta^2 |\mathbf{v}|^2$$

where here, the inner product is

$$\langle \mathbf{v}, \mathbf{u} \rangle \equiv \sum_{j=1}^{n} v_j \overline{u_j}.$$

(8) The discrete Fourier transform maps $\mathbb{C}^n \to \mathbb{C}^n$ as follows.

$$F(\mathbf{x}) = \mathbf{z} \text{ where } z_k = \frac{1}{\sqrt{n}} \sum_{j=0}^{n-1} e^{-i\frac{2\pi}{n}jk} x_j.$$

Show that F^{-1} exists and is given by the formula

$$F^{-1}(\mathbf{z}) = \mathbf{x} \text{ where } x_j = \frac{1}{\sqrt{n}} \sum_{j=0}^{n-1} e^{i\frac{2\pi}{n}jk} z_k$$

Here is one way to approach this problem. Note $\mathbf{z} = U\mathbf{x}$ where

$$U = \frac{1}{\sqrt{n}} \begin{pmatrix} e^{-i\frac{2\pi}{n}0\cdot 0} & e^{-i\frac{2\pi}{n}1\cdot 0} & e^{-i\frac{2\pi}{n}2\cdot 0} & \cdots & e^{-i\frac{2\pi}{n}(n-1)\cdot 0} \\ e^{-i\frac{2\pi}{n}0\cdot 1} & e^{-i\frac{2\pi}{n}1\cdot 1} & e^{-i\frac{2\pi}{n}2\cdot 1} & \cdots & e^{-i\frac{2\pi}{n}(n-1)\cdot 1} \\ e^{-i\frac{2\pi}{n}0\cdot 2} & e^{-i\frac{2\pi}{n}1\cdot 2} & e^{-i\frac{2\pi}{n}2\cdot 2} & \cdots & e^{-i\frac{2\pi}{n}(n-1)\cdot 2} \\ \vdots & \vdots & \vdots & & \vdots \\ e^{-i\frac{2\pi}{n}0\cdot(n-1)} & e^{-i\frac{2\pi}{n}1\cdot(n-1)} & e^{-i\frac{2\pi}{n}2\cdot(n-1)} & \cdots & e^{-i\frac{2\pi}{n}(n-1)\cdot(n-1)} \end{pmatrix}$$

Now argue that U is unitary and use this to establish the result. To show this verify each row has length 1 and the dot product of two different rows gives 0. Now $U_{kj} = e^{-i\frac{2\pi}{n}jk}$, and so $(U^*)_{kj} = e^{i\frac{2\pi}{n}jk}$.

(9) Let A be a complex $m \times n$ matrix. Using the description of the Moore Penrose inverse in terms of the singular value decomposition, show that

$$\lim_{\delta \to 0+} \left(A^* A + \delta I\right)^{-1} A^* = A^+$$

where the convergence happens in the Frobenius norm. Also verify, using the singular value decomposition, that the inverse exists in the above formula. Give an estimate in terms of δ and the singular values of A which will describe how close the approximation on the left is to A^+.

(10) Here is a matrix:

$$\begin{pmatrix} 1 & 2 & 3 \\ 4 & 0 & 6 \\ 7 & 8 & 9 \end{pmatrix}.$$

This matrix is invertible, and its inverse is

$$\begin{pmatrix} -\frac{4}{5} & \frac{1}{10} & \frac{1}{5} \\ \frac{1}{10} & -\frac{1}{5} & \frac{1}{10} \\ \frac{8}{15} & \frac{1}{10} & -\frac{2}{15} \end{pmatrix}.$$

Thus from Problem 3 above, the Moore Penrose inverse is just the ordinary inverse. Use the formula of Problem 9 to obtain an approximation to the Moore Penrose inverse, hence the usual inverse, by assigning $\delta = .001$.

(11) Find an approximation to the Moore Penrose inverse for the matrix

$$\begin{pmatrix} 1 & 2 & 3 & 2 \\ 1 & 0 & 3 & 2 \\ 3 & 2 & 1 & -4 \end{pmatrix}$$

by using δ equal to various small numbers in Problem 9.

(12) Find an approximation to the Moore Penrose inverse for the matrix

$$\begin{pmatrix} 1 & 1 & 3 \\ 2 & 0 & 2 \\ 1 & 1 & 3 \\ 2 & 2 & -4 \end{pmatrix}$$

by using δ equal to various small numbers in Problem 9. Use to obtain an approximation to the best least squares solution to the system

$$\begin{pmatrix} 1 & 1 & 3 \\ 2 & 0 & 2 \\ 1 & 1 & 3 \\ 2 & 0 & 2 \end{pmatrix} \begin{pmatrix} x \\ y \\ z \end{pmatrix} = \begin{pmatrix} 1 \\ 2 \\ 3 \\ 4 \end{pmatrix}$$

Then find all least squares solutions and obtain the one with smallest norm. Compare with what you obtained using the approximate Moore Penrose inverse.

Appendix A

A.1 Using Maple

A.1.1 *Entering a matrix in Maple*

When you open the Maple window this is part of what you see.

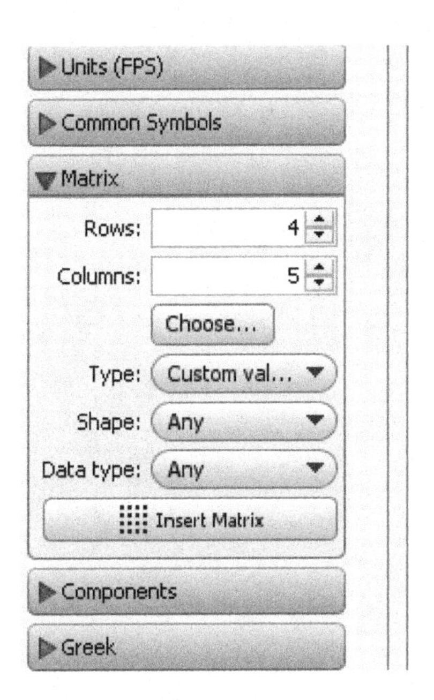

Actually we modified what came up by selecting 4 for four rows and 5 for five columns over on the left. There is much more which appears in the Maple window than what is shown above. The upper left corner of the Maple window will look like the following picture. You may need to move the space bar which is on the right of the left column of stuff up or down to expose that which is relevant to entering a matrix.

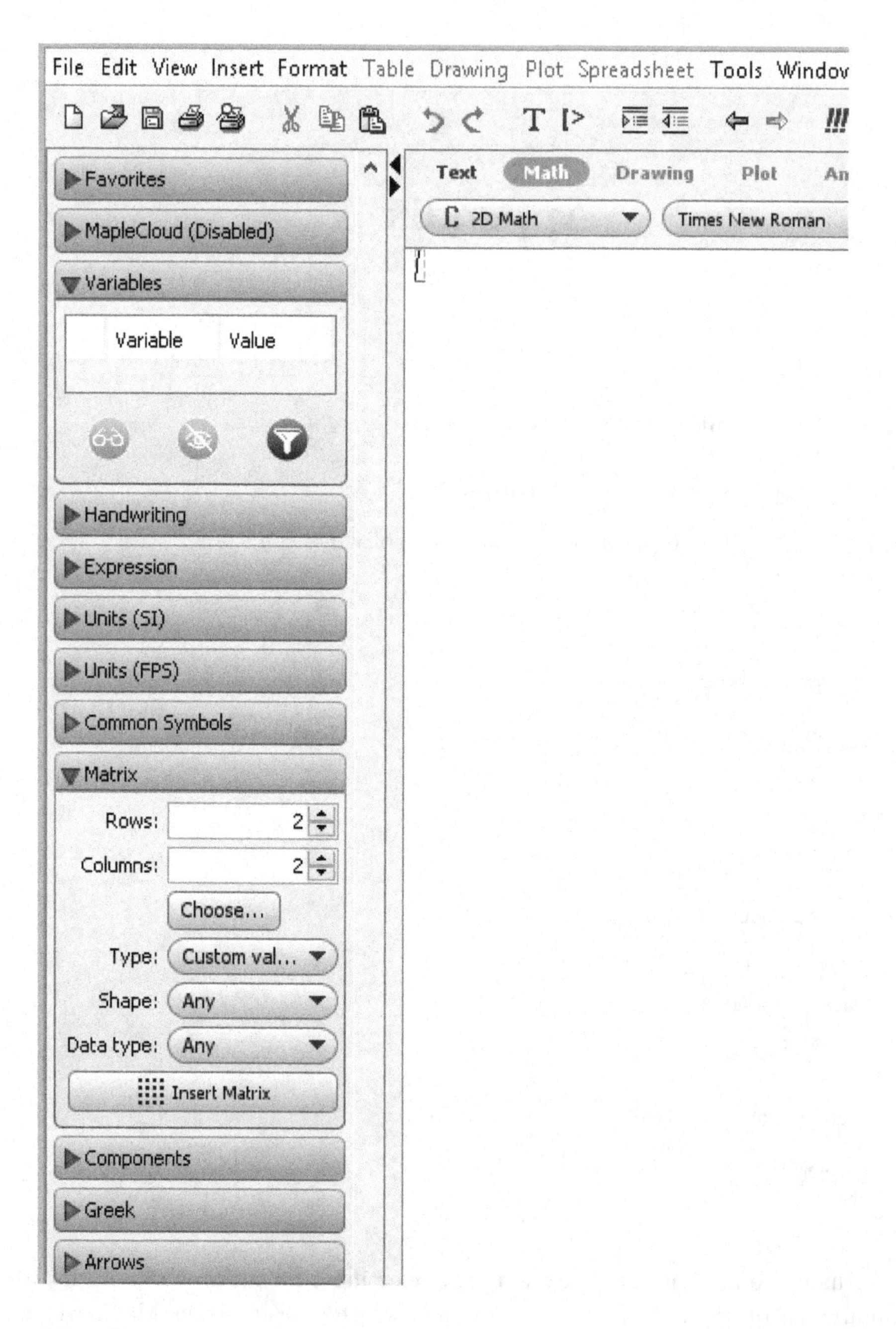

After this came up, we selected four rows and five columns where this is indicated because we wanted to consider a matrix which has four rows and five columns. Then we left clicked on the place where it says: insert matrix. The following is what resulted.

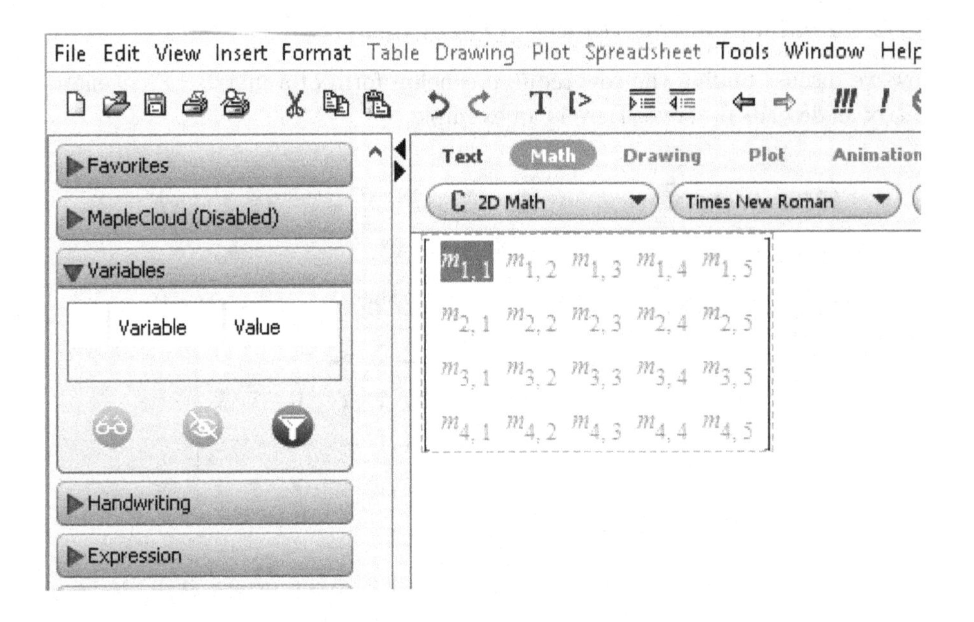

Next we entered the desired matrix by filling in the m_{ij}. This is most easily done by typing in the desired number on the left upper corner and then pressing the tab key. This moves the marker to the right to the next entry and you can type in the next number. Continuing this way, you fill in the entries of the first row. When you press tab again, the marker will jump to the left most entry of the second row and you continue this process till the matrix has been entered. We have done this and obtained the following matrix whose entries are numbers which we picked.

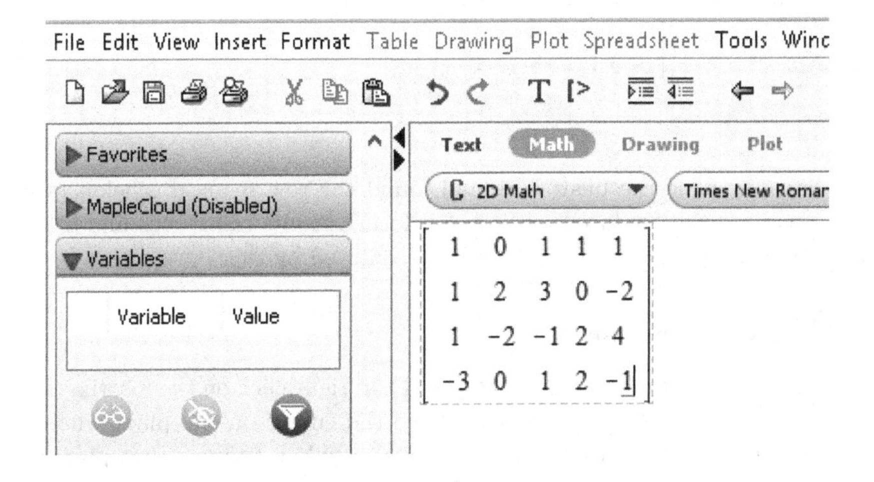

At this point, the matrix has been entered.

A.1.2 *Finding the row reduced echelon form*

Now we discuss finding the row reduced echelon form of a matrix. First enter the matrix as described above. Here is an example.

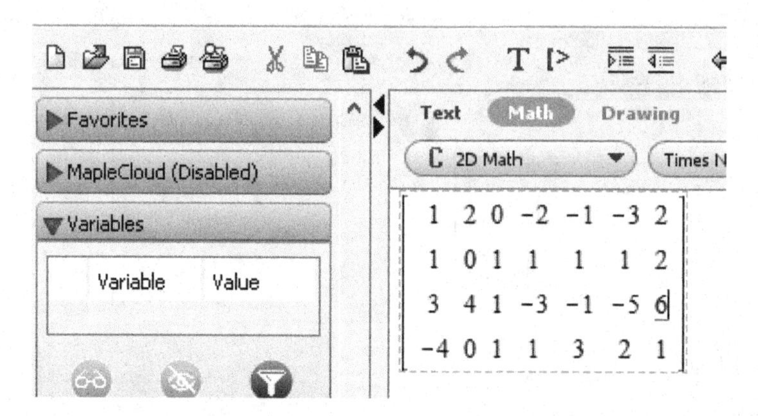

Next place the cursor on the matrix you just entered. Right click. Next place the cursor on Solvers and Forms and then on Row Echelon Form, and then on Reduced. Next left click and this is what results.

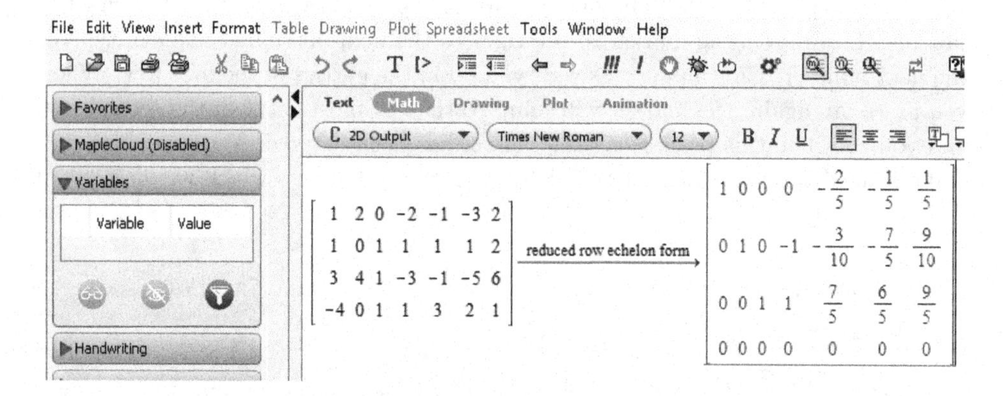

Maple has now done the busy work and found the row reduced echelon form. Note that it calls it the "reduced row echelon form". Some call it this and others call it the row reduced echelon form. Don't be confused by this.

A.1.3 *Finding the inverse*

To find the inverse, enter the matrix as above. Then right click on the matrix. This produces a gray box with many choices. Place the cursor on the place where it says Standard Operations. This produces another gray box and you then place the cursor on the place where it says inverse. Then left click and it produces the inverse for you. The end result will look something like the following.

$$\begin{bmatrix} 3 & 2 & -5 \\ 2 & 7 & 1 \\ 1 & 0 & 5 \end{bmatrix} \xrightarrow{\text{inverse}} \begin{bmatrix} \frac{35}{122} & -\frac{5}{61} & \frac{37}{122} \\ -\frac{9}{122} & \frac{10}{61} & -\frac{13}{122} \\ -\frac{7}{122} & \frac{1}{61} & \frac{17}{122} \end{bmatrix}$$

A.1.4 *Finding a PLU factorization*

Consider the same example used above in illustrating the row reduced echelon form. We find a PLU factorization. Place the cursor on the matrix you just entered. Right click. Next place the cursor on Solvers and Forms and then on LU decomposition, then on Gaussian Elimination and next on Gaussian Elimination (P, L, U) then left click. This is what results.

$$\begin{bmatrix} 1 & 2 & 0 & -2 & -1 & -3 & 2 \\ 1 & 0 & 1 & 1 & 1 & 1 & 2 \\ 3 & 4 & 1 & -3 & -1 & -5 & 6 \\ -4 & 0 & 1 & 1 & 3 & 2 & 1 \end{bmatrix} \xrightarrow{\text{Gaussian elimination (P,L,U)}} \begin{bmatrix} 1 & 0 & 0 & 0 \\ 0 & 1 & 0 & 0 \\ 0 & 0 & 0 & 1 \\ 0 & 0 & 1 & 0 \end{bmatrix}, \begin{bmatrix} 1 & 0 & 0 & 0 \\ 1 & 1 & 0 & 0 \\ -4 & -4 & 1 & 0 \\ 3 & 1 & 0 & 1 \end{bmatrix}, \begin{bmatrix} 1 & 2 & 0 & -2 & -1 & -3 & 2 \\ 0 & -2 & 1 & 3 & 2 & 4 & 0 \\ 0 & 0 & 5 & 5 & 7 & 6 & 9 \\ 0 & 0 & 0 & 0 & 0 & 0 & 0 \end{bmatrix}$$

Note how in this case, there is a permutation matrix so this particular matrix does not have a LU decomposition. If it did, then the first matrix identified as P would be the identity matrix.

A.1.5 *Finding a QR factorization*

To find a QR factorization first enter the matrix. Then right click to bring down a gray box. Next place the cursor on "solvers and forms" and next on "QR Decomposition" and then on "QR Decomposition (Q,R). Then left click. For example, here is the case of a 3×2 matrix.

$$\begin{bmatrix} 2 & 1 \\ 2 & 3 \\ 3 & -3 \end{bmatrix} \xrightarrow{\text{QR decomposition (Q,R)}} \begin{bmatrix} \frac{2}{17}\sqrt{17} & \frac{19}{5474}\sqrt{5474} \\ \frac{2}{17}\sqrt{17} & \frac{53}{5474}\sqrt{5474} \\ \frac{3}{17}\sqrt{17} & -\frac{24}{2737}\sqrt{5474} \end{bmatrix}, \begin{bmatrix} \sqrt{17} & -\frac{1}{17}\sqrt{17} \\ 0 & \frac{1}{17}\sqrt{5474} \end{bmatrix}$$

This is the "thin QR factorization". Note that the matrix on the left is not even square. The columns of this matrix are an orthonormal basis for the two original columns. This is the version of the "QR" factorization which is computed by Maple.

A.1.6 *Finding a singular value decomposition*

First enter the matrix. Next right click on the matrix you have entered. This brings down a gray box. Place the cursor on "Solvers and Forms". Then in the next box, you place it on "Singular Value Decomposition". In the next gray box, you place the cursor on Singular Value Decomposition (U,S,Vt). What you get from this is two square matrices U and $Vt = V^t$ and a column vector in between them consisting of the singular values of the matrix. You will have a description of what is represented on top of the arrow.

$$\begin{pmatrix} 1 & 2 \\ 1 & 1 \\ 0 & 2 \end{pmatrix} \rightarrow \begin{pmatrix} 0.70 & 0.26 & -0.67 \\ 0.4 & 0.63 & 0.67 \\ 0.59 & -0.74 & 0.33 \end{pmatrix}, \begin{pmatrix} 3.2 \\ 0.94 \end{pmatrix}, \begin{pmatrix} 0.35 & 0.94 \\ 0.94 & -0.35 \end{pmatrix}$$

A.1.7 *Generalities*

As you do examples of the above form, you will notice that Maple can do many other computations in the same way. You just place the cursor over a different sequence of items and end up left clicking on the last one. It is very convenient.

Maple is much more powerful than this sketch reveals; see e.g. the book of Heck (2003). We have emphasized the topics most closely related to linear algebra.

A.2 Further Reading

To carry your knowledge of linear algebra further, a good starting point is Horn and Johnson (1990). If you intend to carry out genuine large scale applications, you should have Golub and Van Loan (1983) to hand.

To appreciate the vast scope of applications of linear algebra, and for detailed references to results across a broad spectrum, you should consult the multi-author 'Handbook of Linear Algebra', edited by Leslie Hogben.

You will find references to areas that go well beyond linear algebra, but are naturally suggested by studying it, scattered through this book. Details are collected below under 'References'.

If you would like to trace the historical development of linear algebra, visit the superb MacTutor website:

www-groups.dcs.st-and.ac.uk/ history/

Do not be convinced by this book that matrices in applications are small ($m \times n$ with m, n at most 6). As an antidote, read about the application of linear algebra to Google in K. Bryan and T. Leise, SIAM review vol. 48, pp. 569-581, entitled 'The $25,000,000,000 eigenvalue problem: the linear algebra behind Google'. The relevant $n \times n$ matrix had n about 8×10^9 at the time the article was written in 2005 (it appeared in 2006).

Bibliography

M. Artin, Algebra Pearson. (2011)

P. J. Collins, Differential and Integral Equations, Oxford University Press. (2006)

G. H. Golub and C. F. Van Loan, Matrix Computations, 3rd edn., Johns Hopkins University Press. (1996)

C. H. Edwards, Advanced Calculus of Several Variables, Dover. (1994)

M. D. Greenberg, Advanced Engineering Mathematics, second edition, Prentice Hall. (1998)

G. Hardy, A Course of Pure Mathematics, tenth edition, Cambridge University Press. (1992)

A. Heck, Introduction to Maple, Springer. (2003)

I. N. Herstein, Topics in Algebra, Xerox College Publishing. (1964)

L. Hogben (ed.), Handbook of Linear Algebra, Chapman and Hall. (2007)

R. A. Horn and C. R. Johnson, Matrix Analysis, Cambridge University Press. (1990)

O. D. Johns, Analytical Mechanics for Relativity and Quantum Mechanics. Oxford University Press (2005)

Y. Katznelson, An Introduction to Harmonic Analysis, third edition, Cambridge University Press. (2004)

J. Daniel and B. Nobel, Applied Linear Algebra, Prentice Hall. (1977)

J. Polking, A. Boggess and D. Arnold, Differential Equations with Boundary Value Problems, Pearson Education. (2002)

E. J. Putzer, American Mathematical Monthly, Vol. 73, pp. 2-7. (1966)

W. Rudin, Principles of mathematical analysis, third edition, McGraw Hill. (1976)

E. M. Stein and R. Shakarchi, Fourier Analysis, Princeton University Press. (2003)

J.H. Wilkinson, The Algebraic Eigenvalue Problem, Clarendon Press Oxford. (1965)

Bibliography

Index

Printed in the United States
By Bookmasters